NONMAGNETIC AND MAGNETIC QUANTUM DOTS

Edited by **Vasilios N. Stavrou**

Nonmagnetic and Magnetic Quantum Dots
http://dx.doi.org/10.5772/intechopen.68207
Edited by Vasilios N. Stavrou

Contributors
Wei Ma, Ke Yan, Bei Yan, Akos Nemcsics, Samuel Oluwatobi Oluwafemi, Sundararajan Parani, Ncediwe Tsolekile, Bambesiwe May, K Pandian, Anca Armăşelu, Zhuo Chen, Zequn Chen, Wei Guo, Chuli Sun, Er\'El Granot, Kamla Rawat, Eepsita Priyadarshini, H. B. Bohidar, Ciann-Dong Yang, Jyoti Narayan, Jejiron Baruah, Pablo Zubieta, Yuri Vorobiev, Daniel Olguín, Lynne Macaskie, Angela Murray, John Love, Mark Redwood, Rafael Orozco, Richard Tennant, Frankie Woodhall, Alex Goodridge, Jung Huang

Published by InTech
Janeza Trdine 9, 51000 Rijeka, Croatia

Publishing Process Manager Romina Rovan
Technical Editor SPi Global
Cover InTech Design team

First published April, 2018
Printed in Croatia
Legal deposit, Croatia: National and University Library in Zagreb

Additional hard copies can be obtained from orders@intechopen.com

Nonmagnetic and Magnetic Quantum Dots, Edited by Vasilios N. Stavrou
p. cm.
Print ISBN 978-953-51-3959-1
Online ISBN 978-953-51-3960-7

For EU product safety concerns:
IN TECH d.o.o., Prolaz Marije Krucifikse Kozulić 3, 51000 Rijeka, Croatia,
info@intechopen.com or visit our website at intechopen.com.

Meet the editor

Dr. V. N. Stavrou is currently an adjunct member at the Hellenic Naval Academy, Piraeus, Greece. He received his MSc degree and PhD degree in Theoretical Solid State Physics from the University of Essex in England, in 1995 and in 1999, respectively. He has held postdoctoral positions at the following research institutions: (a) Deutsche Forschungsanstalt fuer Luft und Raumfahrt e.V (German Aerospace Research Center) in Germany; (b) Helsinki University of Technology; (c) State University of New York (SUNY) at Buffalo, USA; and (d) the University of Iowa, USA. He is specifically interested in researching on the electronic, optical, and lattice properties of semiconducting low-dimensional structures (quantum dots and quantum wells, among others). These properties are of special importance for quantum computing architecture and laser technology. He has published his research work in reputable journals related, among others, to quantum computing, spintronics, decoherence in quantum dots, diluted magnetic semiconductors, and phonon models in low-dimensional structures.

Contents

Preface

Quantum dots are mainly made with semiconducting materials in which the carriers are confined in all three dimensions. The carrier confinement makes the quantum dots significant in technology due to highly tunable electronic and optical properties. Their optoelectronic properties depend on the size and the shape of the dots, among other external parameters. Potential applications of these nanostructures include semiconductor lasers, quantum computing, transistors, and sensors, among others. Due to worldwide interest in the domain of technology, several fabrication techniques have been used to control the growth of quantum dots and the quality of the samples. Stranski-Krastanov (SK) random growth in molecular beam epitaxy (MBE) and metalorganic chemical vapor deposition (MOCVD) are the most commonly used growth methods to create high-quality quantum dot samples, in both size and shape. Several advanced computational methods like Monte-Carlo simulations, $k \cdot p$ theory, ab initio calculations, and density functional theory, among others, have been developed to calculate the electronic structure of quantum dots.

In the current book, there is a collection of interesting topics related to quantum dots made with magnetic and nonmagnetic semiconductor materials and their applications. This book is divided into two sections. In Section 1, the chapters are related to nonmagnetic quantum dots and their applications. More specifically, exact models and numerical methods have been presented to describe the analytical solution of the carrier wave functions, the quantum mechanical aspects of quantum dots, and the comparison of the latter to experimental data. Furthermore, methods to produce quantum dots, synthesis techniques of colloidal quantum dots, and applications on sensors and biology, among others, are included in this section. In Section 2, a few topics of magnetic quantum dots and their applications are presented. The section starts with a theoretical model to describe the magnetization dynamics in magnetic quantum dot array and the description of dilute magnetic semiconducting quantum dots and their applications. Additionally, a few applications of magnetic quantum dot in sensors, biology, and medicine are included in Section 2.

As an editor of this book, I would like to thank all the authors for their contribution through the up-to-date research of their high-quality work. Lastly, I would like to express my thanks and gratitude to the InTechOpen team for their support during the preparation of this book.

Dr. Vasilios N. Stavrou
Hellenic Naval Academy
Piraeus, Greece

Section 1

Nonmagnetic Quantum Dots

Chapter 1

Exact Model for Single Atom Transistor

Er'el Granot

Additional information is available at the end of the chapter

http://dx.doi.org/10.5772/intechopen.70445

Abstract

An exact model for a single atom transistor was developed. Using two simplifying assumptions (1) that the device is restricted to a narrow conducting wire and (2) that the atom can be simulated by a point impurity potential, the model can be simplified considerably and an *exact analytical solution* can be derived. Thus, analytical solution is approximated to a close-form solution in three important regimes: at the vicinity of the resonance energy (near the maximum peak), at the vicinity of the inverse resonance, i.e., Fano resonance (near the minimum), and at the threshold energy where a universal transmission pattern appears. Finally, physical values are applied to demonstrate that this device can operate as a transistor, when it is calibrated to work at the vicinity of its maximum and minimum points.

Keywords: quantum dots, quantum point defect, point impurity, quantum transistor, single atom transistor

1. Introduction

In accordance with the rapid growth of calculation power, the transistor dimensions shrink exponentially. Surprisingly, more than 50 years after Gordon Moore made his observation in 1965 (or, more accurately, its revised form a decade later), that the number of transistors on a single chip doubles every couple of years, this observation is still valid [1, 2]. The number of transistors in a chip keeps growing despite the fact that the chip clock speed and its power consumption seem to be stagnated.

To meet the demands of the current trend, the average transistor size should decrease to the dimensions of a single atom, which is the smallest quantum dot, within about a decade.

The ability to move and manipulate single Xenon atoms (in Eigler and Schweizer lab at IBM's Almaden Research Center) in the early 1990s was a great leap in that direction [3].

In the attempts to meet this requirement, scientists already demonstrated the operation on several atoms and dopants [4–9] and even on single atoms [10–12]. The atom can be utilized as a stationary gate [10] or as a dynamic switch [13].

Such a device indeed consists of a single atom, but its conductor leads are of mesoscopic dimensions. Consequently, this is a complicated device to simulate and requires heavy software.

However, since the process is dominated by resonant tunneling, the model can be simplified considerably. In this case, only a single energy level of the quantum dot is relevant to the process, and therefore, it can be simulated by a point defect potential.

A point defect potential has a single eigen bound state, and therefore, it can simulate a quantum dot or a small atom in a relatively narrow spectral domain.

While a delta function can simulate a point defect in one-dimensional (1D) systems, a two-dimensional (2D) delta function cannot scatter and therefore cannot simulate a quantum dot. Azbel suggested to use an Impurity-D-Function (IDF) to simulate point defects in 2D quantum systems [14, 15] (for a comprehensive discussion and derivations, see Ref. [16]).

Several years later, the IDF was utilized in simulations of resonant tunneling through an opaque quantum barrier via a point defect in the presence [17] and absence [18] of a magnetic field. However, in these models, it was taken that there is a degeneracy in the y-direction, i.e., it was assumed that the barrier's transverse dimension is infinite and therefore cannot be applied in a system, where the current is carried by narrow wires (as in modern single atom transistor's devices).

On the other hand, conductance of nanowires with defects, but without a barrier, received lots of attention in the literature, exhibiting a wealth of physical phenomena [4–9, 19–21].

It is the purpose of this chapter to integrate the two, i.e., to formulate a model, which incorporates resonant tunneling via a point defect and wire conductance. That is, both the potential barrier and the impurity are located in the nanowire.

2. The model

The system is illustrated in **Figure 1**. It consists of two semi-infinite conducting wires, which are separated by an insulating gap. Within this gap, there is a quantum dot, which characterizes the resonance atom. To simplify the analysis, it is assumed that the wire boundaries in the y direction are totally reflecting, i.e., the wire is bounded by an infinitely large potential. Moreover, the single atom is modeled by a point defect potential.

It is also taken that this is a 2D model, i.e., there are no variations in the third dimension. This is a good approximation provided the wire is narrower in the z-dimension. Another advantage in constructing the model in 2D is that point impurities potential cannot exist in higher than two dimensions (see Ref. [16]).

http://dx.doi.org/10.5772/intechopen.70445

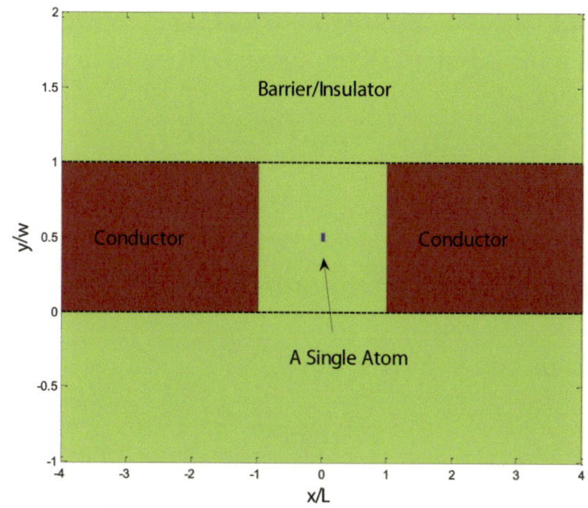

Figure 1. Model schematic.

Therefore, the system can be described by the following stationary Schrödinger equation

$$-\nabla^2\Psi(x,y) + [V(x) + U(y) - D(\mathbf{r}' - \mathbf{r}_0)]\Psi(x,y) = E\Psi(x,y) \tag{1}$$

in which normalized units (where Planck constant is $\hbar = 1$, and the electron's mass is $m = 1/2$) were used. In this equation,

$$U(y) = \begin{cases} 0 & 0 < y < w \\ \infty & else \end{cases} \tag{2}$$

is the boundaries' potential, which confines the dynamics to the wire geometry.

The potential of the gap between the wires is represented by the finite potential barrier,

$$V(x) = \begin{cases} V & |x| < L \\ 0 & else \end{cases} \tag{3}$$

and for the point impurity potential, we use an asymmetric Impurity D Functions (see Refs. [14, 15])

$$D(\mathbf{r}) = \lim_{\rho \to 0} \frac{2\sqrt{\pi}\exp(-y^2/\rho^2)}{\rho \ln(\rho_0/\rho)}\delta(x), \tag{4}$$

where $\mathbf{r} = \hat{x}x + \hat{y}y$, and the location of the point impurity is $\mathbf{r}_0 = \hat{x}x_0 + \hat{y}y_0$. ρ_0 is the de-Broglie wavelength of the impurity's bound eigenstate. The eigenenergy of the bound state of this impurity is

$$E_0 = -\frac{16\exp(-\gamma)}{\rho_0^2} \cong -\frac{8.98}{\rho_0^2} \tag{5}$$

where $\gamma \cong 0.577$ is Euler constant [22].

It should be stressed that this point impurity potential is an excellent approximation to a small quantum dot defect, i.e., a finite but small impurity, with a radius a and potential V_0 provided

$$\rho_0 = 2a \exp\left(\frac{2}{V_0 a^2} + \frac{\gamma}{2}\right). \tag{6}$$

3. Derivation of the exact analytical solution

The solution of Eq. (1) reads (see Refs. [16, 19])

$$\Psi(\mathbf{r}) = \Psi_{inc}(\mathbf{r}) - \frac{G^+(\mathbf{r}, \mathbf{r}_0)\Psi_{inc}(\mathbf{r}_0)}{1 + \int d\mathbf{r}' G^+(\mathbf{r}', \mathbf{r}_0) D(\mathbf{r}' - \mathbf{r}_0)} \int d\mathbf{r}' D(\mathbf{r}' - \mathbf{r}_0) \tag{7}$$

where $\Psi_{inc}(\mathbf{r})$ is the incoming wavefunction, $G^+(\mathbf{r}, \mathbf{r}_0)$ is the outgoing 2D Green function, i.e., $G^+(\mathbf{r}, \mathbf{r}_0)$ is the solution of the partial differential equation

$$-\nabla^2 G^+(\mathbf{r}, \mathbf{r}_0) + [V(x) + U(y) - E] G^+(\mathbf{r}, \mathbf{r}_0) = -\delta(\mathbf{r} - \mathbf{r}_0). \tag{8}$$

Both the incoming wavefunction $\psi_{inc}(\mathbf{r})$ and the Green function can be written as a superposition of the homogenous solution of Eq. (1) $\varphi^{+}_{m,E}(\mathbf{r})$, i.e., solution of the equation where the impurity is absent. These solutions are characterized by two quantum parameters: the energy E and the mode number m, namely

$$-\nabla^2 \phi^{\pm}_{m,E}(\mathbf{r}) + [V(x) + U(y) - E] \phi^{\pm}_{m,E}(\mathbf{r}) = 0 \tag{9}$$

where

$$\phi^{\pm}_{m,E}(\mathbf{r}) = \sqrt{\frac{2}{w}} \sin(m\pi y/w) \chi^{\pm}_{E,m}(x) \tag{10}$$

and $\chi^{\pm}_{E,m}(x)$ are the homogeneous solutions of the 1D equation

$$-\frac{\partial^2}{\partial x^2} \chi^{\pm}_{E,m}(x) + \left[V(x) + (m\pi/w)^2 - E\right] \chi^{\pm}_{E,m}(x) = 0, \tag{11}$$

where the superscript "+" and "−" stand for propagation to the right and to the left respectively.

Similarly, it is convenient to formulate the 2D Green function in terms of the 1D one $[G^{+}_{1D}(x, x'; E)]$:

$$G^+(\mathbf{r}, \mathbf{r}') = \frac{2}{w} \sum_{n=1}^{\infty} \sin(n\pi y/w) \sin(n\pi y'/w) G^{+}_{1D}(x, x'; E_n) \tag{12}$$

where $E_m \equiv E - (m\pi/w)^2$ and $G^+_{1D}(x, x'; E)$ solves the equation

$$-\frac{\partial^2}{\partial x^2} G^+_{1D}(x, x'; E_m) + [V(x) + E_m] G^+_{1D}(x, x'; E_m) = -\delta(x - x') \tag{13}$$

with the boundary condition

$$\frac{\partial}{\partial x} G^+_{1D}(x, x'; E_m) \mp i\sqrt{E_m} G^+_{1D}(x, x'; E_m) = 0 \, \text{for} \, x \to \pm\infty. \tag{14}$$

Therefore,

$$G^+_{1D}(x, x'; E_m) = \begin{cases} \dfrac{\chi^{\pm}_{E,m}(x)/\chi^{\pm}_{E,m}(x_0)}{\chi^{\pm}_{E,m}{}'(x)/\chi^{\pm}_{E,m}(x_0) - \chi^{\pm}_{E,m}{}'(-x)/\chi^{\pm}_{E,m}(-x_0)} & x > x_0 \\ \dfrac{\chi^{\pm}_{E,m}(-x)/\chi^{\pm}_{E,m}(-x_0)}{\chi^{\pm}_{E,m}{}'(x)/\chi^{\pm}_{E,m}(x_0) - \chi^{\pm}_{E,m}{}'(-x)/\chi^{\pm}_{E,m}(-x_0)} & x < x_0 \end{cases} \tag{15}$$

where the tags stand for spatial derivatives.

In the case of a rectangular barrier (in a slightly different writing, see Ref. [23])

$$\chi^+_{k,n}(x) = \begin{cases} \exp(ik_n x) + t_n R_n \exp(-ik_n x) & x < -L \\ t_n C_n \exp(-K_n x) + t_n D_n \exp(K_n x) & |x| < L \\ t_n \exp(ik_n x) & x > L \end{cases} \tag{16}$$

where

$k_m \equiv \sqrt{E_m} = \sqrt{E - (m\pi/w)^2}$ and $K_m \equiv \sqrt{V - E_m} = \sqrt{V - E + (m\pi/w)^2}$;

$$\begin{aligned} t_n &= \frac{\exp(-2ik_n L)}{\cosh(2K_n L) + i(K_n/k_n - k_n/K_n)\sinh(2K_n L)/2} \\ &\cong 2\frac{\exp(-2ik_n L - 2K_n L)}{1 + i(K_n/k_n - k_n/K_n)/2}, \end{aligned} \tag{17}$$

$$C_n = \frac{1}{2}\left(1 - \frac{ik_n}{K_n}\right)\exp(K_n L + ik_n L), \tag{18}$$

$$D_n = \frac{1}{2}\left(1 + \frac{ik_n}{K_n}\right)\exp(-K_n L + ik_n L) \tag{19}$$

and

$$R_n = -\frac{i}{2}\left(\frac{K_n}{k_n} + \frac{k_n}{K_n}\right)\sinh(2K_n L). \tag{20}$$

The general Green function is then

$$G_{1D}^{+}(x, x_0; E_m) \cong \frac{1}{M_n} \begin{cases} \dfrac{\exp(ik_n x)}{C_n \exp(-K_n x_0) + D_n \exp(K_n x_0)} & L < x \\ \dfrac{C_n \exp(-K_n x) + D_n \exp(K_n x)}{C_n \exp(-K_n x_0) + D_n \exp(K_n x_0)} & x_0 < x < L \\ \dfrac{C_n \exp(K_n x) + D_n \exp(-K_n x)}{C_n \exp(K_n x_0) + D_n \exp(-K_n x_0)} & -L < x < x_0 \\ \dfrac{\exp(-ik_n x)}{C_n \exp(K_n x_0) + D_n \exp(-K_n x_0)} & x < -L \end{cases} \tag{21}$$

where

$$M_n = -K_n[\tanh[K_n(L - x_0) + i\theta(k_n)] + \tanh[K_n(L + x_0) + i\theta(k_n)]]$$

and then

$$G_{1D}^{+}(x, x_0; E_m) = \frac{\frac{\tanh[K_m(L-x)+i\theta(k_m)]}{\tanh[K_m(L-x_0)+i\theta(k_m)]}}{-K_m[\tanh[K_m(L - x_0) + i\theta(k_m)] + \tanh[K_m(L + x_0) + i\theta(k_m)]]} \text{ for } |x| < L \tag{22}$$

using

$$\tan\theta(k) = -k/K. \tag{23}$$

Then

$$\begin{aligned} G_{1D}^{+}(x_0, x_0; E_m) &= \frac{1}{-K_m[\tanh[K_m(L - x_0) + i\theta(k_m)] + \tanh[K_m(L + x_0) + i\theta(k_m)]]} \\ &\cong \frac{1}{-2K_m\left[1 - 2\exp[-2K_m L - 2i\theta(k_m)]\cosh[2K_m x_0]\right]} \end{aligned} \tag{24}$$

where the last term is an approximation in the limit of opaque barriers.

When the incoming wavefunction is the mth mode

$$\Psi_{inc}(\mathbf{r}) = \sin\left(\frac{m\pi y}{w}\right)\chi_{E,m}^{+}(x) \tag{25}$$

then, the solution (in all space) reads

$$\begin{aligned} \Psi(\mathbf{r}) = {} & \sin\left(\frac{m\pi y}{w}\right)\chi_{E,m}^{+}(x) \\ & + \frac{\sin\left(\frac{m\pi y_0}{w}\right)\chi_{E,m}^{+}(x_0)\frac{2}{w}\sum_{n=1}^{\infty}\sin\left(\frac{n\pi y}{w}\right)\sin\left(\frac{n\pi y_0}{w}\right)G_{1D}^{+}(x, x_0; E_n)}{\frac{1}{2\pi}\ln\left(\frac{\rho_0}{\rho}\right) + \frac{2}{w}\sum_{n=1}^{\infty}\sin^2\left(\frac{n\pi y_0}{w}\right)G_{1D}^{+}(x_0, x_0; E_n)\exp\left(-\left(\frac{n\pi\rho}{2w}\right)^2\right)} \end{aligned} \tag{26}$$

which can be written as

$$\Psi(x > L, y) = \sum_p \sin\left(\frac{p\pi y}{w}\right)\chi^+_{E,p}(x)$$
$$\times\left\{\delta(p-m) - \frac{\sin\left(\frac{m\pi y_0}{w}\right)\sin\left(\frac{p\pi y_0}{w}\right)\frac{\chi^+_{E,m}(x_0)}{\chi^+_{E,p}(x_0)}\frac{2}{w}G^+_{1D}(x_0, x_0; E_p)}{\frac{1}{2\pi}\ln\left(\frac{\rho_0}{\rho}\right) + \frac{2}{w}\sum_{n=1}^{\infty}\sin^2\left(\frac{n\pi y_0}{w}\right)G^+_{1D}(x_0, x_0; E_n)\exp\left(-\left(\frac{n\pi\rho}{2w}\right)^2\right)}\right\}. \tag{27}$$

In the case where the incoming particle's energy satisfies

$$(\pi/w)^2 < E < (2\pi/w)^2$$

then only a single mode propagates, in which case

$$\Psi(x \to \infty, y) = \sin\left(\frac{\pi y}{w}\right)\chi^+_{E,1}(x)t_{11} \tag{28}$$

where t_{11} is the transmission coefficient to remain at $x \to \infty$ in the first mode, which is

$$t_{11} \equiv 1 - \frac{\sin^2\left(\frac{\pi y_0}{w}\right)\frac{2}{w}G^+_{1D}(x_0, x_0; E_1)}{\frac{1}{2\pi}\ln\left(\frac{\rho_0}{\rho}\right) + \frac{2}{w}\sum_{n=1}^{\infty}\sin^2\left(\frac{n\pi y_0}{w}\right)G^+_{1D}(x_0, x_0; E_n)\exp\left(-\left(\frac{n\pi\rho}{2w}\right)^2\right)} \tag{29}$$

A plot of $T_{11} = |t_{11}|^2$ as a function of the incoming particle's energy is presented in **Figure 2**.

Clearly, a resonance occurs when the real part of the denominator of Eq. (29) vanishes, i.e. when

$$\frac{1}{2\pi}\ln\left(\frac{\rho_0}{\rho}\right) + \frac{2}{w}\sum_{n=1}^{\infty}\sin^2\left(\frac{n\pi y_0}{w}\right)\Re G^+_{1D}(x_0, x_0; E_n)\exp\left(-\left(\frac{n\pi\rho}{2w}\right)^2\right) = 0. \tag{30}$$

In general, it is a complex transcendental equation; however, in case of an opaque barrier, Eq. (24) can be further simplified to

$$G_{1D}(x_0, x_0; E_n) \cong -\frac{1}{2K_n} + i\varepsilon_n \tag{31}$$

when

$$\varepsilon_n \equiv \frac{\exp[-2K_nL]\sin[2\theta(k_n)]\cosh(2K_nx_0)}{K_n} = -2\frac{\exp[-2K_nL]k_n\cosh(2K_nx_0)}{V}, \tag{32}$$

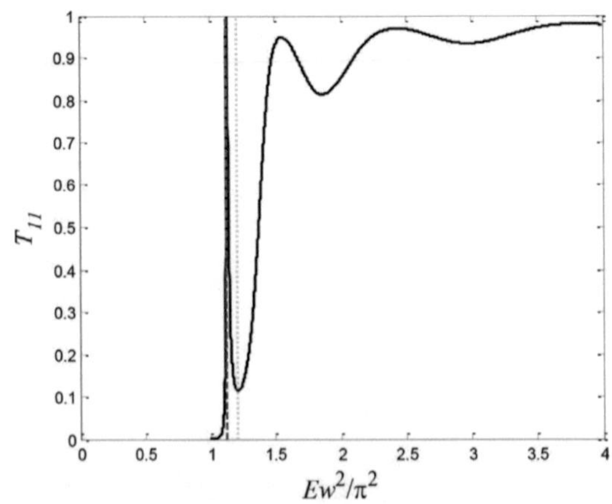

Figure 2. Plot of the $T_{11} = |t_{11}|^2$, i.e., the probability to remain in the first mode of propagation as a function of the normalized energy. The barrier parameters were $L = 2w$ and $V = 2/w^2$, and the defect parameters were $\rho_0 = 300w$, $x_0 = 0$, and $y_0 = w/2$. The dotted line represents the barrier's energy $E_b = V + \pi^2/w^2$, and the dashed line represents the resonance energy E_{res}.

$$\text{and } \sin\left[2\theta(k_1)\right] = \frac{-2kK}{K^2+k^2} = \frac{-2kK}{V}. \tag{33}$$

Then, Eq. (30) can be approximated as

$$\frac{1}{2\pi}\ln\left(\frac{\rho_0}{\rho}\right) - \frac{2}{w}\sum_{n=1}^{\infty}\sin^2\left(\frac{n\pi y_0}{w}\right)\frac{1}{2K_n}\exp\left(-\left(\frac{n\pi\rho}{2w}\right)^2\right) = 0 \tag{34}$$

In the case where the conducting wires is very narrow or the barrier is very high, i.e.,

$$(\pi/w)^2 + V >> E_0 \tag{35}$$

then

$$\frac{1}{2\pi}\ln\left(\frac{\rho_0}{\rho}\right) - \frac{1}{w}\sin^2\left(\frac{\pi y_0}{w}\right)\frac{1}{\sqrt{V-E+(\pi/w)^2}} - \frac{1}{\pi}\sum_{n=2}^{\infty}\sin^2\left(\frac{n\pi y_0}{w}\right)\frac{1}{\sqrt{n^2-1}}\exp\left(-\left(\frac{n\pi\rho}{2w}\right)^2\right) = 0 \tag{36}$$

since

$$\sum_{n=2}^{\infty}\sin^2\left(\frac{n\pi y_0}{w}\right)\frac{1}{\pi\sqrt{n^2-1}}\exp\left(-\left(\frac{n\pi\rho}{2w}\right)^2\right) \cong -\ln(4\rho/w)/2\pi \tag{37}$$

then

$$\frac{1}{2\pi}\ln\left(\frac{4\rho_0}{w}\right) = \frac{1}{w}\sin^2\left(\frac{\pi y_0}{w}\right)\frac{1}{\sqrt{V-E+(\pi/w)^2}} \tag{38}$$

which has a solution provided $4\rho_0 > w$, otherwise the impurity can be regarded as a perturbation and does not carry a resonant level.

When the resonant level exits, then the resonance energy E_R is approximately

$$E_R \cong V + \frac{\pi^2}{w^2}\left\{1 - \sin^4\left(\frac{\pi y_0}{w}\right)\left[\frac{1}{2}\ln\left(\frac{4\rho_0}{w}\right)\right]^{-2}\right\} \tag{39}$$

In **Figure 2**, the resonance energy is presented by a dashed line.

In this approximation,

$$\Psi(\mathbf{r}) \cong \sin\left(\frac{\pi y}{w}\right)\chi_{E,1}^{+}(x)\frac{1}{K_1\frac{w}{2\pi}\ln\left(\frac{4\rho_0}{w}\right)/\sin^2\left(\frac{\pi y_0}{w}\right) - (1 + i2K_1\varepsilon_n)} \tag{40}$$

Since in this regime only, one transverse mode is propagating, the system in practice reduces to a 1D problem, where the 2D impurity can be replaced by a 1D delta function potential

$$V(x) = -\lambda\delta(x) \tag{41}$$

where

$$\lambda = \frac{4\pi}{w\ln(4\rho_0/w)}\sin^2\left(\frac{\pi y_0}{w}\right). \tag{42}$$

Therefore, in the 1D analogy the point potential depends not only on the impurity's de-Broglie wavelength in free space, but on its location (y_0) and the wire's width as well.

In this case, the barrier's transmission can be as high as 1. It depends on the location of the point defect in the horizontal dimension, namely, at the resonance energy

$$\Psi(\mathbf{r}) = i\sin\left(\frac{\pi y}{w}\right)\exp(ik_1x)\frac{\exp(-2ik_1L + i\Xi)}{\cosh(2K_1x_0)} \tag{43}$$

where

$$\tan\Xi = -(K/k - k/K)/2. \tag{44}$$

However, there is a point where a minimum occurs. When the incoming particle's energy satisfies

$$\frac{1}{2\pi}\ln\left(\frac{\rho_0}{\rho}\right) + \frac{2}{w}\sum_{n=2}^{\infty}\sin^2\left(\frac{n\pi y_0}{w}\right)\Re G_{1D}^{+}(x_0, x_0; E_n)\exp\left(-\left(\frac{n\pi\rho}{2w}\right)^2\right) = 0 \tag{45}$$

which at the vicinity of the second mode threshold can be approximated by

$$\frac{1}{2\pi}\ln\left(\frac{\rho_0}{\rho}\right)-\frac{1}{w}\sin^2\left(\frac{2\pi y_0}{w}\right)\frac{1}{\sqrt{V-E+(2\pi/w)^2}}-\frac{1}{\pi}\sum_{n=3}^{\infty}\sin^2\left(\frac{n\pi y_0}{w}\right)\frac{1}{\sqrt{n^2-4}}\exp\left(-\left(\frac{n\pi\rho}{2w}\right)^2\right)=0 \quad (46)$$

or

$$\frac{1}{2\pi}\ln\left(\frac{3.8\rho_0}{w}\right)=\frac{1}{w}\sin^2\left(\frac{2\pi y_0}{w}\right)\frac{1}{\sqrt{V-E+(2\pi/w)^2}}. \quad (47)$$

Again, we see that this equation does not always have a solution. It is required that $3.8\rho_0 > w$, in which case

$$E_{\min}=V+\left(\frac{2\pi}{w}\right)^2\left[1-\sin^4\left(\frac{2\pi y_0}{w}\right)\left[\ln\left(\frac{3.8\rho_0}{w}\right)\right]^{-2}\right] \quad (48)$$

This minimum is presented in **Figure 3** by a dotted line.

In which case, the denominator of Eq. (29) is exactly $\sin^2\left(\frac{\pi y_0}{w}\right)\frac{2}{w}G^+_{1D}(x_0,x_0;E_1)$, and therefore at this point, the transmission is exponentially small, and not zero as in the zero potential case, i.e.,

$$\begin{aligned}\Psi_{\min}(\mathbf{r}) &= \sin\left(\frac{\pi y}{w}\right)\chi^+_{E,1}(x)i\frac{\sin^2(2\pi y_0/w)}{\sin^2(\pi y_0/w)}\frac{\Im G^+_{1D}(x_0,x_0;E_2)}{\Re G^+_{1D}(x_0,x_0;E_1)}\\ &= -\sin\left(\frac{\pi y}{w}\right)\chi^+_{E,1}(x)i\frac{\sin^2(2\pi y_0/w)}{\sin^2(\pi y_0/w)}2K_1\varepsilon_2\end{aligned} \quad (49)$$

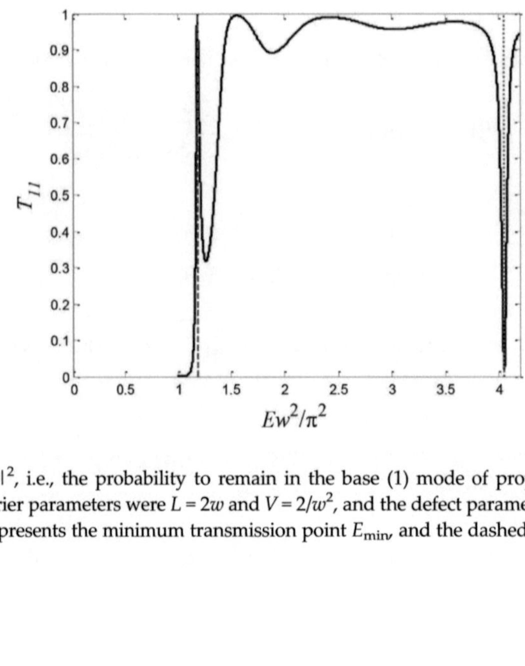

Figure 3. Plot of $T_{11} = |t_{11}|^2$, i.e., the probability to remain in the base (1) mode of propagation as a function of the normalized energy. The barrier parameters were $L = 2w$ and $V = 2/w^2$, and the defect parameters were $\rho_0 = 30w$, $x_0 = 0$, and $y_0 = 0.2w$. The dotted line represents the minimum transmission point E_{min}, and the dashed line represents the resonance energy E_{res}.

which is an exponentially small value. This result agrees with Ref. [24].

It should be stressed, however, that this is a pure 2D phenomenon, which is a consequence of the interaction between the point defect and the wire, and therefore, this minimum disappears in the 1D approximation.

In **Figures 4–9**, a 2D probability density plots (of $|\psi(x, y)|^2$) for various energies are presented.

In **Figure 4**, the energy is too low for the particles to penetrate the barrier, and therefore, almost all of them are reflected from the barrier.

In **Figure 5**, the particle's energy is close to the resonance energy, and therefore, a quasi-bound state is generated at the vicinity of the defect, and the transmission probability is high.

Figures 6 and **7** are examples for local minimum and local maximum respectively.

In **Figure 8**, the particle's energy is close to the minimum (Eq. (48)), which was generated by the interplay between the waveguide and the point defect.

Another important working point is when $K_p = 0$, i.e., $V - E + (p\pi/w)^2 = 0$ and $k_p = \sqrt{V}$.

At this energy, a universal behavior appears. The scattered wavefunction reads

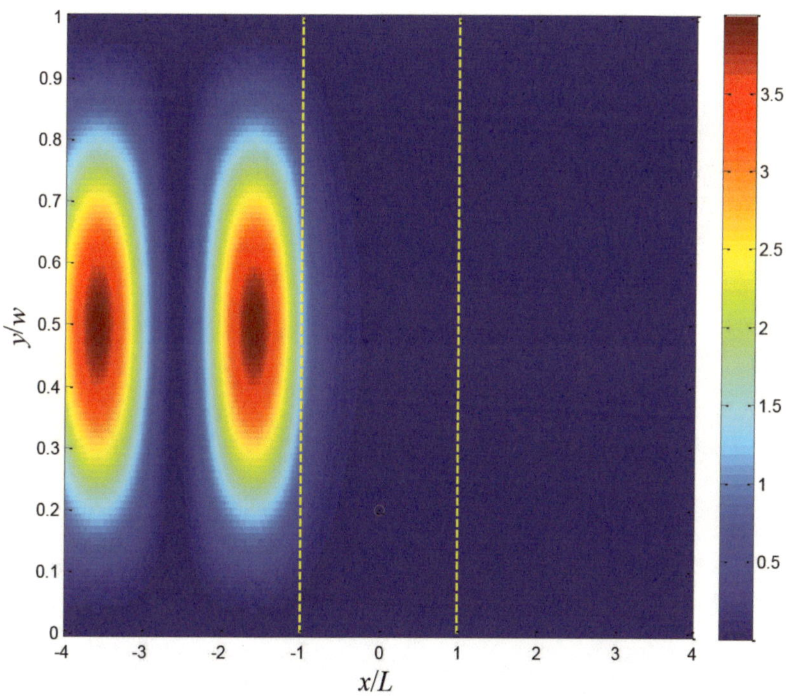

Figure 4. A false colors presentation of the probability density $|\Psi(x, y)|^2$ when the incoming particle's energy is lower than the barrier's height: $E = 10.5w^{-2} < (\pi/w)^2 + V \cong 11.87w^{-2}$. The parameters are same as in **Figure 3**. The dashed lines represent the barrier's boundaries, and the cross at the center of the circle represents the impurity's location.

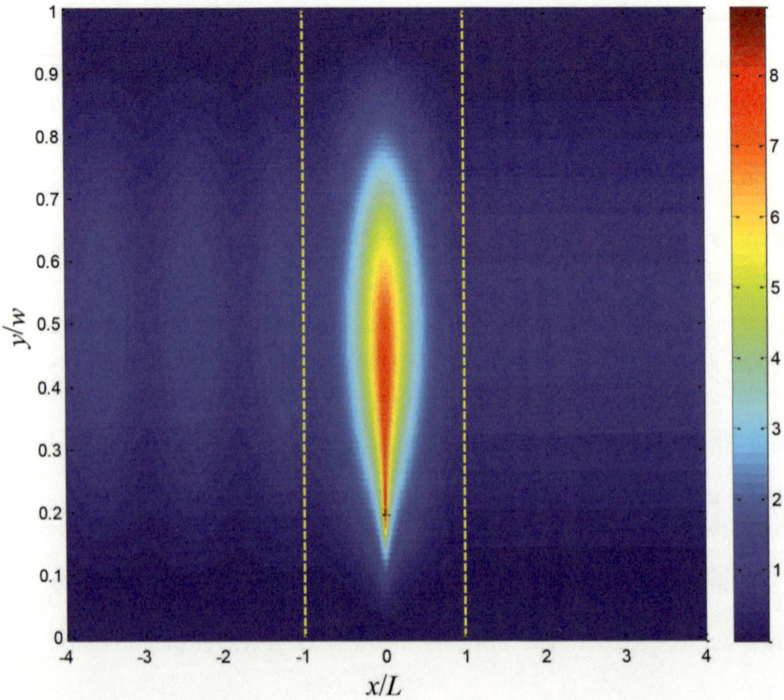

Figure 5. Same as **Figure 4** but when the income particle's energy is close to the resonance energy, i.e. $E = 11.69w^{-2} \cong E_{res}$.

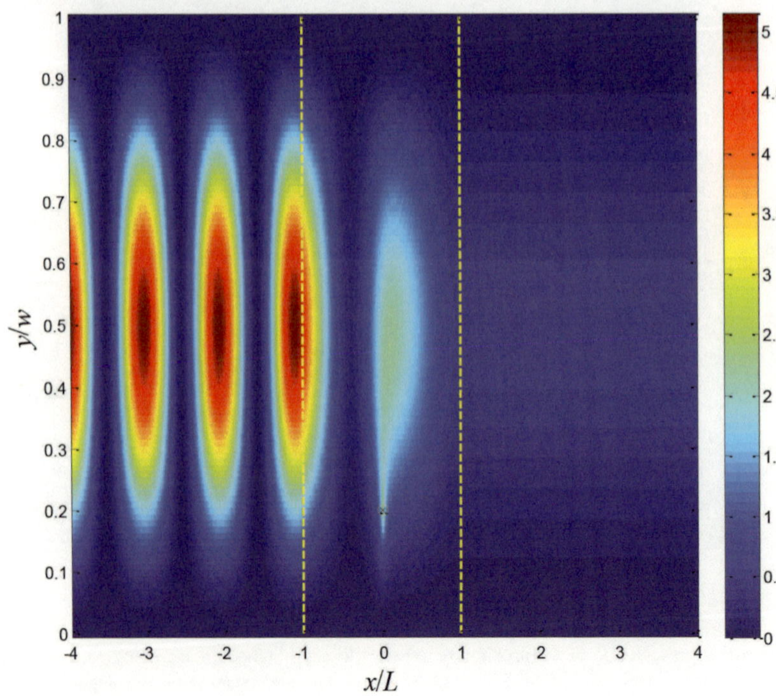

Figure 6. Same as **Figure 4** but when the income particle's energy is close to a local minimum at $E = 12.45w^{-2}$.

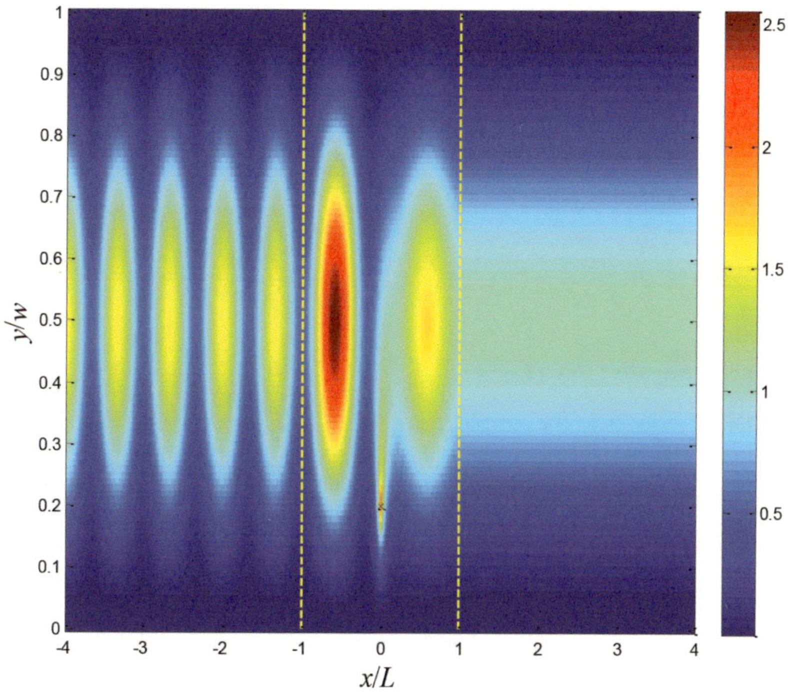

Figure 7. Same as **Figure 4** but when the income particle's energy is close to a local maximum at $E = 15.4w^{-2}$.

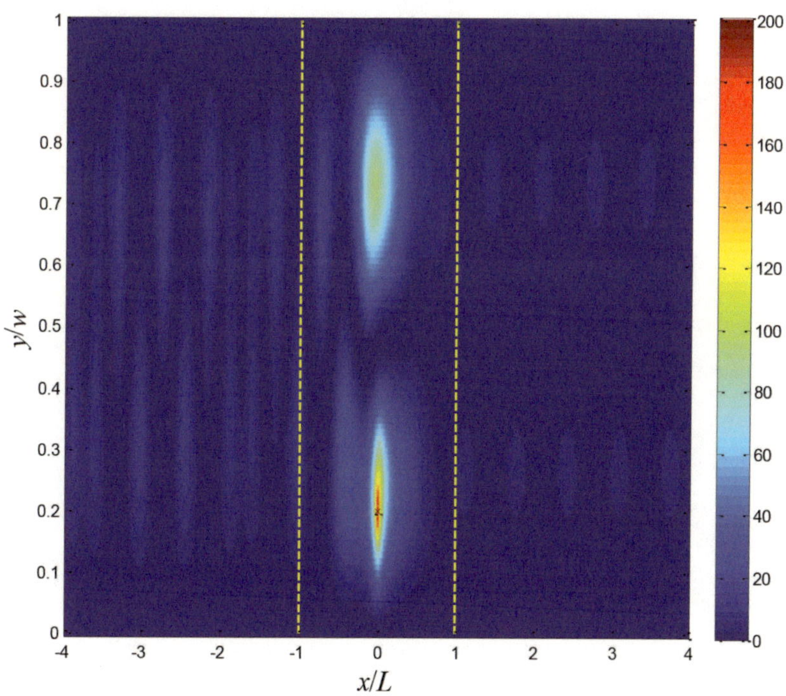

Figure 8. Same as **Figure 4** but when the income particle's energy is close to a local minima at $E = 40.02w^{-2}$.

$$\Psi(\mathbf{r}) = \sin\left(\frac{m\pi y}{w}\right)\chi_{E,m}^{+}(x) + \frac{\sin\left(\frac{m\pi y_0}{w}\right)\chi_{E,m}^{+}(x_0)\sin\left(\frac{p\pi y}{w}\right)G_{1D}^{+}(x,x_0;E_p)}{\sin\left(\frac{p\pi y_0}{w}\right)G_{1D}^{+}(x_0,x_0;E_p)}. \qquad (50)$$

This expression is universal in the sense that it is independent of the point defect potential. It depends only on its location. In case this is a surface defect, i.e., $y_0/w << 1$ then even the dependence on the vertical location vanishes

$$\Psi(\mathbf{r}) = \sin\left(\frac{m\pi y}{w}\right)\chi_{E,m}^{+}(x) + \frac{m}{p}\chi_{E,m}^{+}(x_0)\sin\left(\frac{p\pi y}{w}\right)\frac{G_{1D}^{+}(x,x_0;E_p)}{G_{1D}^{+}(x_0,x_0;E_p)}. \qquad (51)$$

This universality agree with Ref. [25].

For $|x| < L$ Eq. (51) reduces to the simple form

$$\Psi(|x| < L, y) = \sin\left(\frac{m\pi y}{w}\right)\chi_{E,m}^{+}(x) + \frac{\sin\left(\frac{m\pi y_0}{w}\right)}{\sin\left(\frac{p\pi y_0}{w}\right)}\chi_{E,m}^{+}(x_0)\sin\left(\frac{p\pi y}{w}\right) \qquad (52)$$

and in the case of a surface defect, it reduces to even a simpler expression

$$\Psi(|x| < L, y) = \sin\left(\frac{m\pi y}{w}\right)\chi_{E,m}^{+}(x) + \frac{m}{p}\chi_{E,m}^{+}(x_0)\sin\left(\frac{p\pi y}{w}\right) \qquad (53)$$

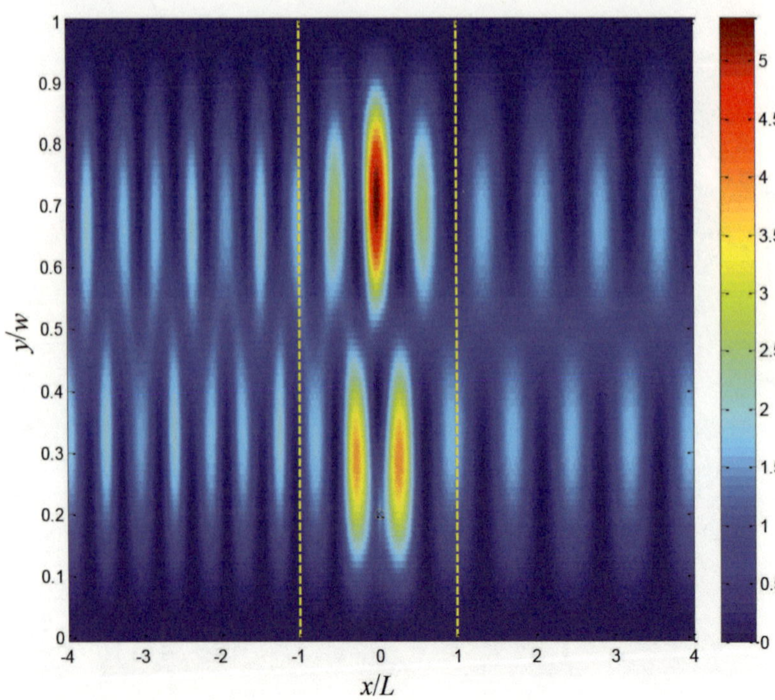

Figure 9. At the transition level $E = V + (\pi/w)^2$, a universal pattern appears.

The fact that the second part is independent of x is also in agreement with Ref. [25].

But unlike Ref. [25], due the barrier, the second mode does propagate, but the expression is still generic (in the sense that it is independent of the impurity's parameter), beyond the barrier it reads

$$\Psi(|x|>L,y)=\sin\left(\frac{m\pi y}{w}\right)\chi^{+}_{E,m}(x)+\frac{m}{p}\chi^{+}_{E,m}(x_0)\sin\left(\frac{p\pi y}{w}\right)\exp\left(i\sqrt{V}(|x|-L)\right). \quad (54)$$

This special universal case is illustrated in **Figure 9**, and it is a manifestation of the effect of Ref. [25], where the footprints of the defect are clearly seen but without any fingerprints. That is, the defect is clearly there, but the scattering is independent of its strength (its eigenenergy).

4. Physical realization and implementation

Let us apply this model to a 1.5-nm wide silicon wire, which is contaminated by a single phosphorous atom. In this case w = 1.5 nm, the phosphorous atom radius is a = 0.098 nm, the effective electron mass in silicon is $m_e \cong 0.2m$. Then the wire transmission (proportional to the device's conductivity in units of e^2/h) as a function of the potential at the atom's center V_0, Eq. (6) (which is proportional to the transistor gate voltage) is plotted in **Figure 10** for two scenarios. In the first scenario, the electron's energy, i.e., the Fermi energy, is E = 0.9 eV and in the second, it is equal to E = 3 eV. In the former scenario, the device works at the vicinity of the quantum dot's resonance, and in the latter, it works at the vicinity of the fano-(anti) resonance.

In both scenarios, a change of about a volt in the gate voltage can change drastically the wire's current. Therefore, it can be implemented as a simplified but rich model for a single atom transistor.

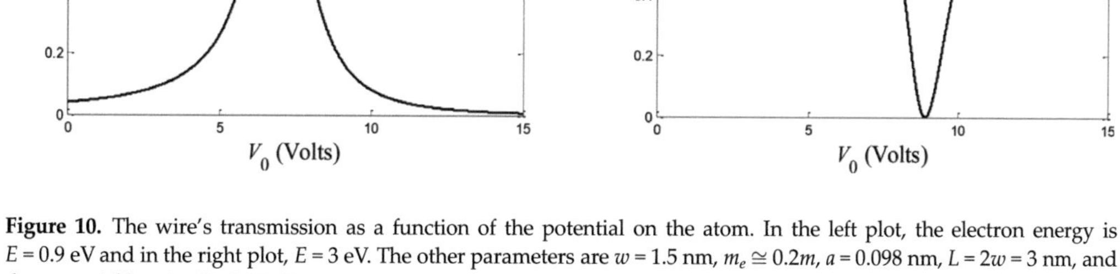

Figure 10. The wire's transmission as a function of the potential on the atom. In the left plot, the electron energy is E = 0.9 eV and in the right plot, E = 3 eV. The other parameters are w = 1.5 nm, $m_e \cong 0.2m$, a = 0.098 nm, $L = 2w$ = 3 nm, and the potential barrier V = 0.15 eV.

Author details

Er'el Granot

Address all correspondence to: erel@ariel.ac.il

Department of Electrical and Electronics Engineering, Ariel University, Ariel, Israel

References

[1] Moore G. Chapter 7: Moore's law at 40. In: Brock D, editor. Understanding Moore's Law: Four Decades of Innovation. Philadelphia, PA: Chemical Heritage Foundation; 2006. pp. 67-84

[2] Takahashi D. Forty Years of Moore's Law. San Jose, CA: Seattle Times; April 18, 2005

[3] Eigler DM, Schweizer EK. Positioning single atoms with a scanning tunnelling microscope. Nature. 1990;**344**:524-526

[4] Koenraad PM, Flatté ME. Single dopants in semiconductors. Nature Materials. 2011;**10**: 91-100

[5] Lansbergen GP, et al. Gate-induced quantum-confinement transition of a single dopant atom in a silicon FinFET. Nature Physics. 2008;**4**:656-661

[6] Calvet LE, Snyder JP, Wernsdorfer W. Excited-state spectroscopy of single Pt atoms in Si. Physical Review B. 2008;**78**:195309

[7] Tan KY, et al. Transport spectroscopy of single phosphorus donors in a silicon nanoscale transistor. Nano Letters. 2010;**10**:11-15

[8] Hollenberg LCL, et al. Charge-based quantum computing using single donors in semiconductors. Physical Review B. 2004;**69**:113301

[9] Schofield SR, et al. Atomically precise placement of single dopants in Si. Physical Review Letters. 2003;**91**:136104

[10] Fuechsle M, Miwa JA, Mahapatra S, Ryu H, Lee S, Warschkow O, Hollenberg LCL, Klimeck G, Simmons MY. A single-atom transistor. Nature Nanotechnology. 2012;**7**:242-246

[11] Xie F-Q, Maul R, Wenzel W, Schn G, Obermair Ch, Schimmel Th. Single-atom transistors: Atomic-scale electronic devices in experiment and simulation. In: International Beilstein Symposium on Functional Nanoscience; Frankfurt am Main, May 2010. pp. 213-228

[12] Fuechsle M, Miwa JA, Mahapatra S, Warschkow O, Hollenberg LCL, Simmons MY. Realisation of a single-atom transistor in silicon. Journal and Proceedings of the Royal Society of New South Wales. 2012;**145**(443 & 444):66-74

[13] Obermair Ch, Xie F-Q, Schimmel Th. The single-atom transistor: Perspectives for quantum electronics on the atomic-scale. Europhysics News. 2010;**41**:25-28

[14] Azbel MY. Variable-range-hopping magnetoresistance. Physical Review B. 1991;**43**:2435

[15] Azbel MY. Quantum particle in a random potential: Implications of an exact solution. Physical Review Letters. 1991;**67**:1787

[16] Granot E. Point scatterers and resonances in low number of dimensions. Physica E. 2006;**31**:13-16

[17] Granot E, Azbel MY. Resonant angular dependence in a weak magnetic field. Journal of Physics: Condensed Matter. 1999;**11**:4031

[18] Granot E, Azbel MY. Resonant tunneling in two dimensions via an impurity. Physical Review B. 1994;**50**:8868

[19] Granot E. Near-threshold-energy conductance of a thin wire. Physical Review B. 1999;**60**:10664

[20] Granot E. Symmetry breaking and current patterns due to a weak imperfection. Physical Review B. 2000;**61**:11078

[21] Weber B, Mahapatra S, Ryu H, Lee S, Fuhrer A, Reusch TCG, Thompson DL, Lee WCT, Klimeck G, Hollenberg LCL, Simmons MY. Ohm's law survives to the atomic scale. Science. 2012;**335**:64

[22] Abramowitz M, Stegun IA. Handbook of Mathematical Functions. New York: Dover Publications; 1972

[23] Merzbacher E. Quantum Mechanics. Hoboken, NJ: Wiley; 1970

[24] Granot E. Universal conductance reduction in a quantum wire. Europhysics Letters. 2004;**68**:860-866

[25] Granot E. Transmission coefficient for a point scatterer at specific energies is affected by the presence of the scatterer but independent of the scatterer's characteristics. Physical Review B. 2005;**71**:035407

Chapter 2

A Quantum Trajectory Interpretation of Magnetic Resistance in Quantum Dots

Ciann-Dong Yang and Shih-Ming Huang

Additional information is available at the end of the chapter

http://dx.doi.org/10.5772/intechopen.74409

Abstract

For a complete description of the electronic motion in a quantum dot, we need a method that can describe not only the trajectory behavior of the electron but also its probabilistic wave behavior. Quantum Hamilton mechanics, which possesses the desired ability of manifesting the wave-particle duality of electrons moving in a quantum dot, is introduced in this chapter to recover the quantum-mechanical meanings of the classical terms such as backscattering and commensurability and to give a quantum-mechanical interpretation of the observed oscillation in the magneto-resistance curve. Solutions of quantum Hamilton equations reveal the existence of electronic standing waves in a quantum dot, whose occurrence is found to be accompanied by a jump in the electronic resistance. The comparison with the experimental data shows that the predicted locations of the resistance jump match closely with the peaks of the measured magneto-resistance.

Keywords: quantum dots, quantum Hamilton mechanics, standing waves, quantum trajectory, magneto-resistance

1. Introduction

As the size of electronic devices is narrowed down to the nanoscale, quantum effects become so prominent that classical mechanics is no longer able to provide an accurate description for electrons moving in nanostructures. However, due to the lack of the sense of trajectory in quantum mechanics, classical or semi-classical mechanics so far has been the sole tool in determining ballistic orbits in quantum dots. Classical orbits satisfying commensurability conditions of geometrical resonances were derived in the literature to determine the magneto-transport behavior of periodic quantum systems. It was reported that the observed regular peaks in the magneto-resistance corresponded to backscattering of commensurate orbits [1], and the critical magnetic fields determined from the backscattering orbits showed an excellent agreement with the observed peak positions in the magneto-resistance curves [2]. A recent study showed that the ballistic motion of electrons within quantum dots can be controlled by an externally applied magnetic field so that the resulting conductance images resemble the classical transmitted and backscattered trajectories [3].

The use of an anisotropic harmonic function, instead of an abrupt hard potential, to describe the confining potential in a quantum dot was shown to be helpful to improve the accuracy of predicting magneto-resistance peaks based on backscattering orbits [4]. Nowadays, the confinement potential forming an electron billiard can be practically patterned to almost arbitrary profile, through which ballistic orbits with chaotic dynamics can be generated to characterize magneto transport [5]. However, the chaotic behavior and its change with magnetic field could not be described in the usual quantum-mechanical picture due to the lack of a trajectory interpretation. Regarding this aspect, the classical description becomes a valued tool for detailed understanding of the transition from low to high magnetic fields in quantum dot arrays [6]. On the other hand, quantum mechanical model for electron billiards was known as quantum billiards [7], in which moving point particles are replaced by waves. Quantum billiards are most convenient for illustrating the phenomenon of Fano interference [8] and its interplay with Aharonov-Bohm interference [9], which otherwise cannot be described by classical methods.

From the existing researches, we have an observation that the ballistic motion in electron billiards was solely described by classical mechanics, while the wave motion in quantum billiards could only be described by quantum mechanics. The aim of this chapter is to give a unified treatment of electron billiards and quantum billiards. We point out that quantum Hamilton mechanics [10, 11] can describe both ballistic motion and wave motion of electrons in a quantum dot to provide us with a quantum commensurability condition to determine backscattering orbits as well as with the wave behavior to characterize the magneto-resistance in a quantum dot.

Quantum Hamilton mechanics is a dynamical realization of quantum mechanics in the complex space [12], under which each quantum operator is realized as a complex function and each wavefunction is represented by a set of complex-valued Hamilton equations of motion. With quantum Hamilton mechanics, we can recover the quantum-mechanical meanings of the classical commensurability condition by showing that there are integral numbers of oscillation in the radial direction, as an electron undergoes a complete angular oscillation around a quantum dot. When the radial and angular dynamics are commensurable, the shape of electronic quantum orbits is found to be stationary like a standing wave. Furthermore, the wave number N, distributed on the circumference of the quantum dot, can be controlled by the applied magnetic field. It will be shown that the classical backscattering orbits discovered in the literature resemble the shape of the quantum standing waves derived here with their wave numbers equal to the numbers of electron's bounces within the quantum dot.

The electronic standing-wave motions considered in this chapter will reveal that a jump of the magneto-resistance in quantum dots is accompanied by a phenomenon of magnetic stagnation, which is a quantum effect that an electron is stagnated or trapped within a quantum dot by an applied magnetic field in such a way that the electron's cyclotron angular velocity is exactly counterbalanced by its quantum angular velocity. We point out that magnetic stagnation is a degenerate case from the electronic standing-wave motion as the wave number N approaches to infinity. The magnetic fields yielding the phenomenon of magnetic stagnation can be determined by the quantum Hamilton equations derived here. Knowing these critical magnetic fields allows us to control the magneto-resistance precisely.

In the following sections, we first introduce quantum Hamilton mechanics and apply it to derive Hamilton equations, which are then used in Section 2 to describe the electronic quantum motions in a quantum dot. By solving the Hamilton equations of motion, Section 3 demonstrates electronic standing-wave motions in various quantum states and characterizes the magnetic field leading to the phenomenon of magnetic stagnation. In Section 4, we show that the magnetic stagnation is the main cause to the resistance oscillation of quantum dots in low magnetic field by comparing the theoretical predictions obtained from Section 3 with the experimental results of the magneto-resistance curve [4, 13].

2. Quantum Hamilton dynamics in a 2D quantum dot

To probe the quantum to classical transition, which involves both classical and quantum features, quantum dots are the most natural systems [14]. Analyzing such systems, we need an approach that can provide both classical and quantum descriptions. Quantum Hamilton mechanics is one of the candidates satisfying this requirement. This chapter will apply quantum Hamilton mechanics to an open quantum dot with circular shape, which is connected to reservoirs with strong coupling. The electronic transport through an open quantum dot can be realized by nano-fabrication techniques as a two-dimensional electron gases system (2DES) at an AlGaAs/GaAs heterostructure, as depicted in **Figure 1**.

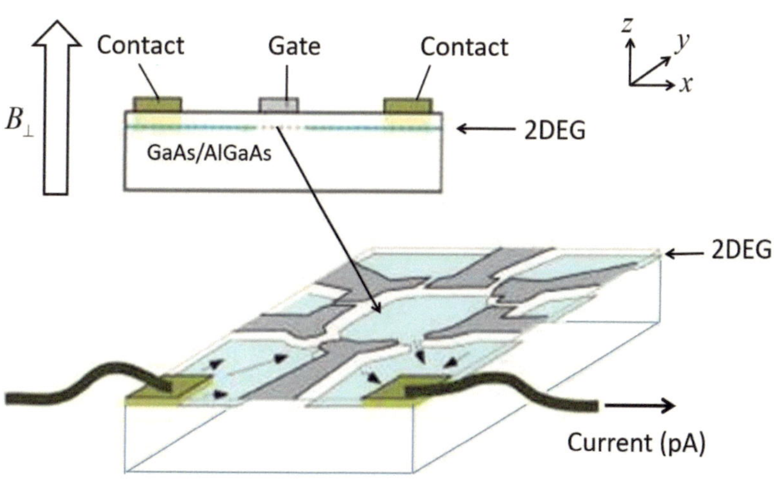

Figure 1. Schematic illustration of a semiconductor heterostructure with a circular dot between the two tunnel barriers.

Under the framework of quantum Hamilton mechanics [10, 12], the equivalent mathematical model of a quantum dot is described as an electron moving in an electromagnetic field with scalar potential V and vector potential $\mathbf{A}$. The related Hamiltonian operator $\widehat{H}$ can be realized as the following complex Hamiltonian function,

$$H(t,\mathbf{q},\mathbf{p}) = \frac{1}{2m}\left(\mathbf{p}+\frac{c}{e}\mathbf{A}\right)\cdot\left(\mathbf{p}+\frac{c}{e}\mathbf{A}\right) + V(t,\mathbf{q}) + \frac{\hbar}{2im}\nabla\cdot\mathbf{p}. \tag{1}$$

We adopt polar coordinates $\mathbf{q}=(r,\theta)$ and momentum $\mathbf{p}=(p_r,p_\theta)$ in the above equation to describe the electronic quantum motion in a 2D circular quantum dot. The resulting complex Hamiltonian Eq. (1) becomes

$$H = \frac{1}{2m^*}\left[\left(p_r+\frac{e}{c}A_r\right)^2 + \frac{1}{r^2}\left(p_\theta+\frac{e}{c}A_\theta\right)^2\right] + V(r,\theta) + \frac{\hbar}{2im^*}\left(\frac{1}{r}p_r + \frac{\partial p_r}{\partial r} + \frac{1}{r^2}\frac{\partial p_\theta}{\partial \theta}\right), \tag{2}$$

where $m^* = 0.067m_e$ is the electron's effective mass in AlGaAs/GaAs heterostructure. The scalar potential $V(r,\theta)$ acts as a confinement potential in the dot, and is modeled by the parabolic function $V = kr^2/2 = m^*\omega_0^2 r^2/2$ to simulate a soft-wall potential. The vector potential $\mathbf{A}$ is determined from the applied magnetic field $\mathbf{B}$ via the relation $\mathbf{B} = \nabla\times\mathbf{A}$. Here, we consider a constant $\mathbf{B}$ along the z direction, which amounts to $A_r = 0$ and $A_\theta = Br^2/2$. Substituting the above assignments of V and $\mathbf{A}$ into the complex Hamiltonian Eq. (2), we obtain

$$H = \frac{1}{2m^*}\left(p_r^2 + \frac{1}{r^2}p_\theta^2\right) + \omega_L p_\theta + \frac{1}{2}m^*\omega^2 r^2 + \frac{\hbar}{2im^*}\left(\frac{1}{r}p_r + \frac{\partial p_r}{\partial r} + \frac{1}{r^2}\frac{\partial p_\theta}{\partial\theta}\right) \tag{3}$$

where $\omega_L = eB/(2m^*c)$ is the Larmor frequency and $\omega = \sqrt{\omega_0^2+\omega_L^2}$ is the composite frequency. The quantum Hamilton-Jacobi equation associated with the Hamiltonian H defined in Eq. (1) reads

$$\frac{\partial S}{\partial t} + H(t,\mathbf{q},\mathbf{p})\bigg|_{p_i=\partial S/\partial q_i} = 0, \tag{4}$$

where S is the quantum action function to be determined. By making use of the substitution

$$p_r = \frac{\partial S}{\partial r},\; p_\theta = \frac{\partial S}{\partial \theta}, \tag{5}$$

the quantum Hamilton-Jacobi Eq. (4) associated with the Hamiltonian in Eq. (3) turns out to be

$$\frac{\partial S}{\partial t} + \frac{1}{2m^*}\left[\left(\frac{\partial S}{\partial r}\right)^2 + \frac{1}{r^2}\left(\frac{\partial S}{\partial\theta}\right)^2\right] + \omega_L\frac{\partial S}{\partial\theta} + \frac{1}{2}m^*\omega^2 r^2 - \frac{i\hbar}{2m^*}\left(\frac{1}{r}\frac{\partial S}{\partial r} + \frac{\partial^2 S}{\partial r^2} + \frac{1}{r^2}\frac{\partial^2 S}{\partial\theta^2}\right) = 0. \tag{6}$$

The recognition of the complex Hamiltonian H in Eqs. (1) and (3) as a complex realization of the Hamiltonian operator $\widehat{H}$ is confirmed by the fact that the quantum Hamilton-Jacobi equation in Eqs. (4) and (6) yields the same Schrodinger equation as constructed from $\widehat{H}$. This can be seen by applying the following transformation

$$S = -i\hbar \ln \Psi, \tag{7}$$

to Eq. (6) to produce the expected Schrodinger equation:

$$i\hbar \frac{\partial \Psi}{\partial t} = -\frac{\hbar^2}{2m^*}\left(\frac{\partial^2 \Psi}{\partial r^2} + \frac{1}{r}\frac{\partial \Psi}{\partial r} + \frac{1}{r^2}\frac{\partial^2 \Psi}{\partial \theta^2}\right) - i\hbar\omega_L \frac{\partial \Psi}{\partial \theta} + \frac{1}{2}m^*\omega^2 r^2 \Psi. \tag{8}$$

Due to the time-independent nature of the applied potentials **A** and V, the wavefunction Ψ in Eq. (8) assumes the following form of solution,

$$\Psi(t, r, \theta) = e^{-iEt/\hbar}\psi(r, \theta), \tag{9}$$

where $\psi(r, \theta)$ satisfies the time-independent Schrodinger equation

$$\hat{H}\psi \triangleq \left[-\frac{\hbar^2}{2m^*}\left(\frac{\partial^2}{\partial r^2} + \frac{1}{r}\frac{\partial}{\partial r} + \frac{1}{r^2}\frac{\partial^2}{\partial \theta^2}\right) - i\hbar\omega_L \frac{\partial}{\partial \theta} + \frac{1}{2}m^*\omega^2 r^2\right]\psi = E\psi. \tag{10}$$

On the other hand, Eq. (3) can be rewritten by using the substitutions Eqs. (5) and (7) as

$$\hat{H}\psi = H\psi, \tag{11}$$

where H and $\hat{H}$ are defined, respectively, by Eqs. (3) and (10). This is a direct proof of the fact that the complex Hamiltonian H is a functional realization of the Hamiltonian operator $\hat{H}$ in a complex space. Indeed, it can be shown [10] that every quantum operator $\hat{A}$ can be realized as a complex function A via the relation $\hat{A}\psi = A\psi$. The combination of Eqs. (10) and (11) reveals the energy conservation law $H = E$, which is a natural result of Hamilton mechanics by noting that the Hamiltonian H in Eq. (3) does not contain time t explicitly and must be a motion constant equal to the system's total energy E.

Upon performing the differentiations $\partial p_r/\partial r$ and $\partial p_\theta/\partial \theta$ involved in Eq. (3), we have to specify in advance the action function S or equivalently the wavefunction ψ via the relation Eq. (7). This requirement makes the complex Hamiltonian H state-dependent. For a given quantum state described by ψ, the complex Hamiltonian H can be expressed explicitly as:

$$H = \frac{1}{2m^*}\left(p_r^2 + \frac{1}{r^2}p_\theta^2\right) + \omega_L p_\theta + \frac{1}{2}m^*\omega^2 r^2 + \frac{\hbar}{2im^*}\left(\frac{1}{r}p_r + \frac{\hbar}{i}\frac{\partial^2 \ln \psi}{\partial r^2} + \frac{1}{r^2}\frac{\hbar}{i}\frac{\partial^2 \ln \psi}{\partial \theta^2}\right). \tag{12}$$

Apart from deriving the Schrodinger equation, the above complex Hamiltonian also gives electronic quantum motions in the state ψ in terms of the Hamilton equations of motion,

$$\frac{dr}{dt} = \frac{\partial H}{\partial p_r} = \frac{1}{m^*}p_r + \frac{\hbar}{2im^*}\frac{1}{r} = \frac{\hbar}{im^*}\frac{\partial \ln \psi}{\partial r} + \frac{\hbar}{2im^*}\frac{1}{r}, \tag{13}$$

$$\frac{d\theta}{dt} = \frac{\partial H}{\partial p_\theta} = \frac{1}{m^* r^2}p_\theta + \omega_L = \frac{\hbar}{im^* r^2}\frac{\partial \ln \psi}{\partial \theta} + \omega_L. \tag{14}$$

The appearance of the imaginary number $i = \sqrt{-1}$ on the right-hand side of the above equations indicates that the quantum trajectory $(r(t), \theta(t))$ has to be defined in the complex space as $(r(t), \theta(t)) = (r_R(t) + ir_I(t), \theta_R(t) + i\theta_I(t))$ to guarantee the solvability of Eqs. (13) and (14). It is just the coupling connection between the real and imaginary parts that gives rise to the quantum phenomena, as we have observed in the real world, such as wave-particle duality [15], tunneling [16], and Heisenberg uncertainty principle [17]. For a given 1D wavefunction $\Psi(t, x)$ expressed in Cartesian coordinates, the complex Hamiltonian Eq. (1) has a simple form:

$$H(t,x,p) = \frac{1}{2m}p^2 + V(t,x) + \frac{\hbar}{2im}\frac{\partial p}{\partial x} = \frac{1}{2m}p^2 + V(t,x) - \frac{\hbar^2}{2m}\frac{\partial^2 \ln \Psi(t,x)}{\partial x^2}. \tag{15}$$

The Hamilton equation for x turns out to be

$$\dot{x} = \frac{\partial H}{\partial p} = \frac{p}{m} = \frac{1}{m}\frac{\partial S}{\partial x} = \frac{\hbar}{im}\frac{\partial \Psi}{\partial x}, \quad x \in \mathbb{C} \tag{16}$$

which can be conceived of as a complex-valued version of Bohmian mechanics [18, 19]. The complex quantum trajectory method based on Eq. (16) has been recently developed into a potential computational tool to analyze wave-packet interference [20] and wave-packet scattering [21].

The wavefunction ψ has to be solved in advance from the Schrodinger Eq. (10), before we determine the electron's quantum trajectory $(r(t), \theta(t))$ from Eqs. (13) and (14). In terms of the dimensionless radial distance $\rho = (\hbar/m\omega)^{-1/2}r$, the eigenvalues $E_{n,l}$ and the related eigenfunction $\psi_{n,l}$ can be solved analytically as [22].

$$E_{n,l} = (2n + |l| + 1)\hbar\omega + l\hbar\omega_L, \tag{17}$$

$$\psi_{n,l}(\rho,\theta) = R_{n,l}(\rho)\Theta_l(\theta) = C_{n,l}e^{-\rho^2/2}\rho^{|l|}L_n^{|l|}(\rho^2)e^{il\theta}, \tag{18}$$

where $n = 0, 1, 2, \cdots$ is the radial quantum number, $l = 0, \pm 1, \pm 2, \cdots$ is the angular quantum number, and $C_{n,l}$ is a normalization factor. The electronic motion in the eigenstate $\psi_{n,l}$ now can be established by integrating Eqs. (13) and (14) with ψ given by Eq. (18):

$$\frac{d\rho}{d\tau} = \frac{1}{i}\left(\frac{d\ln R_{n,l}(\rho)}{d\rho} + \frac{1}{2\rho}\right) \triangleq f_{n,l}(\rho), \tag{19}$$

$$\frac{d\theta}{d\tau} = \frac{l}{\rho^2} + \frac{\omega_L}{\omega}. \tag{20}$$

where the dimensionless time is expressed by $\tau = t\omega$. Eq. (20) indicates that the angular dynamics $\theta(\tau)$ is influenced by the magnetic field B via the relation $\omega_L = eB/(2m^*c)$ and reveals the existence of a critical B such that the Larmor angular velocity ω_L/ω counterbalances the quantum angular velocity l/ρ^2 to yield $d\theta/d\tau = 0$. The stagnation magnetic field denotes the critical B that stagnates the electron with zero angular displacement within a quantum dot. The occurrence of magnetic stagnation retards the electronic transport and causes a jump in resistance. In the following sections, we will characterize the stagnation magnetic field from the equations of motion Eqs. (19) and (20) and verify the consistency between this theoretical prediction and the experimental measurement of resistance.

3. Standing waves and critical magnetic field

The conductivity of a quantum dot depends on how electrons move under the confinement potential within the quantum dot. Eqs. (19) and (20) provides us with all the required information to describe the underlying electronic quantum motion. The radial motion $\rho(\tau)$ described by Eq. (19) and the angular motion $\theta(\tau)$ described by Eq. (20) are, individually, periodic time functions, whose periods, T_ρ and T_θ, can be computed by using the residue theorem. In case that the radial and angular motions are not commensurable, i.e., $T_\rho/T_\theta \notin \mathbb{Q}$, the overall motion is not periodic and the electron's orbit precesses continuously around the periphery of the quantum dot, as shown in **Figure 2a**. By way of this precession orbit, an electron can pass through the quantum dot from the entrance to the exit and contribute to the conductance.

On the other hand, if T_ρ/T_θ is a rational number, the shape of the electron's orbit is stationary like a standing wave, as shown in **Figure 2b**. Except that the orientation of the standing wave happens to align with the direction from the entrance to the exit, as shown in **Figure 2c**, passage through the quantum dot is prohibited, when a standing-wave motion emerges. A standing-wave (non-precessing) orbit has to satisfy the commensurability condition,

$$T_\theta = NT_\rho \tag{21}$$

where N is a positive integer. This condition ensures that when the electron undergoes a complete oscillation in the θ direction, there are integral numbers of oscillation in the r direction. Once electronic standing waves emerge in a quantum dot, the electron after a complete θ revolution will return to the entrance to the quantum dot and consequently contribute to the resistance of the quantum dot.

As shown in **Figure 2d**, the standing-wave motion degenerates into a confined motion such that the electron is trapped into a closed trajectory, in the extreme case $N \to \infty$. When the electron is trapped or stagnated, it is in no way to pass through the quantum dot and causes a remarkable increase in resistance. The special magnetic field corresponding to $N \to \infty$ plays the major role in the magneto-resistance and is to be derived below.

The pattern and the orientation of the standing waves can be controlled by the applied magnetic field B via the relation Eq. (20), which indicates that the angular motion depends on the parameter,

$$\frac{\omega_L}{\omega} = \frac{\omega_L}{\sqrt{\omega_0^2 + \omega_L^2}} = \frac{eB/(2m^*c)}{\sqrt{\omega_0^2 + e^2B^2/(2m^*c)^2}}, \tag{22}$$

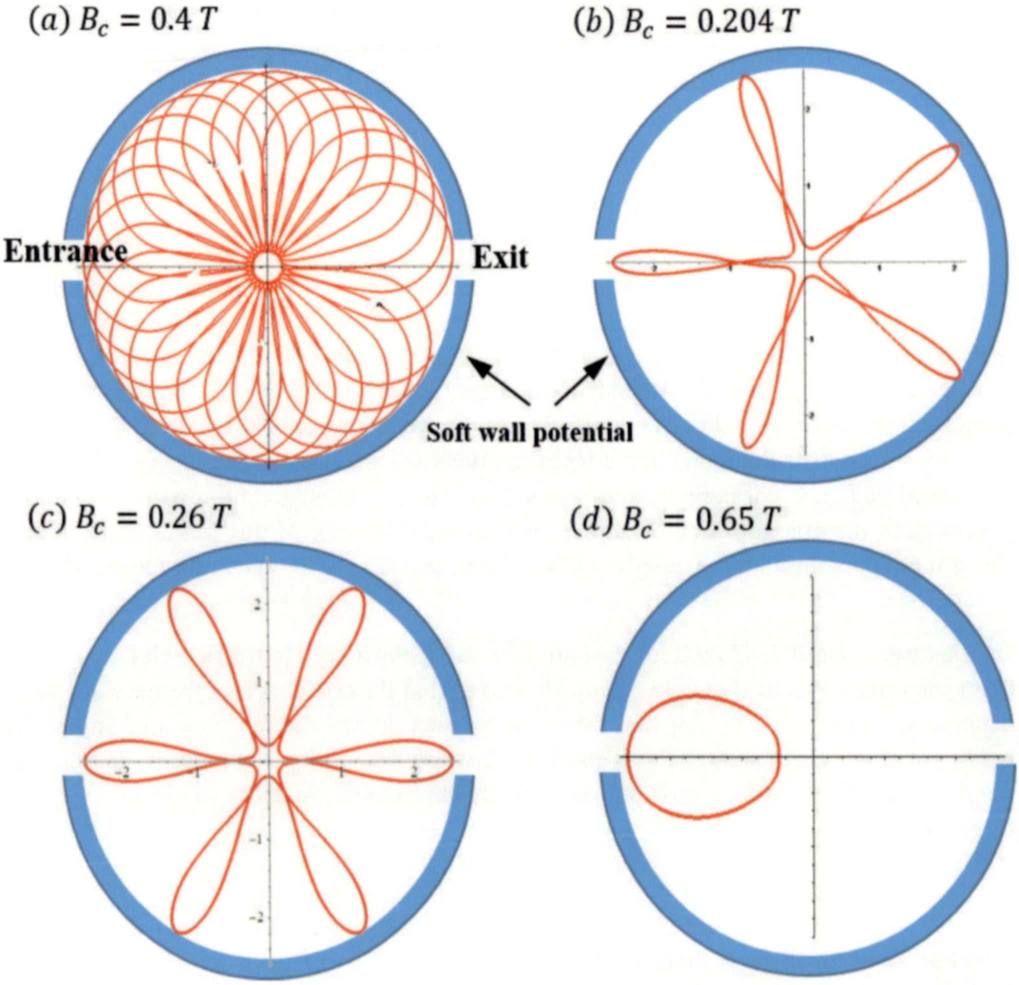

Figure 2. Four types of electronic quantum trajectory in a quantum dot. (a) a magnetic field ($B_c = 0.4T$) not satisfying the commensurability condition Eq. (21) yields precessing trajectories. (b) A critical magnetic field $B_c = 0.204T$ yields standing-wave like quantum trajectory, whose five wavelengths on the circumference do not contact the exit of the quantum dot. (c) A magnetic field $B_c = 0.26T$ yields similar standing-wave trajectory as that in part (b) but with six wavelengths which contact both the entrance and exit of the quantum. (d) At $B_c = 0.65T$, a stagnation magnetic field, the electron is stagnated within an isolated region of θ.

which, in turn, is solely determined by the magnetic field B. We are able to control the resistance of the quantum dot by varying the magnitude of B to satisfy the commensurability condition Eq. (21). Our next issue is to characterize the critical magnetic field B_c that satisfies the commensurability condition Eq. (21). The period T_ρ in Eq. (21) can be evaluated by the radial dynamics Eq. (19). The radial motion $\rho(t)$ is a periodic time function whose trace on the complex ρ plane forms a closed path c_ρ, along which the period T_ρ can be computed as a contour integral as

$$T_\rho = \int d\tau = \oint_{c_\rho} \frac{d\rho}{f_{n,l}(\rho)}. \tag{23}$$

According to the residue theorem, the contour integral in Eq. (23) is equal to $2\pi i$ times the sum of the residues of $1/f_{n,l}(\rho)$ evaluated at its poles within the contour c_ρ, i.e.,

$$T_\rho = \oint_{c_\rho} \frac{d\rho}{f_{n,l}(\rho)} = 2\pi i \sum_k R_k, \tag{24}$$

where R_k is the residue of $1/f_{n,l}(\rho)$ evaluated at its kth pole. Let Ω_k be the set containing all of contours which enclose the same poles of $1/f_{n,l}(\rho)$. According to the residue theorem, the integrals along the contours belonging to the same set Ω_k have the same contour integral, denoted by $T_\rho^{(k)}$. If the number of different ways of pole encirclement is M, we can define M contour sets, $\Omega_1, \Omega_2, \ldots, \Omega_M$, with each contour set corresponding to one particular way of pole encirclement. Along all the possible contours, the period T_ρ defined by Eq. (23) can only have M discrete values, $T_\rho^{(1)}, T_\rho^{(2)}, \ldots, T_\rho^{(M)}$, defined by

$$T_\rho = \oint_{c_\rho} \frac{d\rho}{f_{n,l}(\rho)} = T_\rho^{(k)}, \quad \forall c_\rho \in \Omega_k, \quad k = 1, 2, \cdots, M. \tag{25}$$

The sequence $\left\{T_\rho^{(1)}, T_\rho^{(2)}, \cdots, T_\rho^{(M)}\right\}$ then constitutes a set of quantization levels for the period T_ρ in the quantum state $\psi_{n,l}(\rho, \theta)$.

(A) Standing Wave with $l=0$

In case of $l = 0$, the radial dynamics and azimuth dynamics are decoupled according to Eqs. (19) and (20). A look on the ground state $(n, l) = (0, 0)$ is helpful to understand some common features in the states with $l = 0$. The related wavefunction is given by Eq. (18) as $R_{0,0}(\rho) = e^{-\rho^2/2}$ and $\Theta_0(\theta) = 1$. Substituting this wavefunction into Eqs. (19) and (20) yields the equations of motion for the ground-state electron,

$$\frac{d\rho}{d\tau} = i\frac{2\rho^2 - 1}{2\rho}, \quad \frac{d\theta}{d\tau} = \frac{\omega_L}{\omega}. \tag{26}$$

It appears that that the ground-state electron rotates with a constant angular velocity ω_L/ω around its equilibrium radial position $\rho_{eq} = \sqrt{2}/2$. Therefore, the azimuth period T_θ is simply $2\pi/(\omega_L/\omega)$, and the radial period T_ρ can be computed from Eqs. (24) and (26) as

$$T_\rho = \frac{1}{i}\oint_{c_\rho} \frac{2\rho}{2\rho^2 - 1} d\rho = \pi, \tag{27}$$

where c_ρ is any contour enclosing the pole $\rho_{eq} = \sqrt{2}/2$ on the complex plane of ρ.

The commensurability condition Eq. (21) with the calculated T_ρ and T_θ for the ground state turns out to be

$$\frac{\omega_L}{\omega} = \frac{2}{N}, \quad N = 3, 4, 5, \cdots, \tag{28}$$

where we note $\omega_L/\omega < 1$ from its definition in Eq. (22). The critical magnetic field B_c now can be solved from Eq. (28) as

$$B_c = \frac{B_0}{\sqrt{N^2/4 - 1}}, \quad N = 3, 4, 5, \cdots, \tag{29}$$

where B_0 is the magnetic field whose Larmor frequency ω_L is equal to the natural frequency ω_0 of the harmonic oscillator, i.e., $B_0 = (2m^* c/e)\omega_0$. The relation expressed by Eq. (29) characterizes all the magnetic fields that force the electron to behave like a standing wave in the ground state of a quantum dot.

Regarding excited states, there are multiple periods in the radial motion $\rho(\tau)$ as indicated by Eq. (25). Taking first excited state $(n, l) = (1, 0)$ as an illustrating example, the quantum dynamics is described by

$$\frac{d\rho}{d\tau} = i\frac{2\rho^4 - 11\rho^2 + 6}{2\rho(\rho^2 - 2)} \quad \frac{d\theta}{d\tau} = \frac{\omega_L}{\omega}, \tag{30}$$

which has four equilibrium points at

$$\rho_{eq} = \pm\frac{\sqrt{11 \pm \sqrt{73}}}{2} \tag{31}$$

According to different encirclements of equilibrium points, four sets of complex trajectories $\rho(\tau)$ can be identified as shown in **Figure 3a**, where Ω_1 and Ω_2 denote the sets of all trajectories enclosing only one equilibrium point, Ω_3 denotes the set enclosing two equilibrium points, and Ω_4 denotes the set enclosing all the four equilibrium points.

Corresponding to the four different ways of encirclement, the four quantization levels of T_ρ can be computed from Eq. (24) as

$$T_\rho = \frac{2\pi(73 \pm 3\sqrt{73})}{292}, \pi, 2\pi. \tag{32}$$

The commensurability condition for the occurrence of standing wave in the four contour sets now can be derived from Eq. (21) as

$$\frac{\omega_L}{\omega} = \begin{cases} \dfrac{73 + 3\sqrt{73}}{16N}, N \geq 7, & \rho(\tau) \in \Omega_1 \\ \dfrac{73 - 3\sqrt{73}}{16N}, N \geq 3, & \rho(\tau) \in \Omega_2 \\ \dfrac{2}{N}, \quad N \geq 3, & \rho(\tau) \in \Omega_3 \\ \dfrac{1}{N}, \quad N \geq 2, & \rho(\tau) \in \Omega_4 \end{cases} \tag{33}$$

http://dx.doi.org/10.5772/intechopen.74409

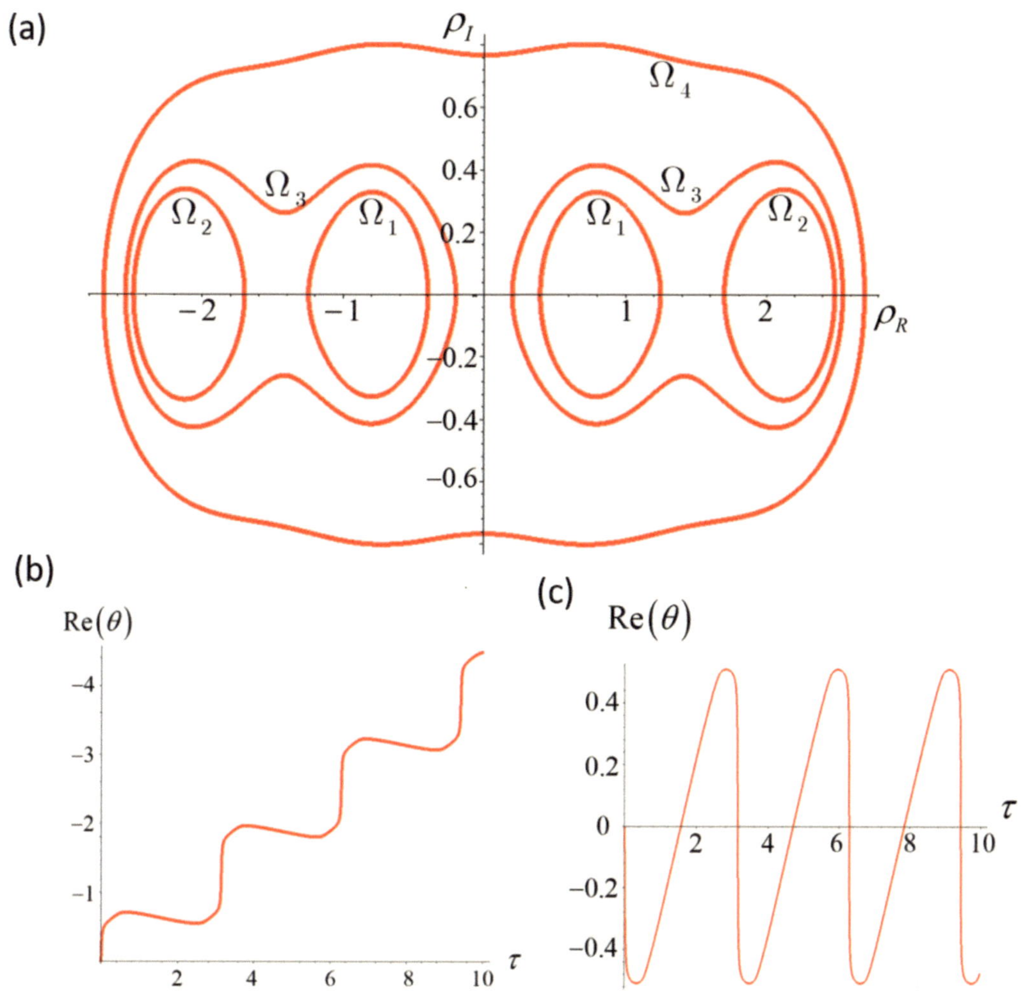

Figure 3. (a) Four sets of complex trajectories $\rho(\tau)$ are identified according to different encirclements of equilibrium points in the state $(n, l) = (1, 0)$. (b) Typical time response of an increasing $\mathrm{Re}(\theta)$ corresponds to the trajectory shown in Figure 2b. (c) Typical time response of an oscillatory $\mathrm{Re}(\theta)$ corresponds to the trajectory shown in Figure 2d.

The related critical magnetic field B_c can be determined by substituting Eq. (33) into Eq. (22). Comparing Eq. (28) with Eq. (33), we can see that the critical B_c, which raises standing waves in the ground state, also raises standing waves in the first excited state. The peaks of the magneto-resistance just concentrate on the dominant critical magnetic field that concurrently produces standing waves in different states.

(B) Standing Wave with $l > 0$.

In the case of $l > 0$, the cyclotron angular velocity ω_L/ω and the quantum angular velocity l/ρ^2 are in the same direction so as to give an intensified resultant $\dot{\theta} = l/\rho^2 + \omega_L/\omega$. The coupling between the azimuth motion $\theta(\tau)$ and the radial dynamics $\rho(\tau)$ makes the evaluation of T_θ more difficult; however, because $\rho(\tau)$ is a periodic function, we can evaluate $\dot{\theta}$ in Eq. (20) by simply replacing l/ρ^2 with its average value $\left(l/\rho^2\right)_{ave}$, if only the period of $\theta(\tau)$ is concerned,

$$T_\theta = \frac{2\pi}{\dot{\theta}_{ave}} = \frac{2\pi}{(l/\rho^2)_{ave} + \omega_L/\omega} = NT_\rho. \tag{34}$$

The time average $(l/\rho^2)_{ave}$ is computed over one period of $\rho(\tau)$ and can be converted into a contour integral along the contour c_ρ traced by $\rho(\tau)$ on the complex plane:

$$\left(\frac{l}{\rho^2}\right)_{ave} = \frac{l}{T_\rho}\int_0^{T_\rho} \frac{d\tau}{\rho^2(\tau)} = \frac{l}{T_\rho}\oint_{c_\rho} \frac{d\rho}{\rho^2 f_{n,l}(\rho)}. \tag{35}$$

Substituting the above $(l/\rho^2)_{ave}$ into Eq. (34), we obtain the critical value of ω_L/ω as

$$\frac{\omega_L}{\omega} = \frac{2\pi}{NT_\rho} - \left(\frac{l}{\rho^2}\right)_{ave}. \tag{36}$$

Due to the constraint $0 \le \omega_L/\omega \le 1$, the admissible integer N lies in the interval

$$\frac{2\pi/T_\rho}{1 + (l/\rho^2)_{ave}} < N \le \frac{2\pi/T_\rho}{(l/\rho^2)_{ave}}, \tag{37}$$

where T_ρ and $(l/\rho^2)_{ave}$ are given by Eqs. (24) and (35), respectively. The admissible range of N is narrowed by increasing angular quantum number l, as can be seen from inequality Eq. (37). There is a maximum allowable l beyond which inequality Eq. (37) has no integer solution and standing-wave motion within the quantum dot disappears. To compare with the quantum state $(n, l) = (1, 0)$ considered previously, let us study the state $(n, l) = (1, 1)$ whose quantum motion is described by

$$\frac{d\rho}{d\tau} = i\frac{2\rho^4 - 11\rho^2 + 6}{2\rho(\rho^2 - 2)}, \quad \frac{d\theta}{d\tau} = \frac{1}{\rho^2} + \frac{\omega_L}{\omega}. \tag{38}$$

The period T_ρ is the same as that derived in Eq. (32), and the period T_θ can be computed by Eq. (34) with $(l/\rho^2)_{ave}$ evaluated by the contour integral Eq. (35) as

$$\left(\frac{1}{\rho^2}\right)_{ave} = \left\{\begin{array}{ll} l(11+\sqrt{73})/12, & \rho(\tau)\in\Omega_1 \\ l(11-\sqrt{73})/12, & \rho(\tau)\in\Omega_2 \\ 2l/3, & \rho(\tau)\in\Omega_3 \cup \Omega_4 \end{array}\right\} \tag{39}$$

Using T_ρ and $(l/\rho^2)_{ave}$ in Eqs. (36) and (37), the critical value of ω_L/ω in the state $(n, l) = (1, 1)$ becomes

$$\frac{\omega_L}{\omega}\left\{\begin{array}{ll} \dfrac{73+3\sqrt{73}}{16N} - \dfrac{11+\sqrt{73}}{12}, N = 3, & \rho(\tau)\in\Omega_1 \\ \dfrac{73-3\sqrt{73}}{16N} - \dfrac{11-\sqrt{73}}{12}, 3 \le N \le 14, & \rho(\tau)\in\Omega_2 \\ 2/N - 2/3, N = 2, 3, & \rho(\tau)\in\Omega_3 \\ 1/N - 2/3, N = 1, & \rho(\tau)\in\Omega_4 \end{array}\right\} \tag{40}$$

The comparison between Eqs. (33) and (40) leads to the observation that the number of the allowed integer N decreases dramatically when l increases from 0 to 1. Since the total different number of N accounts for the number of different ways by which standing wave can be formed, the possibility for the occurrence of standing-wave motion and thus the electronic resistance decreases with increasing angular quantum number l. The main reason is that the increment of the angular velocity $\dot{\theta} = l/\rho^2 + \omega_L/\omega$ with large l accelerates the electron's angular motion around the quantum dot and thus improves the conductance of the quantum dot.

(C) Standing Waves with $l < 0$

In this case, the cyclotron angular velocity ω_L/ω and the quantum angular velocity l/ρ^2 are in opposite directions so as to give a weakened resultant $\dot{\theta} = l/\rho^2 + \omega_L/\omega$. The resultant angular velocity $\dot{\theta}_{ave}$ may be positive, negative or zero, depending on the magnitude of $\left(l/\rho^2\right)_{ave}$, which can be classified into three categories: (1) $\left(l/\rho^2\right)_{ave} \geq 0$, (2) $-1 < \left(l/\rho^2\right)_{ave} < 0$, and (3) $\left(l/\rho^2\right)_{ave} \leq -1$, as listed in **Table 1**.

The cases of $l = 0$ and $l > 0$ considered previously belong to category (1) with $\dot{\theta}_{ave} > 0$, while the case of $l < 0$ belongs to categories (2) and (3). Taking into account the motion with $\dot{\theta}_{ave} < 0$, Eq. (36) now becomes

$$\frac{\omega_L}{\omega} = \pm\frac{2\pi}{NT_\rho} - \left(\frac{l}{\rho^2}\right)_{ave}, \tag{41}$$

where the admissible integer N for the three categories is summarized in **Table 1**.

The critical magnetic field B_c given by Eq. (41) with $\dot{\theta}_{ave} > 0$ and $\dot{\theta}_{ave} < 0$ produces standing-wave motions oscillating, respectively, counterclockwise and clockwise around the quantum dot, as shown in **Figure 2b** and **c**. For an angular quantum number with $-1 < \left(l/\rho^2\right)_{ave} < 0$ in category (2), there exists a special Larmor angular velocity ω_L/ω such that it counterbalances the quantum angular velocity $\left(l/\rho^2\right)_{ave}$ to yield

$$\dot{\theta}_{ave} = \left(l/\rho^2\right)_{ave} + \omega_L/\omega = 0. \tag{42}$$

Range of l	Critical ω_L/ω	Range of integer N
$\left(l/\rho^2\right)_{ave} \geq 0$	$\frac{\omega_L}{\omega} = \frac{2\pi}{NT_\rho} - \left(\frac{l}{\rho^2}\right)_{ave}$	$\frac{2\pi/T_\rho}{1+(l/\rho^2)_{ave}} < N \leq \frac{2\pi/T_\rho}{(l/\rho^2)_{ave}}$
$-1 < \left(l/\rho^2\right)_{ave} < 0$	$\frac{\omega_L}{\omega} = \frac{2\pi}{NT_\rho} - \left(\frac{l}{\rho^2}\right)_{ave}$	$N > \frac{2\pi/T_\rho}{1+(l/\rho^2)_{ave}}$
	$\frac{\omega_L}{\omega} = -\left(\frac{l}{\rho^2}\right)_{ave}$	$N \to \infty$
	$\frac{\omega_L}{\omega} = \frac{-2\pi}{NT_\rho} - \left(\frac{l}{\rho^2}\right)_{ave}$	$N \geq \frac{2\pi/T_\rho}{-(l/\rho^2)_{ave}}$
$\left(l/\rho^2\right)_{ave} \leq -1$	$\frac{\omega_L}{\omega} = \frac{-2\pi}{NT_\rho} - \left(\frac{l}{\rho^2}\right)_{ave}$	$\frac{-2\pi/T_\rho}{(l/\rho^2)_{ave}} \leq N < \frac{-2\pi/T_\rho}{1+(l/\rho^2)_{ave}}$

Table 1. The relation between critical Larmor frequency and angular quantum number l.

The critical B_c satisfying $\dot{\theta}_{ave} = 0$ produces isolated standing waves that form closed trajectories as shown in **Figure 2d**. This critical B_c nullifies the electron's net angular displacement and is called stagnation magnetic field. Because a passage through the quantum dot requires a net angular displacement equal to π, an electron with $\dot{\theta}_{ave} = 0$ is unable to pass the quantum dot and makes no contribution to the conductivity.

In a case study of $l < 0$, we consider the state of $(n, l) = (1, -1)$, whose quantum equations of motion read

$$\frac{d\rho}{d\tau} = i\frac{2\rho^4 - 11\rho^2 + 6}{2\rho(\rho^2 - 2)}, \quad \frac{d\theta}{d\tau} = \frac{-1}{\rho^2} + \frac{\omega_L}{\omega}. \tag{43}$$

The radial trajectories $\rho(\tau)$ are the same as those depicted in **Figure 3a**. Along different sets of radial trajectory, different modes of standing-wave motion are excited. According to the value of $(l/\rho^2)_{ave} = -(1/\rho^2)_{ave}$ given by Eq. (39), it is found that the trajectory set Ω_1 belongs to category (3), while the sets Ω_2, Ω_3 and Ω_4 belong to category (2), as tabulated in **Table 2**.

Typical standing waves in Ω_1, Ω_2 and Ω_3 are shown in **Figure 4** for $N = 5, 7$ and 9. We can see that the geometrical meaning of the integer N defined in Eq. (21) is just the number of electronic waves distributed on the circumference of the quantum dot. Due to $\dot{\theta}_{ave} < 0$ in Ω_1 trajectory set, as indicated in **Table 2**, the mean rotation direction of the electron in Ω_1 is clockwise. Because $\dot{\theta}_{ave}$ merely denotes the mean angular velocity, locally we may have $\dot{\theta} > 0$ during some short periods in which the electron rotates in an opposite direction as shown in **Figure 3b**.

In the state ψ, we have two stagnation frequencies at $\omega_L/\omega = (11 - \sqrt{73})/12$ and $\omega_L/\omega = 2/3$. In the presence of magnetic stagnation, Larmor angular velocity ω_L/ω is counterbalanced by

Set	Frequency range	Critical frequency	Integer N
Ω_1	$0 \le \frac{\omega_L}{\omega} < 1$	$\frac{\omega_L}{\omega} = -\frac{73+3\sqrt{73}}{16N} + \frac{11+\sqrt{73}}{12}$	$4 \le N \le 9$
Ω_2	$\frac{11-\sqrt{73}}{12} \le \frac{\omega_L}{\omega} < 1$	$\frac{\omega_L}{\omega} = \frac{73-3\sqrt{73}}{16N} + \frac{11-\sqrt{73}}{12}$	$N \ge 4$
	$\frac{\omega_L}{\omega} = \frac{11-\sqrt{73}}{12}$	$\frac{\omega_L}{\omega} = \frac{11-\sqrt{73}}{12}$	$N \to \infty$
	$0 \le \frac{\omega_L}{\omega} < \frac{11-\sqrt{73}}{12}$	$\frac{\omega_L}{\omega} = -\frac{73-3\sqrt{73}}{16N} + \frac{11-\sqrt{73}}{12}$	$N \ge 15$
Ω_3	$\frac{\omega_L}{\omega} > \frac{2}{3}$	$\frac{\omega_L}{\omega} = \frac{2}{N} + \frac{2}{3}$	$N \ge 7$
	$\frac{\omega_L}{\omega} = \frac{2}{3}$	$\frac{\omega_L}{\omega} = \frac{2}{3}$	$N \to \infty$
	$0 \le \frac{\omega_L}{\omega} < \frac{2}{3}$	$\frac{\omega_L}{\omega} = -\frac{2}{N} + \frac{2}{3}$	$N \ge 3$
Ω_4	$\frac{\omega_L}{\omega} > \frac{2}{3}$	$\frac{\omega_L}{\omega} = \frac{1}{N} + \frac{2}{3}$	$N \ge 4$
	$\frac{\omega_L}{\omega} = \frac{2}{3}$	$\frac{\omega_L}{\omega} = \frac{2}{3}$	$N \to \infty$
	$0 \le \frac{\omega_L}{\omega} < \frac{2}{3}$	$\frac{\omega_L}{\omega} = -\frac{1}{N} + \frac{2}{3}$	$N \ge 2$

Table 2. Distribution of the critical frequencies in the state $(n, l) = (1, -1)$.

http://dx.doi.org/10.5772/intechopen.74409

the quantum angular velocity $\left(l/\rho^2\right)_{ave}$ such that the electron's net angular displacement $\Delta\theta_{ave}$ is zero and the electron is stagnated within the quantum dot. The instantaneous dynamics of $\rho(\tau)$ and $\theta(\tau)$ are solved from Eq. (43) at the stagnation frequency $\omega_L/\omega = 2/3$ and the results are shown in **Figure 3c**. As expected, the net change of $\theta(\tau)$ is zero after a period of oscillation. The projection of the computed complex trajectory on the real $x - y$ plane is a closed path as illustrated in **Figure 2d**. This closed path produced by magnetic stagnation isolates the electron from the exit of the quantum dot and is the main cause of electronic resistance.

Apart from the consequence of $\dot{\theta}_{ave} = 0$, the effect of magnetic stagnation is also reflected in the wave number N. From Eq. (21), the relation between $\dot{\theta}_{ave}$ and N can be expressed by

$$N = \frac{2\pi/T_\rho}{\dot{\theta}_{ave}} = \frac{2\pi/T_\rho}{\left(l/\rho^2\right)_{ave} + \omega_L/\omega} \tag{44}$$

There are infinitely many wavelengths distributed on the circumference of the quantum dot, as $\dot{\theta}_{ave}$ approaches to zero. The variation of the wave number N with respect to the critical Larmor frequency ω_L/ω for the quantum state $(n, l) = (1, -1)$ is demonstrated in **Figure 5a**. A prominent change of N appears in the vicinity of the two stagnation frequencies $\omega_L/\omega = \left(11 - \sqrt{73}\right)/12$ and 2/3, at which the wave number N approaches to infinity. These two stagnation frequencies

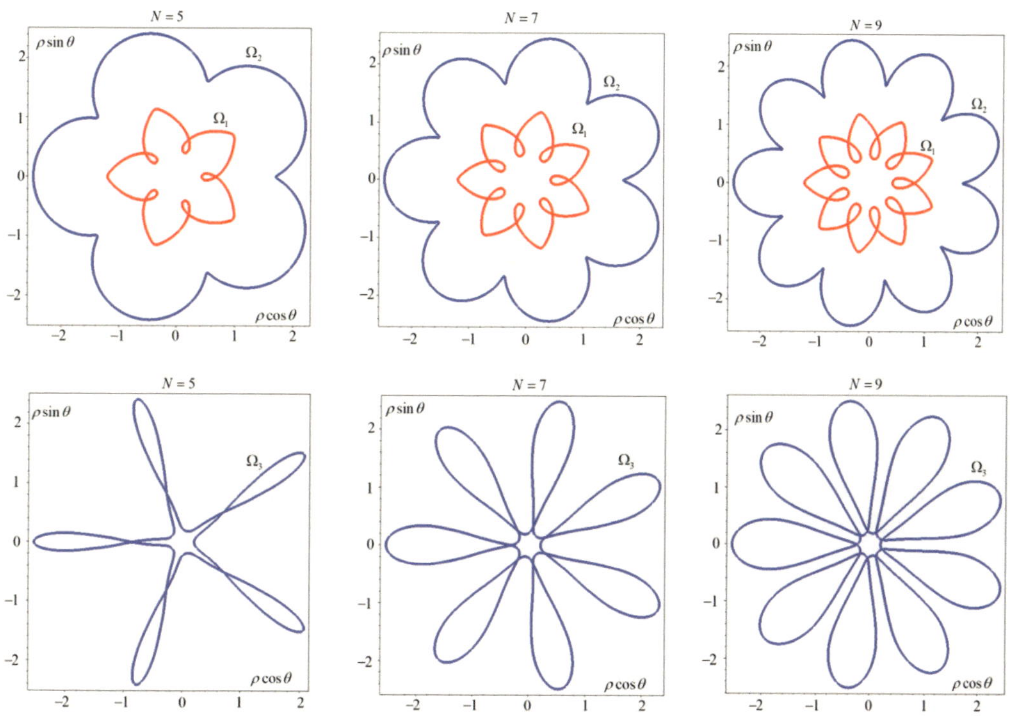

Figure 4. Typical standing-wave motions in the state $(n, l) = (1, -1)$ with wave number $N = 5$,7 and 9. The trajectory sets Ω_1, Ω_2, and Ω_3 refer to the three sets of radial trajectory defined in **Figure 3a**.

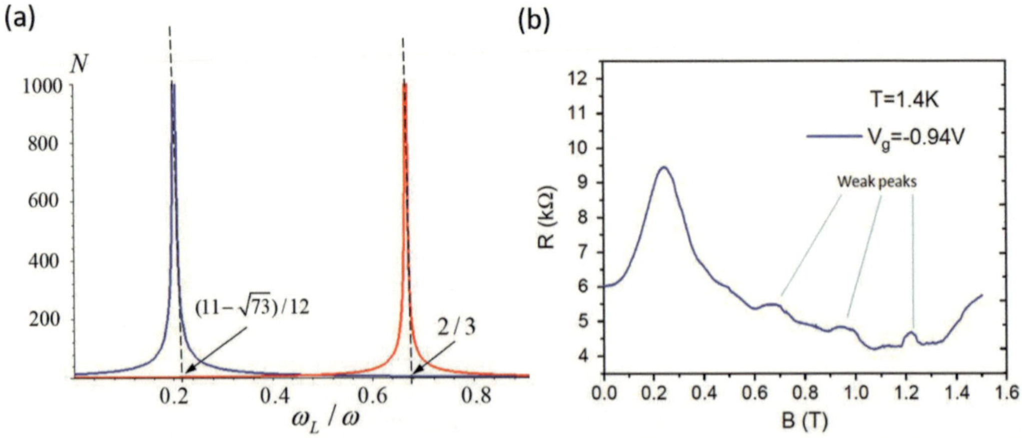

Figure 5. (a) The variation of wave number N with respect to the Larmor frequency ω_L/ω in the quantum state $(n, l) = (1, -1)$. (b) The two stagnation frequencies, $\omega_L/\omega = (11 - \sqrt{73})/12$ and 2/3, coincide with the two peaks of the experimental curve of resistance.

coincide with the locations of the resistance peaks by comparing with the experimental results as shown in **Figure 5b**.

4. Experimental verification

This section will compare the above theoretical predictions with the existing experimental data [4, 13] to confirm the fact that the effect of magnetic stagnation is the main cause to the resistance oscillation of quantum dots in low magnetic field. The experiment was performed in an AlGaAs/GaAs heterostructure with a carrier concentration $n_e = 2.5 \times 10^{11}\text{cm}^{-2}$. Resistance was measured at temperature $T = 1.4K$ using a sensitive lock-in amplifier at currents of typically 1 nA and a frequency of 12 Hz. The resulting resistance measurement in the range of low magnetic field $B \leq 1.3$ is depicted in **Figure 5b** showing a strong peak located around $B = 0.22$T and three weak peaks at $B = 0.65$T, 0.97T, and 1.21T.

Thus far, our analysis on quantum trajectory focuses on some specific states. In order to know the influence of the applied magnetic field on the resistance, we have to consider all the possible quantum states occupied in the device. At temperature $T = 1.4K$, where the resistance is measured, the possible states to be occupied can be estimated by the Fermi-Dirac distribution,

$$f(E) = \frac{1}{1 + e^{\left(E_{n,l} - E_F\right)/k_B T}}, \tag{45}$$

where $E_{n,l}$ is the energy level given by Eq. (17), and $E_F = 8.5$ meV is the Fermi energy of the AlGaAs/GaAs heterostructure. All the possibly occupied states and their associated stagnation frequencies are listed in **Table 3**.

ω_L/ω		n				
		0	**1**	**2**	**3**	**4**
l	0	0	0	0	0	0
	−1	2/3	2/3, 0.205	2/3, 0.373, 0.119	2/3, 0.52, 0.19	2/3, 0.252, 0.124
	−2	4/5	4/5, 0.316	4/5, 0.543, 0.2	4/5, 0.73, 0.31, 0.146	0.4, 0.213, 0.115
	−3	6/7	6/7, 0.391	6/7, 0.64, 0.26	0.39, 0.196	*
	−4	8/9	8/9, 0.445	0.7, 0.3, 8/9	0.454, 0.237	*
	−5	*	0.486	0.75, 0.347	*	*
	−6	*	0.52	0.78, 0.38	*	*
	−7	*	0.546	*	*	*
	−8	*	0.57	*	*	*

Table 3. Stagnation frequencies ω_L/ω evaluated in the quantum states (n, l) at T = 1.4 K.

An incident electron subjected to an applied magnetic field B may enter any one of the occupied states listed in **Table 3**. The electronic resistance induced by B depends on the global transportation behavior across the quantum dot through all the allowable states. Magnetic stagnation slows down the electron's angular rate and retards the passage of the electron. The angular motion is fully retarded and the electron is trapped in the quantum dot without contribution to the conductance, as $\dot{\theta}_{ave} \to 0$. To quantify the effect of magnetic stagnation, we define a magneto-stagnation function as following

$$S(B) = -\sum_{n,l} \ln\left|\dot{\theta}_{ave}(n, l, B)\right| = -\sum_{n,l} \ln\left|\frac{\omega_L}{\omega} + \left(\frac{l}{\rho^2}\right)_{ave}\right|_{n,l} \tag{46}$$

where the summation is taken over all the states listed in **Table 3**. The expression of ω_L/ω as a function of B has already been given by Eq. (22). Upon comparing the prediction of Eq. (46) with the experimental results, we evaluate the constants in ω_L/ω according to the experimental setup [4, 13], which gave $\hbar\omega_c = \hbar eB/m = 1.76B\ meV$and $\hbar\omega_0 = \hbar\sqrt{k/m} = 0.64meV$. Using these data in Eq. (22) yields

$$\frac{\omega_L}{\omega} = \frac{0.88B}{\sqrt{0.4096 + 0.7744B^2}} \tag{47}$$

which in turn is substituted into Eq. (46) to express the magneto-stagnation function $S(B)$ as an explicit function of B.

The electron's total angular velocity via all admissible quantum states at $T = 1.4K$ can be estimated by the function $S(B)$. Because of $S(B) \to \infty$ as $\dot{\theta}_{ave} \to 0$, a large value of $S(B)$ implies that there is a high resistance to the electron's angular movability. Accordingly, $S(B)$ can be reasonably treated as an alternative description of electronic resistance. **Figure 5a** and **b** illustrates the first evidence of this correspondence. At the two stagnation frequencies $\omega_L/\omega = \left(11 - \sqrt{73}\right)/12$ and $2/3$, corresponding to the two peaks of the resistance curve around $B = 0.22\text{T}$ and $B = 0.65\text{T}$, $S(B)$ approaches to infinity, even though only the state $(n, l) = (1, -1)$ is considered in **Figure 5a**.

If magnetic stagnation takes place simultaneously in many states, its effect will be amplified. Stagnation frequencies such as $\omega_L/\omega = 2/3$, $4/5$, and $6/7$ appear concurrently in different quantum states, as can be seen from **Table 3**. Because the stagnation function considers the superposition of $\ln\left|\dot{\theta}_{ave}(n, l, B)\right|$ coming from all the allowable states, the value of $S(B)$ is intensified at such stagnation frequency concurring in different states. According to the conversion formula Eq. (47), the magnetic field relating to the stagnation frequencies $\omega_L/\omega = 2/3$, $4/5$, and $6/7$ is found to be $B = 0.65\text{T}$, 0.97T, and 1.21T, respectively, which are just the locations of the three weak peaks of the resistance curve as shown in **Figure 5b**.

Figure 6 demonstrates the strong correspondence between the stagnation function $S(B)$ and the resistance curve, where the resistance curve is superposed on the gray-level plot of the stagnation function $S(B)$ with the intensity of darkness representing the magnitude of $S(B)$. As can be seen, the gray-level distribution matches closely with the resistance curve and in that the dark bands of $S(B)$ correctly locate the peaks of the resistance. The gray-level plot of $S(B)$ has several narrow dark bands and one broad dark band. The narrow dark bands come from the isolated stagnation frequencies at $\omega_L/\omega = 2/3$, $4/5$, and $6/7$, and their locations coincide with the three weak peaks of the resistance curve. The broad dark band of $S(B)$ covers the neighborhood of the strong peak of the resistance curve, which is formed by a series of closely distributed stagnation frequencies centered at $\omega_L/\omega = 0.29$, or equivalently,

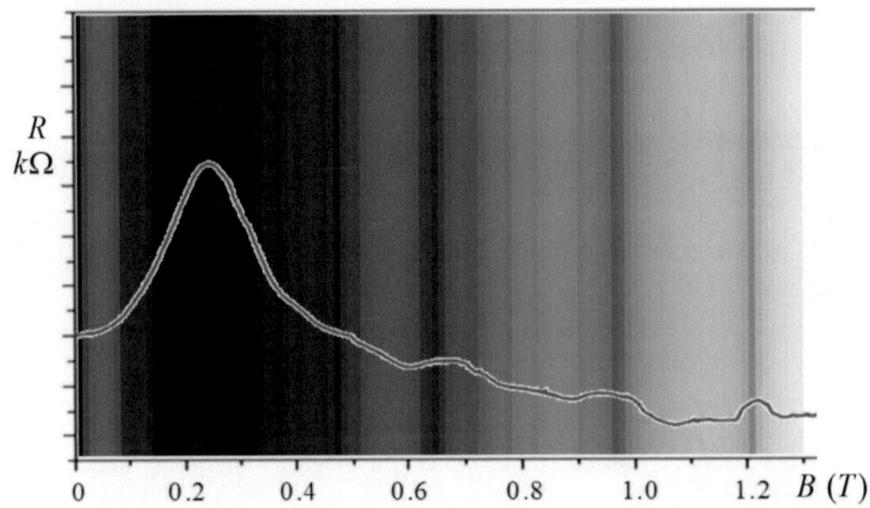

Figure 6. A gray-level plot of the stagnation function $S(B)$ with the darkness intensity representing the value of $S(B)$ is compared with the resistance curve. The resistance curve [4, 13] has a strong peak located around $B = 0.22\,\text{T}$ and three weak peaks at $B = 0.65\,\text{T}$, $0.97\,\text{T}$, and $1.21\,\text{T}$. It appears that the locations of the three narrow dark bands coincide with the three weak peaks of the resistance curve, while the broad dark band covers the neighborhood of the strong peak of the resistance curve.

at $B = 0.22\,\mathrm{T}$ from Eq. (47). The agreement between the experimental data of magneto-resistance and the magneto-stagnation function $S(B)$ constructed from the quantum Hamilton dynamics Eqs. (19) and (20) is not surprising, if we recall that Eqs. (19) and (20) is fully determined by the wavefunction $\psi_{n,l}(r, \theta)$, which is responsible for the observed magneto-resistance in quantum dots.

5. Conclusions

Parallel to the existing probabilistic description for a quantum dot by a probability density function $\psi^*\psi$, this chapter considered an alternative trajectory description according to a dynamic representation of ψ constructed from quantum Hamilton mechanics. The equivalence between a given wavefunction $\psi(x)$ and its dynamic representation $\dot{x} = f(x)$ ensures that the various quantum properties possessed by ψ also manifest in its dynamic representation. The established Hamilton dynamics for a quantum dot predicts that there are special magnetic fields, which can trap electrons within the quantum dot and cause a significant raise in the resistance. The comparison with experimental data validates this theoretical prediction. Apart from the magneto-transport considered in this chapter, many other features of a quantum dot, which were studied previously from a probabilistic perspective based on ψ, now can be reexamined from a trajectory viewpoint based on the dynamic representation of ψ proposed here.

Author details

Ciann-Dong Yang* and Shih-Ming Huang

*Address all correspondence to: cdyang@mail.ncku.edu.tw

Department of Aeronautics and Astronautics, National Cheng Kung University, Tainan, Taiwan

References

[1] Ochiai Y, Widjaja AW, Sasaki N, Yamamoto K, Akis R, Ferry DK, Bird JP, Ishibashi K, Aoyagi Y, Sugano T. Backscattering of ballistic electrons in a corrugated-gate quantum wire. Physical Review B. 1997;**56**:1073-1076

[2] Lin LH, Aoki N, Nakao K, Ishibashi K, Aoyagi Y, Sugano T, Holmberg N, Vasileska D, Akis R, Bird JP, Ferry DK, Ochiai Y. Magneto-transport in corrugated quantum wires. Physica E: Low-dimensional Systems and Nanostructures. 2000;**7**:750-755

[3] Aoki N, Brunner R, Burke AM, Akis R, Meisels R, Ferry DK, Ochiai Y. Direct imaging of electron states in open quantum dots. Physical Review Letters. 2012;**108**:136804

[4] Brunner R, Meisels R, Kuchar F, ElHassan M, Bird JP, Ishibashi K. Investigations of backscattering peaks and of the nature of the confining potential in open quantum dots. Physica E: Low-dimensional Systems and Nanostructures. 2004;**21**:491-495

[5] Brunner R, Meisels R, Kuchar F, Akis R, Ferry DK, Bird JP. Classical and quantum dynamics in an array of electron billiards. Physica E: Low-dimensional Systems and Nanostructures. 2008;**40**:1315-1318

[6] Brunner R, Meisels R, Kuchar F, Akis A, Ferry DK, Bird JP. Magneto-transport in open quantum dot arrays at the transition from low to high magnetic field: Regularity and chaos. International Journal of Modern Physics B. 2007;**21**:1288-1296

[7] Morfonios CV, Schmelcher P. Control of Magnetotransport in Qunatum Billiards. Switzerland: Springer International Publishing; 2017

[8] Fransson J, Kang M, Yoon Y, Xiao S, Ochiai Y, Reno J, Aoki N, Bird JP. Tuning the Fano resonance with an intruder continuum. Nano Letters. 2014;**14**:788-793

[9] Poniedziałek MR, Szafran B. Multisubband transport and magnetic deflection of Fermi electron trajectories in three terminal junctions and rings. Journal of Physics. Condensed Matter. 2012;**24**:085801

[10] Yang CD. Quantum Hamilton mechanics Hamilton equations of quantum motion, origin of quantum operators, and proof of quantization axiom. Annals of Physics. 2006;**321**: 2876-2926

[11] Yang CD. Complex mechanics. In: Progress in Nonlinear Science. Vol. 1. Hong Kong: Asian Academic Publisher; 2010

[12] Yang CD. Modeling quantum harmonic oscillator in complex domain. Chaos, Solitons, & Fractals. 2006;**30**:342-362

[13] Elhassan M, Akis R, Bird JP, Ferry DK, Ida T, Ishibashi K. Magnetically induced Bragg scattering of electrons in quantum-dot crystals. Physical Review B. 2004;**70**:205341

[14] Ferry DK, Burke AM, Akis R, Brunner R, Day TE, Meisels R, Kuchar F, Bird JP, Bennett BR. Open quantum dots-probing the quantum to classical transition. Semiconductor Science and Technology. 2011;**26**:043001

[15] Yang CD. Wave-particle duality in complex space. Ann. Physics. 2005;**319**:444-470

[16] Yang CD. Complex tunneling dynamics. Chaos,Solitons and Fractals. 2007;**32**:312-345

[17] Yang CD. Trajectory interpretation of the uncertainty principle in 1D systems using complex Bohmian mechanics. Physics Letters A. 2008;**372**:6240-6253

[18] Bohm D. A suggested interpretation of the quantum theory in terms of hidden variables. Physical Review. 1952;**85**:166-193

[19] Holland PR. The Quantum Theory of Motion. Cambridge: Cambridge University Press; 1993

[20] Chou CC, Sanz AS, Miret-Artes S, Wyatt RE. Hydrodynamic view of wave-packet interference: Quantum caves. Physical Review Letters. 2009;**102**:250401

[21] Wyatt RE, Rowland BA. Computational investigation of wave packet scattering in the complex plane: Propagation on a grid. Journal of Chemical Theory and Computation. 2009;**5**:443-451

[22] Darwin CG. The diamagnetism of the free electron. Mathematical Proceedings of the Cambridge Philosophical Society. 1931;**27**:86-90

Chapter 3

Droplet Epitaxy as a Tool for the QD-Based Circuit Realization

Ákos Nemcsics

Additional information is available at the end of the chapter

http://dx.doi.org/10.5772/intechopen.70613

Abstract

The chapter describes a novel technology, called droplet epitaxy, in the view point of quantum-circuit realization. This technology is useful when quantum dots are to be produced, of different shape and size in various densities. There are self-assembling methods to achieve spatial ordering or spatial positioning. Out of some of the possible applications as an example, the register and cellular automata circuit will be described.

Keywords: droplet epitaxy, quantum dot, self-assembling, lateral alignment, vertical stacking

1. Introduction

The most frequently quoted integration tendency in microelectronics is covered by the so-called Moor's law, which predicts the growth of the component concentration on microchips, doubling in every 2 years and forecasting further miniaturization. The CMOS technology itself is approaching its theoretical limits. Further limitations are also caused by quantum effects and certain anomalies in the technology in materials science. The spread in size, when CMOS technology is approaching the nano-region, represents further problems in microchip design. These difficulties make us wander about the next step in microchip technology, which would follow the present CMOS technology. The answer is hidden either in the promising state of spinotronics, an electronics based on graphene, or in circuits based on Josephson junction [1–4]. Quantum dots (QDs) or groups of QDs are also possible candidates of a new technology for the creation of electronic circuitry (**Figure 1A**). Nanotechnology based on GaAs and related compounds are also the most likely candidates for the development of new technology.

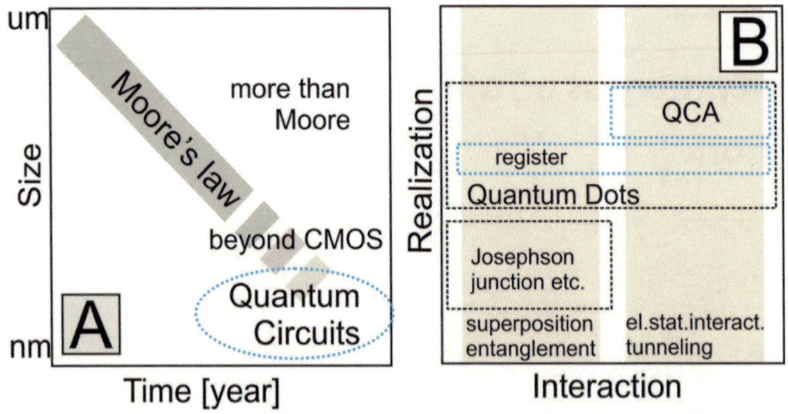

Figure 1. (A) Quantum circuit is a possible solution beyond CMOS technology; (B) realization of quantum circuits such as register or quantum dot cellular automata (QCA). The quantum dot-based realization can bridge the bit- and qubit-based circuits.

The work of quantum circuits (like quantum computing) is based on quantum mechanical phenomena, so the realization must be in a nanometer scale. In this field, one of the promising candidates is the QD-based technology. The QD-based computing technology fundamentally differs from earlier systems. Conventional digital computing technology uses voltage values to represent binary states. By contrast, QD-based computing system uses the position of electrons in QDs to represent binary states. Here, we can distinguish two main types according to the interactions. One of them utilizes superposition and entanglement, and another one utilizes electrostatic interaction and tunneling. For the computation, the first one uses the so-called qbits.

The quantum computer uses the quantum states to encode and process information. The unit of quantum information is the qubit, which can be shown as a two-stage system such as a QD. Opposite to some classical object, a quantum system can exist not only in the ground state $|0>$ or the excited state $|1>$, but in some linear superposition of these two stages. The possibility of the handling of these stages provides the main advantage of quantum computing [5]. One type uses the charge of an electron to form a qubit. The qubit realization is possible by single or two electron QDs. In close neighbor, two semiconducting QDs can be coupled with each other. They spatially confine an individual charge carrier in a discrete energy level, interact quantum mechanically with each other. The ordered QD pair ensemble system offers the potential of implementing tunable qubit arrays. The utilization of the ordering of charge-coupled QDs enables to realize also quantum circuits with utilization of classical bit. One of them is called as computational register and the other one is memory register, respectively [6, 7]. One of the main tasks of the quantum computer is the encoding of the qubit. The QD-based circuits can bridge the qubit- and bit-based circuits.

In this chapter, a very novel technology, called droplet epitaxy (DE), will be discussed in the applicational view point. What kind of possibilities can be served by DE for the technology of quantum circuitry? This method is useful when QDs are to be produced, of different shape and size in various densities. This technology is used already, for boosting efficiency, in device technology, such as lasers, LEDs, and solar cells. Their accurate positioning in complex structures like nano-sized circuits is very important. The lithographic direct processing used in microelectronics cannot be applied anymore; instead, the material's self-assembling properties is to be used, which is an inherent feature of every substance used.

There are various self-assembling methods to achieve spatial arranging or positioning. The nucleation of the nano-structure, which can be induced by local stress field, is to be used for locating laterally or vertically some nano-objects. This method, called controlled self-assembling, has three different forms. The first forms self-contained objects like QDs, QD pairs, or QD clovers (four-coupled QDs) by manipulating the technological parameters. The second uses the natural features (steps of monolayers (MLs), or dislocations, etc.) to induce the required ordering by self-assembling effect. The third induces the required order by applying artificial influence on the process, like for instance focused ion beam (FIB) or creation of nanoholes (NHs). The combination of these methods can provide possibility to create complex nano-structures. This is called hierarchical self-organization, which provides potentional creation of quantum circuits.

The chapter is organized in the following way. In the first parts, following "Introduction" section, we briefly describe the technique of DE. The following part describes the recent opportunities of the self-organizing-based creation of DE nanostructures. The last part describes two possibilities of applications: as an example, the QD register and circuit of quantum cellular automata will be discussed. The purpose of this paper is to speak to people engaged in circuit research with the aim of bringing together material scientists and circuit designers onto a common platform in order to overcome the problem of the present restrictions in further miniaturization.

2. Fundamentals of droplet-epitaxial technology

For the fabrication of QDs and other zero-dimensional nano-structures, various techniques have been developed. The molecular beam epitaxy (MBE) is the most advanced technology in this area for nano-structure preparation. For a long time, the only known method for the production of epitaxially grown zero-dimensional structures was the strain-induced method, based on lattice mismatch in Stranski-Krastanov growth mode [8–12]. InAs-based QDs on GaAs surface are the archetypal system. The driving force of the self-organized QD formation is the strain energy induced by the lattice mismatch, which in approximately 7% of the case the conditions restrict the material choice. Two groups of shape formations, like pyramids and domes, can be created with defect-free QD transformations.

The DE is a viable alternative technology to the production of strain-driven QDs [13–20]. Here, the material choice is not restricted by the lattice mismatch condition, which is a further advantage to a process, based on the strain-induced growth mode. DE also makes possible the fabrication of strain-free QDs and other nano-structures. This shape diversity of the produced nano-structures makes it advantageous in applications. The technology used for the growth governs the size, shape, and the elementary distribution of the developed structures. These physical parameters are very important in applications.

In DE applications, GaAs and related substances will be used as sample materials. That case, the clustering on the surface is carried out with the help of Volmer-Weber growth mode. This is a common idea, based on the splitting of the III- and V-column material supply, during the MBE growth (**Figure 2A**). The QD preparation consists of two main parts such as the formation of metallic nano-sized droplet on the surface and its crystallization. Here, the QD preparation consists of two main parts such as the formation of metallic nano-sized droplet on the surface and its crystallization with the help of the non-metallic component of the compound semiconductor [20]. In this way, not only conventional-shaped QDs but ring-like or double-ring-like zero-dimensional nano-structures can be created. Further possible nano-structures are the filled nano-hole and QD pairs or other ensembles, depending on growth parameters (**Figure 2B**). It must be noted that this DE technique is entirely compatible with the MBE technology. This attribute makes possible to combine the DE method with the other conventional MBE processes.

A typical QD preparation is illustrated in the following [21]: at first, on GaAs (001) wafer, an $Al_{0.3}Ga_{0.7}As$ layer is grown. After the layer preparation, the sample is cooled to 200°C. Following this, Ga (θ = 3.75 ML) is deposited with the flux of 0.75 ML/s without any arsenic flux. After the Ga deposition, a 60-s waiting time comes. The annealing is carried out at a temperature of 350°C and at an As pressure of 5×10^{-5} Torr. The process of GaAs crystallization starts at the edge of the droplet, initialized by the three-phase line at this point, serving as discontinuity for the crystal seeding. Although, in principle, interaction can take place at any point of the droplet, due to the thermal movement, the atoms, arriving to the edge, will start the seeding of the crystallization process.

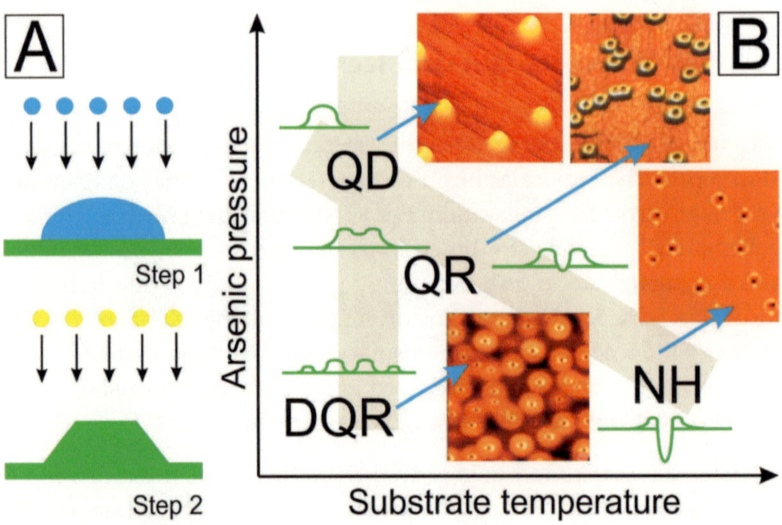

Figure 2. (A) The droplet epitaxial nano-structure production consists of two basic growth sequences; (B) versatile shaped nano-object can be created depending on the technological parameters (where QR is quantum ring, DQR is double quantum ring, and NH is nano-hole).

DE formation of ring-like QDs is similar as the previous description earlier, but the technological parameters are somewhat different; however, the AlGaAs layer preparation process is the same [22]. After that, the sample is cooled to 300°C. On the surface, Ga is deposited as described before. The same Ga is deposited with the flux of 0.19 ML/s without any arsenic flux. During the annealing, the temperature remained the same (300°C), but the arsenic pressure changed to 4×10^{-6} Torr. During the nano-structure formation, diffusion of the constituents has an important role.

A further recent method for the fabrication of strain-free QDs is the filling of nano-holes [23]. The nano-hole is created by localized thermal etching by liquid metallic droplet, and the created nano-hole is filled subsequently. A localized thermal etching takes place at conventional MBE growth temperatures, and we expect only very low level of crystal defects. The nano-holes are created in a self-organized fashion by local material removal. For inverted QD fabrication, nano-holes are generated by using Al droplets on AlAs surface. Following that, the holes are filled with GaAs to form QDs of controllable height. The nano-hole filling is carried out with GaAs in pulsed mode. The creation of QDs occurs with an inverted technology (**Figure 3**).

QD pairs can be prepared on AlGaAs surface by using the anisotropy of the (001)-oriented surface [25]. There are two known preparational processes. One of them is carried out under lower temperature, with a fewer amounts of deposited MLs. The other one is prepared under higher temperature at a higher amount of deposited Ga. In the first case, AlGaAs with an Al content of 0.27% is grown on GaAs (001) surface. Following that, Ga droplets are created at 330°C temperature on the substrate. The crystallization occurs at 200°C, under strict control of the arsenic flux. The resulting structure basically consists of two QDs aligned in the $[0\bar{1}1]$ crystallographic direction. In the other technology also, AlGaAs surface is being used. At 550°C substrate temperature, a large amount of Ga is deposited, to create droplets on the surface.

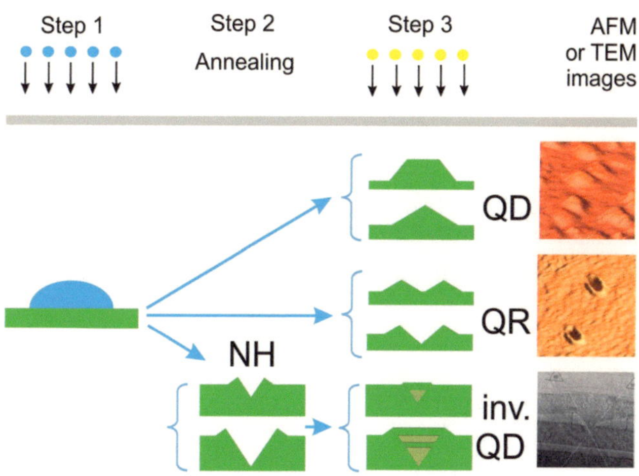

Figure 3. According to the technological parameters, the initial metallic droplet can lead to various zero-dimensional semiconductor nano-structures (where QR is quantum ring, NH is nano-hole, inv.QD is QD produced by nano-hole filling: inverted QD technology) (the AFM and TEM pictures originate from Refs. [22, 24], respectively).

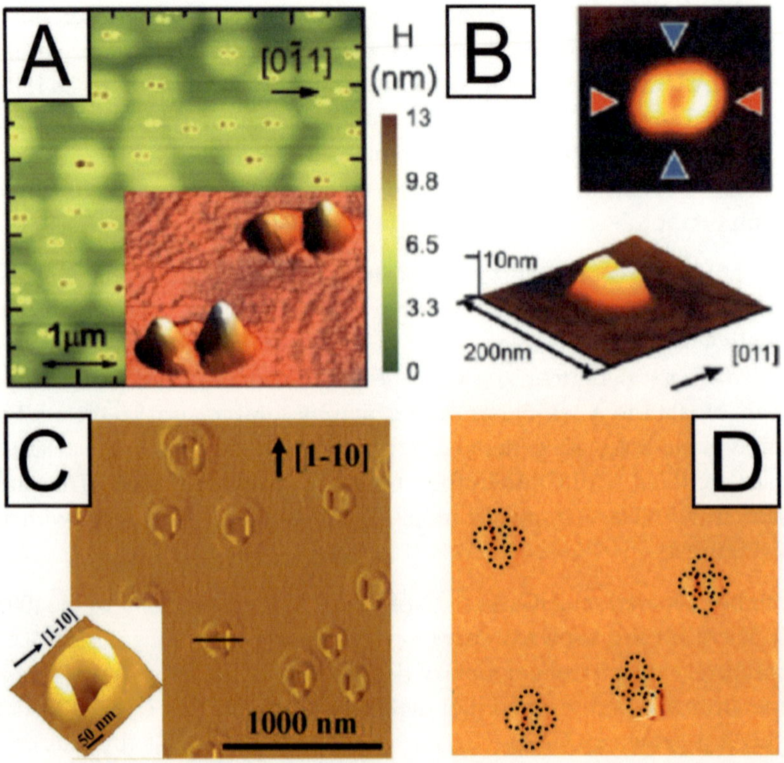

Figure 4. (A) and (B) AFM picture of QD pairs; (C) and (D) QD clovers (the AFM pictures originate from Refs. [24–26], respectively).

The structure is crystallized by an accurate control of the flux. The resulting dots are rather large. The individual pairs have an interdot distance of about 130 nm and are aligned along the [0$\bar{1}$1] direction. QD pairs are shown in **Figure 4A** and **B**.

Nano-objects consisting of four parts can also be grown by DE. The structure is a split-ring formation. A typical technological process, when the samples are grown on GaAs (001) substrates, is as follows. First, $In_{0.15}Ga_{0.85}$ of 20 ML is deposited with a rate of 1 ML/s at 360°C. Then, the formed droplets are exposed to arsenic beam for 5 min at a temperature of 200°C to crystallize the nano-droplets. Following this, the substrate temperature is raised to 450°C for the regrowth process with a growth rate of 0.05 ML/s. The structures are shown in **Figure 4C** and **D** [26, 27].

3. Ordered nano-structures

The self-assembling ordered QDs can be linearly, circularly, and also vertically alignmented. The most promising method for achieving long-range laterally ordered self-assembled QDs is the combination of substrate pre-patterning and self-assembled growth. The pre-patterning can be carried out by using the naturally occurring anomalies on the crystalline surface or can be made artificially by external influences. There are three kinds of linear alignmentation methods (**Figure 5**). One of them is the surface cross-hatch-induced mode, and the other kind is the alignmentation created by ML step. These are utilization of naturally formed surface effects. The third one is a fully artificial method, where the alignmentation is induced by ion beam-created surface damage.

The dislocations, generated at the substrate/layer junction, show themselves on the surface as ridges and troughs. At a sufficiently high density of dislocations, the development of misfit dislocation network shows up at this junction. Such a network, consisting of two arrays of single dislocations with alternating glide planes, will result in a quadratic surface structure. The dislocation network shows itself at the surface, which is called as a cross-hatch pattern. This pattern coexists with the crystallites, giving the possibility of using the interplay between the two strain-relief mechanisms for self-ordering of QDs.

The cross-hatch pattern creation has been already demonstrated in different material systems such as $In_xGa_{1-x}As/GaAs$ [28], $In_xGa_{1-x}As/InP$ [29], and $Si_{1-x}Ge_x/Si$ [30], attributed to misfit dislocations and glides. Its production is as follows. Self-assembled InAs QDs are grown on cross-hatched surface, consisting of 50 nm $In_{0.15}Ga_{0.85}As$ layers on GaAs (0 0 1) substrate. The lattice-mismatched $In_{0.15}Ga_{0.85}As$ layer is left growing well beyond the critical layer thickness for the formation of misfit dislocation in order to form long orthogonal cross-hatch pattern oriented along the [1 1 0] and [0$\bar{1}$1] crystalline directions (**Figure 5A**). On top of the cross-hatched surface, InAs layer growth at a low growth rate of 0.01 ML/s and at a thickness of 0.8 ML originates spontaneous QD formation. It was found that the

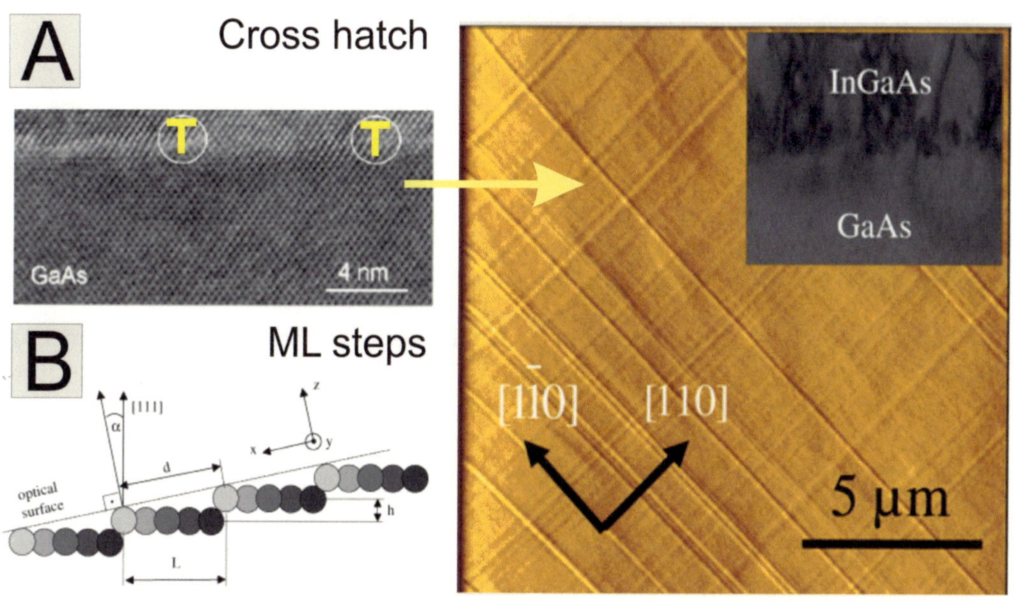

Figure 5. Linearly aligned QDs; (A) the QD alignment is induced by cross-hatch (B) and by monolayer (ML) steps, (C) and by ion-induced surface damage (the AFM and SEM pictures originate from Refs. [31, 39], respectively).

substrate temperature reduction immediately after the QD formation will result in a majority of QDs alignment on the cross-hatch pattern. With a short growth interruption, duration of 30 s, before reducing the substrate temperature, the QDs will form, almost exclusively on the cross-hatches and the surface formation is named QD hatches. Exceeding the optimum interruption time will result in inhomogeneous, sparsely connected QD hatches, possibly due to desorption of In atoms [31].

The second possibility to self-aligned QD ordering in a crystalline layer uses step bunching of preexisting ML steps on the miscut (0 0 1) substrate (**Figure 5B**). Crystallite ordering on vicinal-oriented (0 0 1) surface is guided by spontaneously formed step-bunched ripple patterns. The ripple distance and orientation can be engineered by varying the polar and azimuthal miscut directions of the substrate [32].

The focused ion beam bombardment is a widely used technique for surface preparation and nano-patterning for the fabrication of self-assembling nanostructures such as nano-ripples, nano-needles, nano-holes, and also QDs. FIB-induced self-assembly of ordered nano-structures has been reported on metals, semiconductors, and insulators as well [32–38].

Ordered Ga nano-droplets can be self-assembled under ion beam bombardment at off-normal incidence [39]. The homogeneity, size, and density of Ga nano-droplets can be controlled by the incident ion beam angle. The beam current also plays a crucial role in the self-ordering of Ga nano-droplets. It has been found that the droplets exhibit a similar droplet size but higher density and better homogeneity with an increased current of ion beam. Compared to the destructive formation of nano-droplets by direct ion beam bombardment, the controllable assembly of nano-droplets on intact surfaces can be used as templates for DE fabrication of arranged semiconductor nano-structures (**Figure 5C**).

The start of circularly aligned QD molecule can be initialized by a droplet edge (**Figure 6A**) or by a rim of nano-holes (**Figure 6B**). It is a simple method of preparing ring-shape InP nano-structures on $In_{0.49}Ga_{0.51}P$ by using DE. The surface morphology of the structure depends strongly on the ML of In. For instance, the ring-shaped nano-structure is formed at 1.6-ML In thickness. The ring-shaped QD molecule is formed when the deposited ML of In is less than 3.2 ML. It has been found that the density, height, and average number of QD per molecule are dependent on the In MLs and on its deposition rate [40].

A relatively simple way to fabricate vertical QD molecules is to grow stacks of QDs (**Figure 7A**). It is known that the surface strain field modulation from a buried island layer influences the island nucleation in the next layer and this leads to a spontaneous vertical alignment [41, 42]. The electronic coupling between vertically aligned QDs has been demonstrated earlier [40, 43–48]. Further vertically stacked QD ensembles can be created by sequentially filled nano-hole. First, Al or Ga droplets are created on the AlGaAs surface. After them an annealing appears, where the substrate temperature is ranged between 550 and 650°C, the arsenic pressure is under 10^{-7} Torr (**Figure 7B**). During this annealing, the initial droplet transforms into a nano-hole surrounded by a protrusion. The nano-hole is filled by pulsed mode. The filling consists of 0.5-s GaAs deposition followed by a 30-s pause. The stacked QDs are separated by an AlGaAs barrier layer deposition [49].

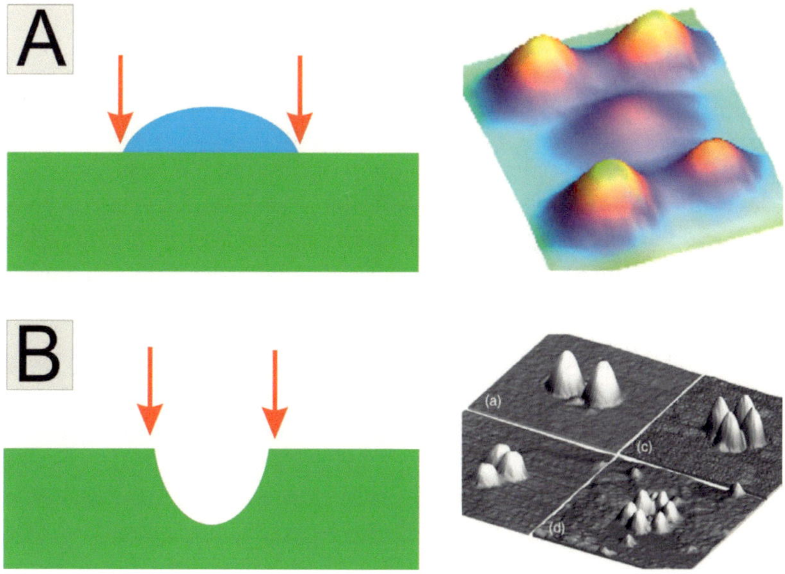

Figure 6. Circularly aligned QDs (QD molecule); (A) the QD nucleation is at droplet edge (B) and at the rim of hole opening. The arrows indicate the seeding places (the AFM pictures originate from Refs. [46, 47], respectively).

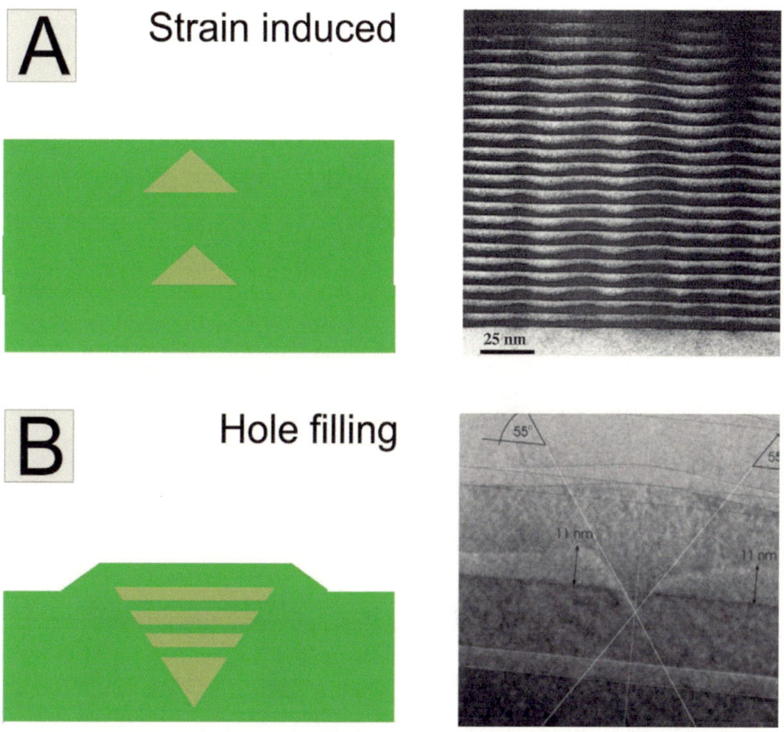

Figure 7. The vertical alignment; (A) the vertical stacking is induced by strain, (B) vertically coupled QDs by nano-hole (NH) filling (the TEM pictures originate from Refs. [24, 48], respectively).

4. Applications in quantum circuitry

In this chapter, we discuss two types of circuits composed from aligned QDs. One of them is the linearly aligned register. A QD register for quantum computing can be realized by uniformly aligned QDs or by QD pairs with the help of directed DE assembly [50]. A possible realization can be the following. The linearly aligned GaAs QDs is created on an AlGaAs surface. This structure is embedded by a barrier material of AlGaAs. When the cover layer few MLs only then the subsequently deposited metallic droplets are positioned most likely by the QD sites below (**Figure 8**).

The second discussed structure is the QD cellular automata, which was firstly proposed in the beginning of 1990s [51]. The QD-based cellular automaton is one of the most promising device structures in the future [52–55]. The circuit consists of coupled QD array to realize Boolean logic functions [9] and to perform useful computations. Two main advantages of QD cellular automata are the exceptionally high logic integration derived from the small QD size, and the low power consumption. QD cellular automata can be used to implement complex digital circuits by properly arranged QD clovers. Such circuits are, for example, full adder, multiplexer, programmable logic array, multivibrator or can be also designed memory circuits, such as quantum dot cellular automatic random access memory and serial memory. The basic building block of QD cellular automata device named cell is presented in **Figure 9A**. QD cellular automata unit cell consists of four QDs in a square array coupled by tunnel barriers, and two

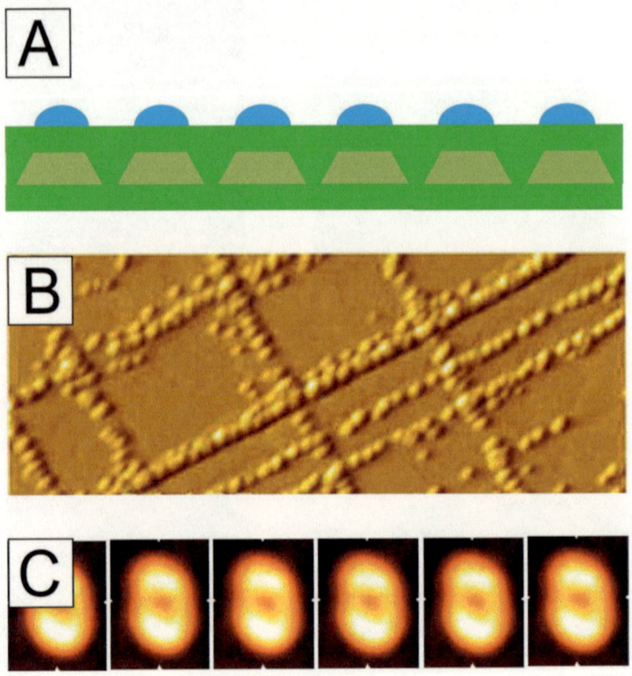

Figure 8. Realization of QD register: (A) cross section of literally aligned QD series with vertically positioned self-assembling metallic clusters; (B) along cross-hatch-aligned QDs; (C) the alignmentation of QD pairs is also possible.

electrons are injected into the cell (**Figure 9B**). Due to Coulombic repulsion, the two electrons reside in opposite corners representing two polarizations. Some basic elements for QD cellular automata logic implementation are wire, inverter, and majority voter [52–60].

The DE-grown QDs as building elements for quantum computing were first proposed in 2009 [61, 62]. The alignmentation of the QD clovers can be realized like a single dot, which can lead to the wire implementation. Here, the linear inhomogeneity of the surface can be utilized. For the gate realization, these inhomogeneities for directed assembly must be generated artificially. The QD cellular automata can be realized in more levels (**Figure 9C** and **D**). The couplings between the circuits on the adjacent levels can be carried out with vertical alignment.

For any required operation, it is very important to determine the optimal size of the QDs and their distances from each other. Not only the size but the working temperature is also important. At the realization, it is important to take into consideration that the switching fidelity increases with decreasing temperature [63]. It is predicted that the density of the QD-based circuits could exceed the device density of 10^{12} cm^{-2} and the operating speed could reach the frequency of THz region [64]. The clocking in THz region can also be realized with the help of DE. One of the effective ways to generate THz pulses is realized by near-infrared femtosecond laser irradiation on semiconductor/metal surfaces with the help of plasmon enhancement [65–67]. The DE is an appropriate technology to create such positioned semiconductor and metallic nano-particles. The structure can be realized by DE-grown self-alignment of QD molecules and metallic nano-particle [25, 68]. Promising perspective is provided with a recent result to the realization of the nano-positioned metallic nano-particle on QD molecule, which can be useful not only at the THz clocking but also at QD register, too [69].

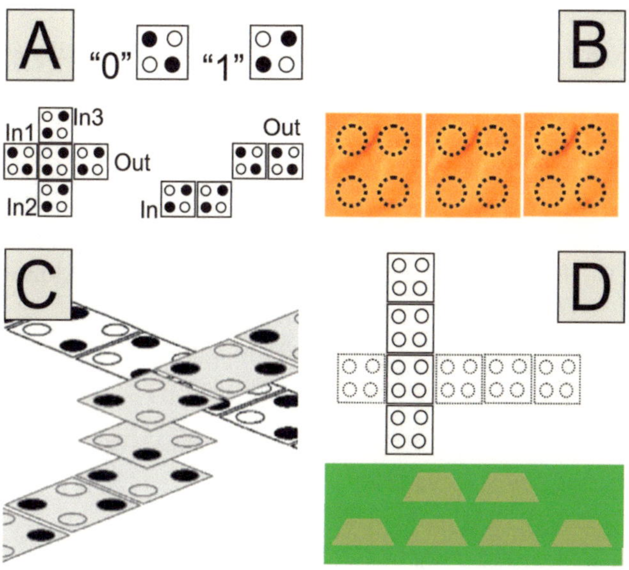

Figure 9. Realization of QD clover-based QD cellular automata; (A) the basic unit and the majority gate and inverter gate of the QD cellular automata; (B) the realization of QD cellular automata by QD.

5. Conclusion

There are still a number of open scientific problems awaiting a solution. For the perfect operation of the circuits, the optimal QDs and their distances from each other must be determined. It is a fact that the sizes and the shape of the QD are not independent from their elementary density, which finally determines the distances among the QDs. The task is rather complex. If we can understand the details of the evolution mechanism of the DE-grown nano-structures, we can approach the technological solution of the circuit formation. It is another possibility to take into account the technological capability at the circuit design, which increases the importance of the mutually common thinking among different professionals. Lately, the number of published papers in this area has increased drastically, which is an encouraging sign for the possible technological solution.

Acknowledgements

This work was supported party by NKFI-OTKA-114457(FemtoTera) and partly by OE(KVK and ADTI) research grants, which are acknowledged.

Author details

Ákos Nemcsics

Address all correspondence to: nemcsics.akos@kvk.uni-obuda.hu

Institute for Microelectronics and Technology, Obuda University, Budapest, Hungary

References

[1] Carballo JA, Chan WTJ, Gargini PA, Kahng AB, Nath S. In: Proceedings of 32nd IEEE International Conference on Computer Design (ICCD); 19-22 October 2014; Seoul; 2014. pp. 139-146

[2] Csurgay ÁI, Csaba Gy, Porod W. ECCTD'01—European Conference on Circuit Theory and Design; 28-31 August 2001; Espoo, Finland; 2001. p.I-9

[3] Csurgay ÁI. On circuit models for quantum-classical networks. International Journal of Circuit Theory and Applications. 2007;**35**:471-484

[4] Csurgay ÁI, Porod W. Circuit models for arrays of nanoelectronic resonator appearances of discrete breathers. International Journal of Circuit Theory and Applications. 2007;**35**:305-313

[5] Deutsch D. Quantum theory, the Church-Turing principle and the universal quantum computer. Proceedings of the Royal Society of London A. 1985;**97**:400

[6] Vedral V. Introduction to Quantum Information Science. Oxford: Oxford University Press; 2006

[7] Stolze J, Suter D. Quantum Computing – A Short Course from Theory to Experiment. Wiley-VCH: Weinheim; 2004

[8] Leonard D, Krisnamurthy M, Reaves CM, Denbaas SP, Petroff PM. Direct formation of quantum-sized dots from uniform coherent islands of TnGaAs on GaAs surfaces. Applied Physics Letters. 1993;**63**:3203

[9] Bressler-Hill V, Varma S, Lorke A, Nosho BZ, Petroff PM, Weinberg WH. Island scaling in strained heteroepitaxy: InAs/GaAs (001). Physical Review Letters. 1995;**75**:3209

[10] Yang W, Lee H, Johnson TJ, Sercel PC, Normann AG. Electronic structure of self-organized InAs/GaAs quantum dots bounded by {136} facets. Physical Review B. 2000; **61**:2784

[11] Stier O, Grundmann M, Bimberg D. Electronic and optical properties of strained quantum dots modelled by 8-band k.p theory. Physical Review B. 1999;**59**:5688

[12] Heyn C, Bolz A, Maltezopulos T, Johnson RL, Hansen W. Intermixing in self-assembled InAs quantum dot formation. Journal of Crystal Growth. 2005;**278**:46

[13] Koguchi N, Takahashi S, Chikyow T. New MBE growth method for InSb quantum well boxes. Journal of Crystal Growth. 1991;**111**:688

[14] Koguchi N, Ishige K. Growth of GaAs epitaxial microcrystals on an S-terminated GaAs substrate by succesive irradiation of Ga and As molecular beams. Japanese Journal of Applied Physics. 1993;**32**:2052

[15] Mano T, Kuroda T, Mitsuishi K, Yamagiwa M, Guo XJ, Furuya K, Sakoda K, Koguchi N. Ring-shaped GaAs quantum dot laser grown by droplet epitaxy: Effects of post-growth annealing on structural and optical properties. Journal of Crystal Growth. 2007; **301-302**:740

[16] Mano T, Tsukamoto S, Fujioka H, Oshima M, Koguchi N. Indium segregation in the fabrication of heterogeneous droplet epitaxy. Journal of Crystal Growth. 2001; **227-228**:1069

[17] Gong Z, Nin ZC, Huang SS, Fang ZD, Sun BQ, Xia JB. Formation of GaAs/AlGaAs and InGaAs/GaAs nanorings by droplet molecular beam epitaxy. Applied Physics Letters. 2005; **87**:093116

[18] Kuroda T, Mano T, Ochiai T, Sanguinetti S, Sakoad K, Kigo G, Koguchi N. Optical transitions in quantum ring complexes. Physical Review B. 2005;**72**:205301

[19] Hwang S, Nin Z, Fang Z, Ni H, Gong Z, Xia JB. Complex quantum ring structures formed by droplet epitaxy. Applied Physical Letters. 2006;**89**:031921

[20] Nemcsics Á. Stavrou V. (Ed.) Quantum Dots Prepared by Droplet Epitaxial Method in Quantum Dots—Theory and Applications. InTech Publishing; 2015

[21] Heyn C, Stemmann A, Schramm A, Welsch H, Hansen W, Nemcsics Á. Faceting during during GaAs quantum dot self-assembly by droplet epitaxy. Applied Physical Letters. 2007;**90**:203105

[22] Nemcsics Á, Heyn C, Stemmann A, Schramm A, Welsch H, Hansen W. The RHEED tracking of the droplet epitaxial grown quantum dot and ring structures. Materials Science and Engineering B. 2009;**165**:118

[23] Heyn C, Stemmann A, Köppen T, Strelow C, Kipp T, Grave M, Mendach S, Hansen W. Highly uniform and strain-free GaAs quantum dots fabricated by filling of self-assembled nanoholes. Applied Physics Letters. 2009;**94**:183113

[24] Nemcsics Á, Heyn C, Tóth L, Dobos L, Stemmann A, Hansen W. Cross sectional transmission electron microscopy of GaAs quantum dots fabricated by filling of droplet-etched nanoholes. Journal of Crystal Growth. 2011;**335**:58

[25] Yamagiwa M, Mano T, Kuroda T, Tateno T, Sakoda K, Kido G, Koguchi N, Minami F. Self-assembly of laterally aligned GaAs quantum dot pairs. Applied Physics Letters. 2006;**89**:113115

[26] Boonpeng P, Jevasuwan W, Suraprapapich S, Ratanathammaphan S, Panyakeow S. Quadra-quantum dots grown on quantum rings having square-shaped holes: Basic nanostructure for quantum dot cellular automata application. Microelectronic Engineering. 2009;**86**:853-856

[27] Nemcsics Á, Balázs J, Pődör B, Makai J, Szemann A. Photoluminescence studies of GaAs quantum dots and quantum rings. Physica Status Solidi C. 2011;**8**:2826

[28] Rajan K, Devine R, Moore WT, Magine P. Dislocation structure in InGaAs/GaAs strained-layer superlattices. Journal of Applied Physics. 1987;**62**:1713

[29] Dumont H, Auvray L, Dazord J, Souliere V, Monteil Y, Bouix J. Strain-induced surface morphology of slightly mismatched InGaAs films grown on vicinal (100) InP substrates. Journal of Applied Physics. 1999;**85**:7185

[30] Shiryaev SY, Jensen F, Hansen JL, Petersen JW, Larsen AN. Nanoscale structuring by misfit dislocations in SiGe/Si epitaxial systems. Physical Review Letters. 1997;**78**:503

[31] Thet CC, Panyakeow S, Kanjanachuchai S. Growth of InAs quantum-dot hatches on InGaAs/GaAs cross-hatch virtual substrates. Microelectronic Engineering. 2007;**84**: 1562-1565

[32] Rusponi S, Boragno C, Valbusa U. Ripple structure on Ag(110) surface induced by ion sputtering. Physical Review Letters. 1997;(14):2795-2798

[33] Wei Q, Zhou X, Joshi B, Chen Y, Li KD, Wei Q, Sun K, Wang L. Self-assembly of ordered semiconductor nanoholes by ion beam sputtering. Advanced Materials. 2009;(28): 2865-2869

[34] Rose F, Fujita H, Kawakatsu H. Real-time observation of FIB-created dots and ripples on GaAs. Nanotechnology. 2008;**19**:874-880

[35] Du Y, Atha S, Hull R, Groves J, Lyubinetsky I, Baer D. Focused-ion-beam directed self-assembly of Cu2O islands on SrTiO3(100). Applied Physical Letters. 2004;**84**:5213-5215

[36] Rusponi S, Costantini G, Boragno C, Valbusa U. Scaling law of the ripple morphology on Cu(110). Physical Review Letters. 1998;(81):4184-4187

[37] Lian J, Wang L, Sun X, Yu Q, Ewing RC. Patterning metallic nanostructures by ion-beam-induced dewetting and rayleigh instability. Nano Letters. 2006;(5):1047-1052

[38] Chason E, Mayer T, Kellerman B, McIlroy D, Howard A. Roughnening instability and evolution of the Ge(001) surface during ion sputtering. Physical Review Letters. 1994;(72):3040-3043

[39] Xu X, Wu J, Wang X, Zhang M, Li J, Shi Z, Li H, Zhou Z, Ji H, Niu XM, Wang Z. Ion-beam-directed self-ordering of Ga nanodroplets on GaAs surfaces. Nanoscale Research Letters. 2016;**11**:38

[40] Jevasuwan W, Panyakeow S, Ratanathammaphan S. In-droplet-induced formation of InP nanostructures by solid-source molecular beam epitaxy. Microelectronic Engineering. 2007; **84**:1548-1551

[41] Xie Q, Madhukar A, Chen P, Kobayashi NP. Vertically self-organized InAs Quantum Box Islands on GaAs (100). Physical Review Letters. 1995;**75**:2542

[42] Schmidt OG, Eberl K. Multiple layers on self-assembled Ge/Si islands: Photoluminescence, strain fields, material interdiffusion, and island formation. Physical Review B. 2000; **61**:13721

[43] Krenner HJ, Sabathil M, Clark EC, Kress A, Schuh D, Bichler M, Abstreiter G, Finley JJ. Direct observation of controlled coupling in an individual quantum dot molecule. Physical Review Letters. 2005;**94**:057402-1-4

[44] Ortner G, Bayer M, Lyanda-Geller Y, Reinecke TL, Kress A, Reithmaier JP, Forchel A. Control of vertically coupled InGaAs/GaAs quantum dots with electric field. Physical Review Letters. 2005;**94**:157401-1-4

[45] Borri P, Langbein W, Woggon U, Schwab M, Bayer M, Fafard S, Wasilewski Z, Hawrylak P. Exciton dephasing in quantum dot molecules. Physical Review Letters. 2003;**91**:267401-1-4

[46] Sablon KA, Lee JH, Wang ZM, Shultz JH, Salamo GJ. Configuration control of quantum dot molecules by droplet epitaxy. Applied Physics Letters. 2008;**92**:203106

[47] Rastelli A, Songmuang R, Kiravittaya S, Schmidt OG. In: Schmidt OG, editor. Lateral Alignment of Epitaxial Quantum Dots. Springer; 2007. p. 103

[48] Grundmann M, Ledentsov NN, Kirstaedter N, Alferov ZI. Semiconductor quantum dots for application in diode lasers. Thin Solid Films. 1998;**318**:83-87

[49] Heyn C, Sonnenberg D, Hansen W. In: Wang ZM, editor. Nanodroplets, Lecture Notes in Nanoscale Science and Technology. 2013;**18**:363

[50] Balandin A, Jin G, Wang KL. Issues of practical realization of a quantum dot register for quantum computing. Journal of Electronic Materials. 2000;**29**:549-553

[51] Lent CS, Tougaw PD. Lines of interacting quantum-dot cells: Binary wire. Journal of Applied Physics. 1993;**74**:6227

[52] Orlov AO, Amlani I, Bernstein GH, Lent CS, Snider GL. Realization of a functional cell for quantum-dot cellular automata. Science. 1997;**277**:928

[53] Oya T, Asai FT, Amemiya Y. Design of generalized pipeline cellular array in quantum-dot cellular automata. IEEE Transactions on Nanotechnology. 2003;**2**:15

[54] Ganesh EN. Power analysis of quantum cellular automata circuits. Procedia Materials Science. 2015;**10**:381-394

[55] Sen B, Nag A, De A, Sikdar BK. Towards the hierachial design of multilayer QCA logic circuit. Journal of Computational Science. 2015;**11**:233-244

[56] Choi M, Patitz Z, Jin B, Tao F, Park N, Choi M. Designing layout-timing independent quantum-dot cellular automata circuits by global assynchrony. Journal of Systems Architecture. 2007;**53**:551-567

[57] Rao NG, Srikanth PC, Sharan P. A novel quantum dot cellular automata for 4 bit code converters. Optik. 2016;**127**:4246-4249

[58] Mohammadi M, Mohammadi M, Gorgin S. An efficient design of full adder in quantum-dot cellular automata technology. Microelectronics Journal. 2016;**50**:35-43

[59] Sheikhfaal S, Angizi S, Sarmadi S, Moaiyeri MH, Sayedsalehi S. Desugning efficient QCA logical circuits with power dissipation analysis. Microelectronics Journal. 2015; **46**:462-471

[60] Gladshtein M. Design and simulation of novel adder/subctarctor on quantum-dot cellular automata: Radical departure from Boolean logic circuits. Microelectronics Journal. 2013;**44**:545-552

[61] Sablon KA. Towards quantum computing: A hybrid approach that will unleash a plethorn of new QD nanostructures, bringing us a step further to laterally coupled QDs. Nanoscale Research Letters. 2009;**4**:1254

[62] Panyakeow S. Quantum nanostructures by droplet epitaxy. Engineering Journal. 2009; **13**:51

[63] Tiihonen J, Schramm A, Kylänpää I, Rantala TT. Exact modeling of finite temperature and quantum-dot cellular automata. Journal of Physics D. 2016;**49**:1

[64] Takoori MB, Momenzadeh M, Huang J, Lombardi F. Design and performance evaluation of approximate floating point multipliers. Proceedings of 22nd IEEE VLSI Test Symposium. 2004. p. 291

[65] Zhang XC, BB H, Darrow JT, Anton DH. Generation of femtosecond electromagnetic pulses from semiconductor surfaces. Applied Physical Letters. 1990;**56**:1011

[66] Ramakrishnan G, Kumar N, Ramanandan GKP, Adam AJL, Hendrix RWA, Planken PCM. Plasmon enhanced terahertz emission from semiconductor/metal interface. Applied Physical Letters. 2014;**104**:071104

[67] Carreno F, Antón MA, Melle S, Calderón OG, Cabrera-Granado E. Plasmon enhanced terahertz emission in self-assembled quantum dots by femtosecond pulses. Journal of Applied Physics. 2014;**115**:064304

[68] Urbanczyk A, Hamhnis GJ, Nötzel R. Site-controlled Ag nanocrystals grown by molecular beam epitaxy – Towards plasmonic integration technology. Applied Physical Letters. 2010;**97**:043105

[69] Elborg M, Noda T, Mano T, Kuroda T, Sakoda K. Self-assembly of vertically aligned quantum ring-dot structure by Multiple Droplet Epitaxy. Journal of Crystal Growth. 2017;**477**:239-242

Chapter 4

Colloidal III–V Nitride Quantum Dots

Zequn Chen, Chuli Sun, Wei Guo and Zhuo Chen

Additional information is available at the end of the chapter

http://dx.doi.org/10.5772/intechopen.70844

Abstract

Colloidal quantum dots (QDs) have attracted intense attention in both fundamental studies and practical applications. To date, the size, morphology, and composition-controlled syntheses have been successfully achieved in II–VI semiconductor nanocrystals. Recently, III-nitride semiconductor quantum dots have begun to draw significant interest due to their promising applications in solid-state lighting, lasing technologies, and optoelectronic devices. The quality of nitride nanocrystals is, however, dramatically lower than that of II–VI semiconductor nanocrystals. In this review, the recent development in the synthesis techniques and properties of colloidal III–V nitride quantum dots as well as their applications are introduced.

Keywords: colloidal synthesis, III–V nitride, quantum dots, semiconductor, optoelectronic properties

1. Introduction

Due to the uniquely tunable electronic structure and low-cost synthesis in a controllable way, colloidal quantum dots (QDs) have attracted intense attention in both fundamental studies and practical applications [1–5], such as solar cell, quantum dot light-emitting diode, and spectrometer. Usually, semiconductor quantum dot properties can be varied by their size, composition, morphology, and phase structure. To date, with the rapid development of synthesis technique, the size, morphology, and composition-controlled syntheses of colloidal quantum dots have been successfully achieved [6–9].

III–V semiconductors are crystalline binary compounds formed by combining metallic elements from group III and nonmetallic elements from group V of the periodic table [10]. In the III–V nitride, the wurtzite phase is the stable form and they have direct bandgaps ranging from 0.7 eV for InN, to 3.4 eV for GaN, and to 6.2 eV for AlN. They can combine with each other to form alloys with bandgaps value from 0.7 to 6.2 eV, covering a wide range of spectra from ultraviolet (UV) to infrared region, exhibiting large potential applications for electronic and optoelectronic devices.

Recently, III-nitride semiconductor (such as GaN, InN, and AlN) quantum dots (QDs) have begun to draw significant interest due to their promising applications in solid-state lighting, lasing technologies, and optoelectronic devices. The quality of nitride nanocrystals is, however, dramatically lower than that of II–VI semiconductor nanocrystals. For synthesis of III–V nitride quantum dots with uniform distribution, it is important to have a very fast nucleation and relatively slow growth process, which requires the growth unit concentration to reach high super-saturation level. However, such condition is very difficult to achieve for nitride quantum dots due to their strong covalent bonding and lack of suitable precursors. Although during the past few decades, various methods and precursors have been studied, the effective reaction with a control over group III elements and nitrogen in the solution has still remained difficult. Herein, we review the recent development in the synthesis techniques and properties of colloidal III–V nitride quantum dots as well as their applications. Meanwhile, the overview will partially involve the development and improvement of III-nitride QDs grown by vapour-phase methods. More detailed introduction of colloidal III–V nitride quantum dots, that is, GaN, InN, and their alloys, are presented in the following discussion.

1.1. Gallium nitride quantum dots

GaN is a technologically important direct semiconductor for development of short-wavelength optoelectronic devices, high-speed microwave device, and high-density integrated circuit [11–13]. Its bandgap is 3.4 eV. GaN also have chemical and radiation resistance, and is therefore being considered as a stable photocatalyst in photoelectrochemical (PEC) cells for the production of fuels [14]. Colloidal QDs made from this material are expected to comprise good thermal, chemical, and radiation stability with the excellent optical properties. Therefore, since Xie [15] group succeeded in preparing GaN nanoparticles by simple inorganic reactions at 300°C, considerable efforts have been made towards the solution-based synthesis of GaN QDs at low temperature and the understanding of optical and electronic properties. In the past 20 years, many researchers have prepared zero-dimensional (0D) GaN nanostructures by top-down approach. Various methods based on a bottom-up approach, like solvothermal methods, thermal decomposition, and so on, have been used to synthesize 0D GaN QDs. However, for the wet-chemical approach, controlling the size of 0D GaN QDs and even their optical and electrical properties remains a significant challenge.

1.2. Indium nitride quantum dots

Among III–V nitrides, indium nitride is of relatively low-thermal stability. For example, the InN thin film conducts thermal decomposition under dinitrogen desorption at 500–550°C [16]. InN is one of the least studied materials in the III–V compounds. Previously, it was believed that the fundamental bandgap of InN was 1.9 eV, until much more recent studies on higher quality films of InN have clearly shown that the true value of its direct bandgap is at 0.7 eV [17], making it a very promising compound for optoelectronic applications. Recently, with the further study of the group III nitride, the semiconductor properties of InN have attracted increasing attention. It has a good property on electron transport, making it face the huge application on high-speed electronic device due to its small-effective mass [18]. InN is also considered as an excellent material for low-cost, high-efficiency solar cell, photomask, light-emitting diodes, laser diodes, sensors, and THz radiation [19–22]. In addition, InN is considered as a promising candidate for biological imaging and in vivo medical applications because of its nontoxicity and its infrared emission in the optically transparent region of water and blood [23].

With the increasing importance on InN, how to obtain the InN with high quality is also attracting much attention. Although several synthesis methods, such as solvothermal methods [24, 25], sputtering [26], and molecular beam epitaxy (MBE) [27], were developed to prepare 0D InN QDs, more effort needs to be devoted to improve the quality of InN QDs with a controlled way.

1.3. Indium gallium nitride quantum dots

Group III nitrides have a high light-emitting efficiency due to its direct bandgap as well as high radiative transition rate. GaN and InN can form component continuous solid solution and superlattice, like InGaN. The alloy's bandgap can be tuned by controlling the ratio of GaN/InN and with the increasing composition of In, the VBM of the InGaN increases in energy almost linearly [28]. In addition, due to the quantum size effect, the bandgap of the InGaN can be further tuned by changing the size and shape of the QDs, so that these semiconductors can be used for red to ultraviolet emitting devices [29].

Unlike the synthesis methods of GaN and InN, the way to prepare InGaN quantum dots are mainly top-down approach, such as plasma-assisted molecular beam epitaxy (PA-MBE) [30], metal-organic vapour phase epitaxy (MOVPE) [31], and metal-organic chemical vapour deposition (MOCVD) [32].

There are abundant papers and books to review the development of 1D and 2D III–V nitride and other nanostructures. However, there are few reviews on the research status of the colloidal III–V nitride quantum dots. This research field is important for fundamental science and technology and is growing fast. In this review, we focus on recent progresses in the synthesis, crystal structure, and optoelectronic properties of colloidal III–V nitride quantum dots.

2. Crystal structure

There are three common crystal structures shared by the group III nitrides, namely, the wurtzite, zinc blende, and rocksalt structures. Usually, the thermodynamically stable structures are wurtzite for bulk AlN, GaN, and InN. The large difference in electronegativity between the group III and group V elements (Al = 1.18, Ga = 1.13, In = 0.99, N = 3.1) results in very strong chemical bonds within the III-nitride material system, which not only is at the origin of most of the exceptional III-nitride physical properties (listed in **Table 1**), but also greatly hinder the low-temperature solution synthesis of III–V nitrides [33]. For the growth of III–V nitride QDs, vapour deposition based on substrate is a more popular approach.

	AlN	GaN	InN
Lattice constant, a (Å)	3.112	3.189	3.545
Lattice constant, c (Å)	4.982	5.186	5.703
Thermal expansion coefficient α_a (10^{-6} K^{-1})	5.27 (20–800°C)	4.3 (17–477°C)	5.6 (280°C)
Thermal expansion coefficient α_c (10^{-6} K^{-1})	4.15 (20–800°C)	4.0 (20–800°C)	3.8 (280°C)
Electron effective mass, m_e (m_0)		0.2	0.11
Hole effective mass, m_h (m_0)		0.8	0.5 (mhh) 0.17 (mlh)
Refractive index, n	2.2 (0.60 μm) 2.5 (0.23 μm)	2.35 (1.0 μm) 2.60 (0.38 μm)	2.56 (1.0 μm) 3.12 (0.66 μm)
ε (0)	9.14	10.4 (E‖c) 9.5 (E⊥c)	
ε (∞)	4.84	5.8 (E‖c) 5.4 (E⊥c)	9.3
Thermal conductivity, (κher/cm K)	2.0	1.7–1.8	
Melting point (°C)	2000	>1700	1100
ΔG^0 (kcal/mol)	−68.2	−33.0	−23.0
Heat capacity, C_p (cal/mol K)	7.6	9.7	10.0

Table 1. Physical properties of III–V nitride semiconductors [33].

3. Colloidal III–V nitride quantum dots

3.1. Syntheses of colloidal nitride quantum dots

So far, many efforts have been devoted to synthesize colloidal nitride quantum dots by the solution-phase routes. Over the past 20 years, many groups have prepared the colloidal III–V nitride quantum dots by various solution-based methods. These methods can be classified into the following approaches, namely solvothermal [24, 25, 34], hydrothermal [35], and thermal decomposition of single [36, 37] and two precursors [38]. The hydrothermal process refers to the reaction of reactants and water in a pressurized reaction environment to form nanoparticles. The process of forming of nanocrystals undergoes two stages, dissolution and crystallization. In the primary reaction, the aggregation and binding of the precursor particles are destroyed, making the particles dissolute in the hydrothermal solvent, transporting into the solution in the form of ions or ionic groups, and crystallizing the crystalline grain after nucleation. Solvothermal method was developed on the basis of hydrothermal method, which uses the organic solvent as solution, replacing water. In this way, some compounds sensitive to water (reacting with water, hydrolysis, resolution, or instability) like III–V group semiconductors, carbides, fluorides, and so on, can be prepared. LaMer and Dinegra thought that the preparation of monodisperse nanocluster needed a transient and discrete nucleation process, and then controlled the crystal nucleus growth slowly [39]. Putting the reactants rapidly into the container makes the precursor concentration higher than the nucleation threshold value. A short period of sudden nucleation can reduce saturation. As long as the rate of the consuming concentration for nanocrystal growth reaction is less than the rate of precursor injection, there are no new nanocrystals formed. Therefore, the size of the particle distributions mostly depends on the time from nucleation to growth. This method has some advantages that cannot be replaced by other methods: (1) uniform morphology; (2) narrow size distribution; and (3) high crystallinity due to relatively high reaction temperature.

3.1.1. *Gallium nitride*

The preparation of colloidal GaN quantum dots via thermal decomposition commonly requires the suitable Ga and N resources as the precursors. As the growth temperature increases, the precursors will be decomposed rapidly, react, and then generate lots of small nanometal clusters. Finally, the GaN nanoparticles are formed after the further growth and the size of the particles depends on the reaction time and temperature. The precursor contains polymeric gallium imide [40, 41] and gallium cupferron with hexamethyldisilazane, [42] etc. The selection of the precursor is very important for successful growth of colloidal GaN quantum dots. Janik and Wells [41] prepared powders of mixed hexagonal/cubic nanocrystals of GaN by deamination of polymeric gallium imide ($\{Ga(NH)3/2\}_n$) at 210°C. Their work showed that nanosize GaN could be prepared by polymeric gallium imide ($\{Ga(NH)3/2\}_n$) at high temperature, due to the lack of any organic substituents in the precursor, which made $\{Ga(NH)3/2\}_n$ a good candidate for the generation of carbon-free GaN. However, it is unfortunate that these methods for the size control were limited and did not allow the nanocrystals to be dispersed in solvents to form transparent solutions of QDs suitable for optical measurements. One important factor in the synthesis of GaN is the purity of the final product. Carbon is usually left on QD surfaces after pyrolysis and it is difficult to remove. The elemental analyses of the GaN QDs showed 2.49–3.65% carbon content. Although the GaN QDs obtained by the above methods were of poor quality, it opened a window for researchers to find better ways to obtain the high-quality GaN QDs.

According to the previous reports, Mićić et al. [40] still used {Ga(NH)3/2}n to prepare the GaN but they added trioctylamine (TOA) and hexadecylamine (HDA) during the heating process. The HDA could improve hydrophobicity of the GaN surface because of its less sterical hinderance and much dense surface cap. Mićić also showed that TOA/HDA decreased carbon adsorption on the QD particles and after purification yields a white colloidal solution. A transmission electron microscopy (TEM) image of the GaN quantum dots is given in **Figure 1**, which shows the spherical GaN QDs with diameter ranging from 23 to 45 Å. High-resolution micrographs show <111> lattice fringes in some particles have the proper orientation for observing fringes (**Figure 1**, bottom right panels). The particle size was estimated by simply counting the lattice fringes for each particle (interplanar spacing for <111> GaN = 2.52 Å); and the average diameter is 30 Å ± 40%. The bottom left panel in **Figure 1** shows the electron diffraction pattern of the GaN nanocrystals.

The polymer is not a good solvent for controlling the size and the high-yield production of QDs. In the polymer solvent, during the nucleation process, due to the melt in the solvent, dispersibility is not good, the nanoclusters may be happened to aggregate. To solve this

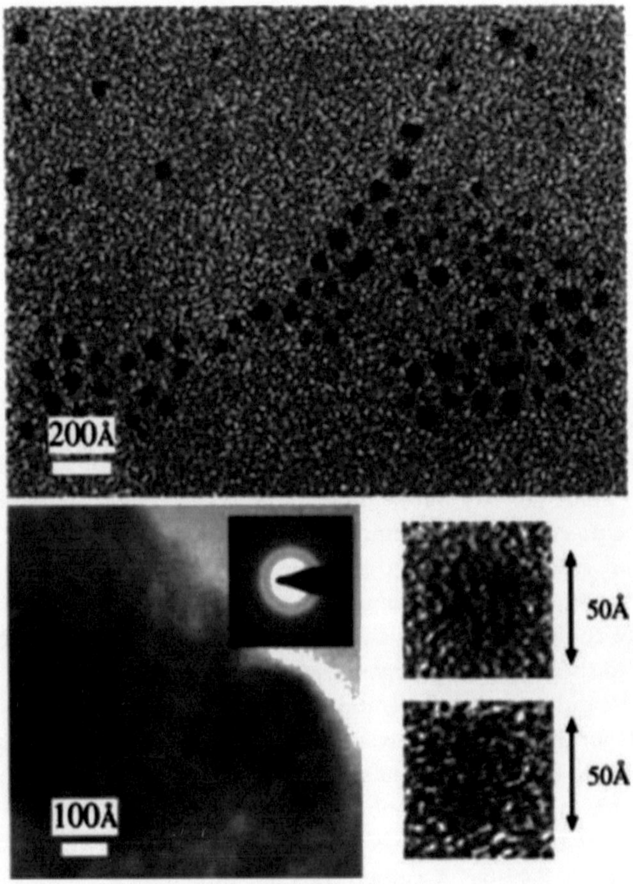

Figure 1. TEM image of GaN QDs taken in bright field. Top panel shows low magnification of QDs and some linear alignment. Bottom two right panels show high magnification and lattice fringes of QD oriented with the <111> axis in the plane of the micrograph. Bottom left panel shows electron diffraction pattern of GaN QDs indicating zinc-blende structure [40].

problem, Pan et al. [43] found that GaN could be obtained by dimeric amidogallium precursor ($Ga_2[N(CH_3)_2]_6$) through pyrolysis without the need for the polymeric intermediate. In this way, not only it produced colloidal GaN quantum dots, but also offered the possibilities of controlling the dots' size. In addition, the gaseous ammonia needed in the nucleation process was cancelled. Colloidal GaN QDs was got by transmission electron microscopy (TEM) imaging. **Figure 2** shows a TEM image of the GaN nanoparticles. Several spherical particles with diameters of 2–4 nm are shown in the **Figure 2**. Although the particle size distribution obtained here is not comparable with those obtained in the highly optimized II–VI group systems, it is believed that this reaction will be improved through optimizing the reaction conditions. However, the as-prepared samples' crystallinity is poor, which can be confirmed by the TEM images. No lattice fringes were observed in HRTEM images. HAD may has a contribution for the pyrolysis reaction. When HDA was eliminated from the reaction mixture,

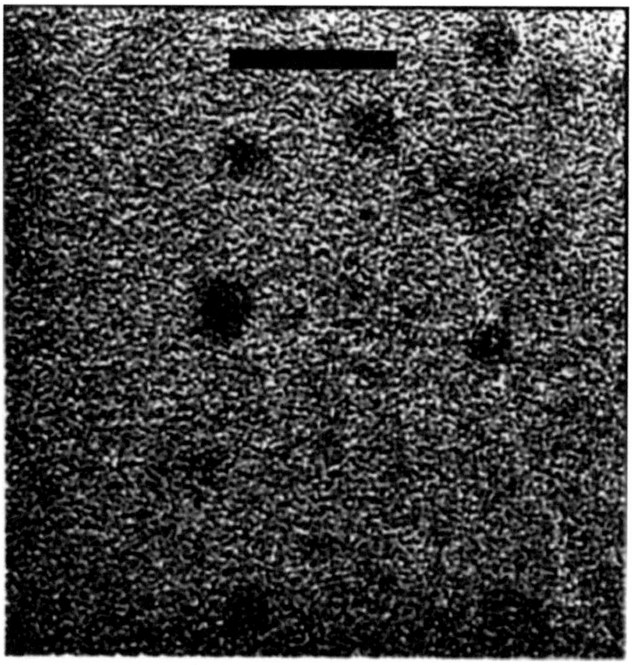

Figure 2. TEM image of GaN nanoparticles obtained from pyrolysis of $Ga_2[N(CH3)_2]_6$. Scale bar is 10 nm [43].

no GaN was produced. Elemental analysis on the GaN product prepared in the presence of HDA revealed a Ga/N mass ratio of 4.88:1 (theoretical 4.98:1), indicative of nearly stoichiometric GaN.

The sample also had a high-carbon content (C/N) of 2.27:1, indicating that carbon was incorporated into the particles. Certainly, it was consistent with capping of nanosize particles by capping ligands, such as HDA or TOA that contains long aliphatic chains.

Generally, to get the GaN, the post-treatment temperature is at least 500°C. Nitrides of lanthanide and transition metals (M) can be prepared by solid reaction at the temperature ranging from 600 to 1000°C. The chemical equation is as follows:

$$MCl_n + Li_3N \rightarrow MN + LiCl \quad (1)$$

Xie et al. [15] reported that the crystalline GaN particles could be synthesized by simple inorganic reactions at temperature of 300°C in the autoclave (no capping ligand in the whole preparation process). They used $GaCl_3$ and Li_3N as gallium and nitrogen precursors in the liquid, respectively, and the reaction equation is:

$$GaCl_3 + Li_3N \rightarrow GaN + 3LiCl \quad (2)$$

These crystallites of GaN have an average size of 32 nm and display a uniform shape (**Figure 3A**). The images of GaN particles were observed by High-resolution electron microscopy (HREM). In

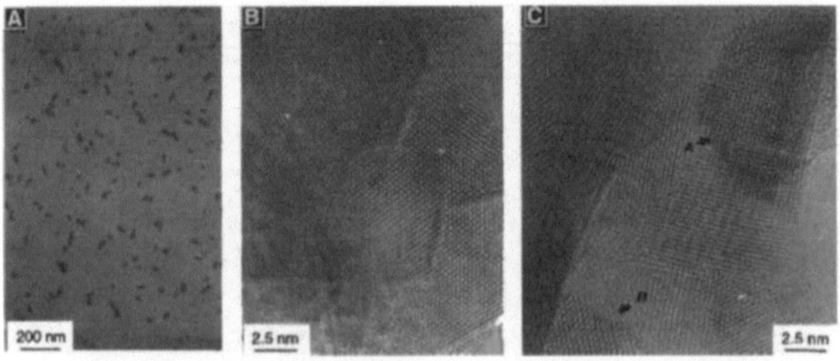

Figure 3. (A) A TEM micrograph of nanocrystalline GaN. (B and C) HREM images of nanocrystalline GaN: (B) lattice fringes of (001) plane in GaN with a wurtzite structure and (C) lattice fringes of (100) and (110) planes in GaN (marked A and B, respectively) with a rocksalt structure [15].

Figure 3B, the (001) lattice fringes of GaN in the wurtzite structure appear frequently, indicating the preferential orientation of the plate-like GaN particles. The areas marked by arrowheads (A and B) in **Figure 3C** represent a typical structural image of [100] and [110] orientations, respectively, of GaN in the rocksalt structure.

3.1.2. Indium nitride

Among the nitride, InN is the most unstable, which decomposes above 500°C [44]. Hence, it is difficult to prepare InN crystalline. Historically, polycrystalline indium nitride was synthesized by radio frequency sputtering [28], which results in high free-electron concentration, significant oxygen contamination, and an absorption edge at about 1.9 eV [45–47]. Recently, high-quality crystalline InN has been grown by molecular beam epitaxy (MBE) [27]. The typically observed bandgap of high-quality wurtzite-InN grown by MBE is around 0.65–0.7 eV [34, 48]. However, the quality of indium nitride samples grown by low-cost solution or other vapour methods has still remained challenging. Therefore, there still exists a huge challenge to synthesize high-quality InN nanocrystals using low-cost solution or vapour methods. So far, Xiao et al. [24] adopted the solvothermal using $NaNH_2$ and In_2S_3 as novel nitrogen and indium sources to prepare the indium nitride at 180–200 °C with the particle size ranging from 10 to 30 nm, and the reaction equation is:

$$In_2S_3 + 6NaNH_2 \rightarrow 2InN + 3Na_2S + 4NH_3 \quad (3)$$

Hsieh [25] also used solvothermal to prepare the InN NCs with an average diameter of 6.2 ± 2.0 nm utilizing $InBr_3$ and $NaNH_2$ in a low temperature, ambient pressure, and liquid-phase condition.

Generally, ammonia is used as nitrogen source in the vapour-phase growth process. The relative high growth temperature in vapour-phase methods can help nucleation overcome the reaction difficulty encountered by solution methods. However, unlike the solution-based methods (where the nucleation and growth process could be separated by choosing appropriate ligands and solvents), the vapour-phase methods usually trigger off nonuniform nanocrystal morphology due to poor control in the nucleation process. Moreover, the vapour-phase methods often encounter the aggregation problem due to the large surface energy of nanocrystals. It is concluded that each of the aforementioned methods for synthesis of monodisperse InN nanocrystals faces significant challenges in achieving well-defined size and shape. Our group [49] had addressed these critical issues by exploiting a new synthesis approach that resulted in monodisperse InN nanocrystals with uniform size and morphology and superior optical quality, and first successfully prepared the cubic InN nanocrystals with aforementioned advantages, by combining solution- and vapour-phase methods under silica shell confinement (SVSC), as schematically shown in **Figure 4(a)**.

In this method, the In_2O_3 nanocrystals with well-defined size and morphology were first synthesized by a solution-based method. The In_2O_3 nanocrystals were coated by silica shell before nitridation, this is because silica is inert and can be easily removed by HF acid. The obtained In_2O_3@SiO_2 nanopowders were put into a tube furnace. After being purged with NH_3 gas for 20 min, the furnace was heated to 500–700°C and kept for 5 h under NH_3 flow at 300 ml/min. Finally, large-scale InN@SiO_2nanocrystals with uniform size and morphology were obtained through the SVSC route. After removing the silica shell, InN NCs can be dispersed into DI water and then transferred to various nonpolar organic solvents by phase transfer, as shown in **Figures 4(b)** and **(c)**.

Figure 5(a) and **(b)** shows transmission electron microscopy (TEM) and high-resolution transmission electron microscopy (HRTEM) images of the InN nanocrystals. The indium nitride nanocrystals with nearly monodisperse spherical shape can be observed from the TEM images. The diameter distribution of the synthesized indium nitride nanocrystals can be revealed from the **Figure 5(c)**, revealing a fairly uniform size distribution of the InN nanocrystals from 5.0 to

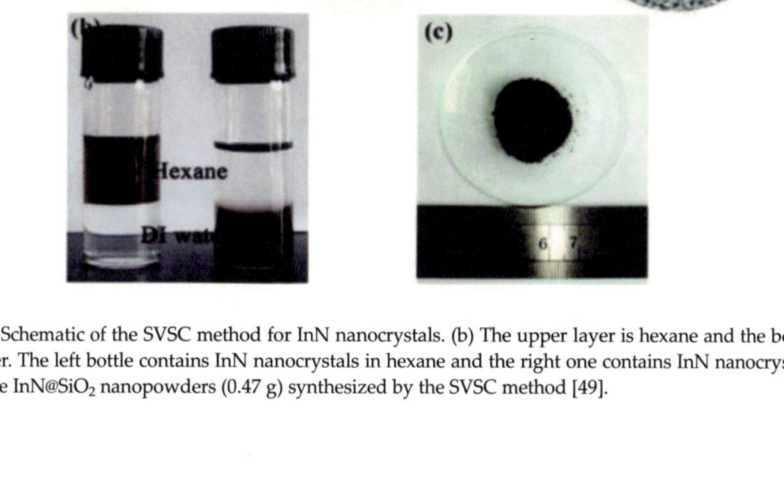

Figure 4. (a) Schematic of the SVSC method for InN nanocrystals. (b) The upper layer is hexane and the bottom layer is distilled water. The left bottle contains InN nanocrystals in hexane and the right one contains InN nanocrystals in water. (c) Large-scale InN@SiO_2 nanopowders (0.47 g) synthesized by the SVSC method [49].

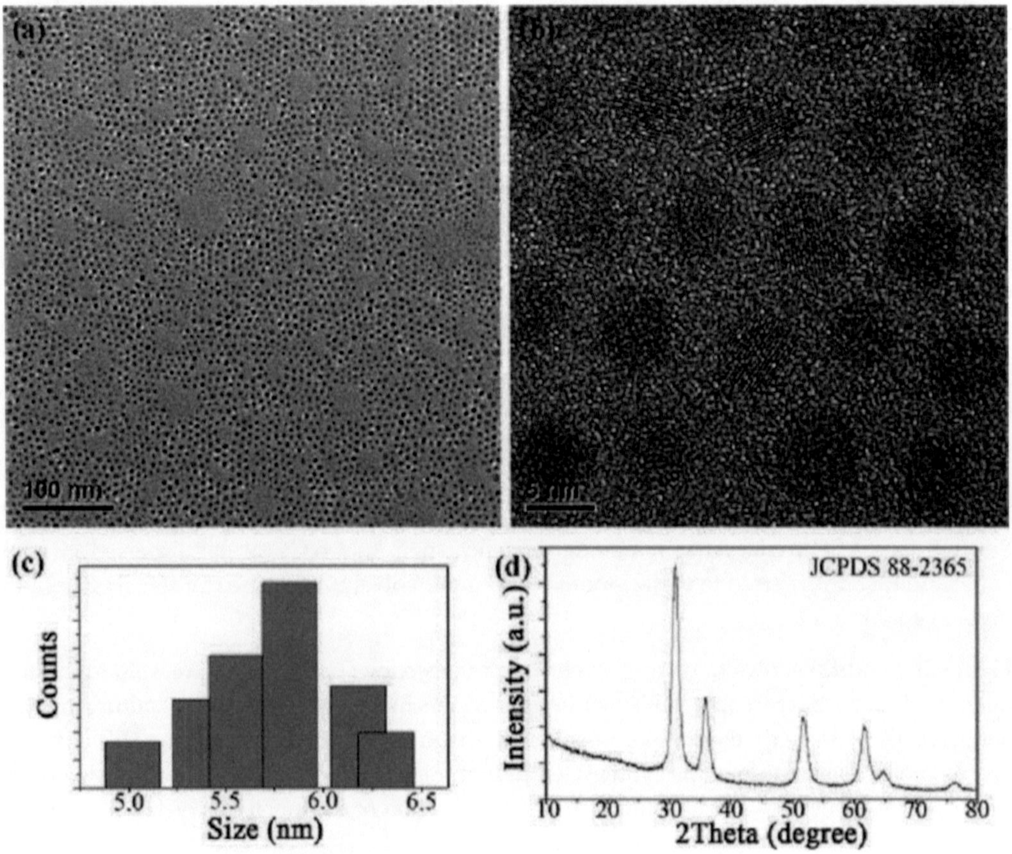

Figure 5. (a) Low-resolution TEM; (b) HRTEM images of the InN nanocrystals; (c) size distribution of the InN nanocrystals; and (d) XRD spectrum of the InN@SiO_2nanocrystals obtained at 550°C [49].

6.3 nm. The average diameter of the InN nanocrystals is calculated to be 5.7 ± 0.6 nm, after counting about 200 nanocrystals. **Figure 5d** shows the X-ray diffraction (XRD) pattern of the InN nanocrystals. All the peaks can be matched with the cubic InN (JCPDS No. 88-2365) except the peak at $2\theta = 22.5°$, which corresponds to the amorphous silica. No peaks of In_2O_3 were observed in the XRD pattern, indicating that all In_2O_3 nanocrystals had been converted to InN. As illustrated in **Figure 6**, the InN@SiO_2 nanocrystals in the form of powders (not in the solution) exhibited infrared PL at room temperature. The fluctuation of PL signal at ~1900 nm could be due to water absorption and might not originate from the sample. The PL spectrum is characterized by the presence of three distinct emission peaks.

3.1.3. *Indium gallium nitride*

Xiao et al. [50] demonstrated for the first time, a new route that quantum size effect can be used to prepare the epitaxial nanostructures with great significance for the achievement of a broad range of future nanoelectronic and nanophotonic devices. The process is quantum size-controlled photoelectrochemical (QSC-PEC) etching. Quantum dots' bandgap depends on the

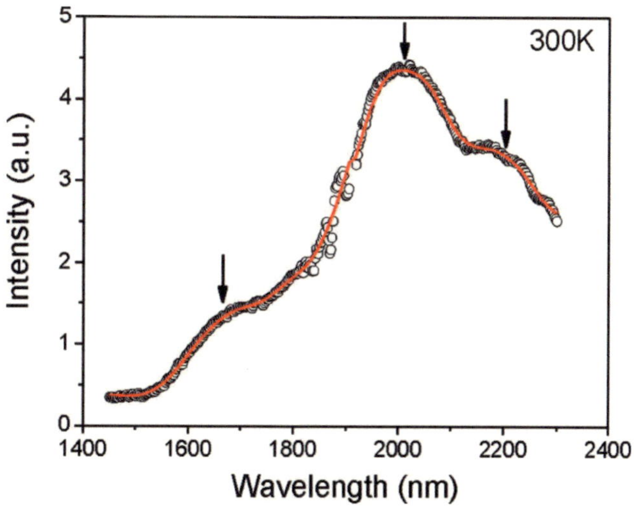

Figure 6. Photoluminescence spectrum of InN@SiO_2 nanocrystals at room temperature [49].

nanostructure size. Therefore, they used QSC-PEC etching to fabricate InGaN QDs of controlled size starting from the InGaN thin film. They used H_2SO_4 aqueous solution as the electrolyte and a tunable, relatively narrow band laser source as photoexcitation. The sample consisted of $In_{0.13}Ga_{0.87}N$ films (3–20 nm) grown on c-plane GaN/sapphire. They discussed the influence of solution pH during quantum size-controlled PEC etch process. [51] When the solution pH lies between 5 and 11, both Ga- and In-oxides are formed at the surface. Etching rates are very low and InGaN QDs are not formed. In the dark etching of InGaN at pH above 5, the above situation may also occur. However, when the solution pH is below 3, oxide-free QDs with self-terminated sizes can be successfully realized. In strongly acidic solutions, the oxides are not formed during the PEC etching process, due to all the oxide productions can be dissolved by the electrolyte. Therefore, PEC etching can be used to prepare InGaN QDs. Meanwhile, there are other methods of preparing InGaN quantum dot, such as plasma-assisted molecular beam epitaxy (PA-MBE) [30], metal-organic vapour phase epitaxy (MOVPE) [31], and metal-organic chemical vapour deposition (MOCVD) [32].

4. Summary and future directions

The field of colloidal III–V nitride quantum dots has been constantly gaining interest among science and engineering communities during the past decades. In this chapter, we have summarized the research status and progress in this field, including their preparation techniques and optoelectronic properties. Although much progress has already been made in the field of colloidal III–V nitride quantum dots, significant challenges involving the fabrication of quantum dots with uniform morphology and size and control of the electronic and optical properties in terms of composition and structure remain to be solved. Once high-quality colloidal III–V nitride quantum dots are synthesized successfully, more new discoveries and applications will be exploited, such as solid-state lighting, lasing technologies, and optoelectronic devices, as well as the booming quantum photonics technology.

Acknowledgements

This work was financially supported by the National Natural Science Foundation of China (Grant No. 51472031, 51102017 and 21503014).

Author details

Zequn Chen[1], Chuli Sun[2], Wei Guo[2] and Zhuo Chen[1]*

*Address all correspondence to: zchen@bit.edu.cn

1 Department of Materials Physics and Chemistry, Beijing Key Laboratory of Construction Tailorable Advanced Functional Materials and Green Applications, School of Materials Science and Engineering, Beijing Institute of Technology Institution, Beijing, P.R. China

2 Department of Physics, Beijing Institute of Technology, Beijing, P.R. China

References

[1] Jie B, Moungi GB. A colloidal quantum dot spectrometer. Nature. 2015;**523**:67-70. DOI: 10.1038/nature14576

[2] Tang J, Sargent EH. Infrared colloidal quantum dots for photovoltaics: Fundamentals and recent progress. Advanced Materials. 2011;**23**:12-29. DOI: 10.1002/adma.201001491

[3] Yin Y, Alivisatos AP. Colloidal nanocrystal synthesis and the organic-inorganic interface. Nature. 2005;**437**:664-670. DOI: 10.1038/nature04165

[4] Colvin VL, Schlamp MC, Alivisatos AP. Light-emitteing-diodes made from cadmium selenide nanocrystals and a semiconducting polymer. Nature. 1994;**370**:354-357. DOI: 10. 1038/370354a0

[5] Jeong HK, Jae HL, Myeong JP, Seong HL, Kook HC, Chang HL. High-power genuine ultraviolet light-emitting diodes based on colloidal nanocrystal quantum dots. Nano Letters. 2015;**15**:3793-3799. DOI: 10.1021/acs.nanolett.5b00392

[6] Yu WW, Qu LH, Guo WZ, Peng XG. Experimental determination of the extinction coefficient of CdTe, CdSe, and CdS nanocrystals. Chemistry of Materials. 2003;**15**:2854-2860. DOI: 10.1021/cm034081k

[7] Moreels I, Lambert K, De Muynck D, Vanhaecke F, Poelman D, Martins JC, Allan G, Hens Z. Composition and size-dependent extinction coefficient of colloidal PbSe quantum dots. Chemistry of Materials. 2007;**19**:6101-6106. DOI: 10.1021/cm071410q

[8] Xia YN, Xiong YJ, Lim B, Skrabalak SE. Shape-controlled synthesis of metal nanocrystals: Simple chemistry meets complex physics. Angewandte Chemie International Edition. 2009;**48**:60-103. DOI: 10.1002/anie.200802248

[9] Puntes VF, Krishnan KM, Alivisatos AP. Colloidal nanocrystal shape and size control: The case of cobalt. Science. 2001;**291**:1019-1020. DOI: 10.1126/science.1057553

[10] Reiss P, Carriere M, Lincheneau C, Vaure L, Tamang S. Synthesis of semiconductor nanocrystals, focusing on nontoxic and earth-abundant materials. Chemical Review. 2016;**116**:10731-10819. DOI: 10.1021/acs.chemrev.6b00116

[11] Nakamura S, Pearton SJ, Fasol G. The Blue Laser Diode: The Complete Story. 2nd ed. New York: Springer-Verlag; 2000. 104 p. DOI: 10.1007/978-3-662-04156-7

[12] Wu Y, Jacob-mitos M, Moore ML, Heikman S. A 97.8% efficient GaN HEMT boost converter with 300-W output power at 1 MHz. IEEE Electron Device Letters. 2008;**29**:824-826. DOI: 10.1109/LED.2008.2000921

[13] Nakamura S, Faso G. The Blue Laser Diode. The Complete Story. Berlin: Springer; 1999. 72 p. DOI: 10.1007/978-3-662-04156-7

[14] Kocha SS, Peterson MW, Arent DJ, Redwing JM, Tischler MA, Turner JA. Electrochemical investigation of the gallium nitride-aqueous electrolyte interface. Journal of the Electrochemical Society. 1995;**142**:L238-L240. DOI: 10.1149/1.2048511

[15] Xie Y, Qian Y, Wang W, Zhang S, Zhang Y. A benzene-thermal synthetic route to nanocrystalline GaN. Science. 1996;**272**:1926-1927. DOI: 10.1126/science.272.5270.1926

[16] Guo Q, Kato O, Yoshida A. Thermal stability of indium nitride single crystal films. Journal of Applied Physics. 1993;**73**:7969-7971. DOI: 10.1063/1.353906

[17] Yasushi N, Yoshiki S, Tomohiro Y. RF-molecular beam epitaxy growth and properties of InN and related alloys. The Japan Society of Applied Physics. 2003;**42**:2549-2559. DOI: 10.1143/JJAP.42.2549

[18] Faso GS, Nakamura S. The Blue Laser Diode: GaN Based Light Emitters and Lasers. Berlin, Hong Kong: Springer; 1997. 15 p. DOI: 10.1007/978-3-662-03462-0

[19] Starikov E, Shiktorov P, Gruninskis V. Monte Carlo calculations of THz generation in wide gap semiconductors. Physica B. 2002;**341**:171-175. DOI: 10.1016/S0921-4526(01) 01374-6

[20] Neff H, Semchinova OK, AMN L, Filimonov A, Holzhueter G. Photovoltaic properties and technological aspectsof $In_{1-x}Ga_xN$/Si, Ge ($0 < x < 0.6$) heterojunction solar cells. Solar Energy Materials & Solar Cells. 2006;**90**:982-997. DOI: 10.1016/j.solmat.2005.06.002

[21] Nguyen HPT, Chang Y-L, et al. InN p-i-n nanowire solar cells on Si. IEEE Journal of Selected Topics in Quantum Electronics. 2011;**17**:1062-1069. DOI: 10.1109/JSTQE.2010. 2082505

[22] Wu JQ. When group-III nitrides go infrared: New properties and perspectives. Journal of Applied Physics. 2009;**106**:011101-011128. DOI: 10.1063/1.3155798

[23] Michalet X, Pinaud FF, Bentolila LA, et al. Quantum dots for live cells, in vivo imaging, and diagnostics. Science. 2005;**307**:538-544. DOI: 10.1126/science.1104274

[24] Xiao J, Xie Y, Tang R, Luo W. Benzene thermal conversion to nanocrystalline indium nitride from sulfide at low temperature. Inorganic Chemistry. 2003;**42**:107-111. DOI: 10.1021/ic0258330

[25] Hsieh JC, Yun DS, Hu E, Belcher AM. Ambient pressure, low-temperature synthesis and characterization of colloidal InN nanocrystals. Journal of Materials Chemistry. 2010; **20**:1435-1437. DOI: 10.1039/b922196d

[26] Hovel HJ, Cuomo JJ. Electrical and optical properties of rf-sputtered GaN and InN. Applied Physics Letters. 1972;**20**:71-73. DOI: 10.1063/1.1654051

[27] Chang Y-L, Mi Z, Li F. Photoluminescence properties of a nearly intrinsic single InN nanowire. Advanced Functional Materials. 2010;**20**:4146-4151. DOI: 10.1002/adfm.201000739

[28] Moses PG, Van de Walle CG. Band bowing and band alignment in InGaN alloys. Applied Physica Letters. 2010;**96**:021908-021910. DOI: 10.1063/1.3291055

[29] Nakamura S. The roles of structural imperfections in InGaN-based blue light-emitting diodes and laser diodes. Science. 1998;**281**:956-961. DOI: 10.1126/science.281.5379.956

[30] Deshpande S, Frost T, Yan LF, et al. Formation and nature of InGaN quantum dots in GaN nanowires. Nano Letters. 2015;**15**:1647-1653. DOI: 10.1021/nl5041989

[31] Tsai HL, Wang TY, Yang JR, et al. Observation of ultrahigh density InGaN quantum dots. Journal of Applied Physics. 2007;**102**:013521-013524. DOI: 10.1063/1.2745848

[32] Zhang L, Teng C-H, Hill TA, et al. Single photon emission from site-controlled InGaN/GaN quantum dots. Applied Physica Letters. 2013;**103**:192114-192118. DOI: 10.1063/1.4830000

[33] Mohamed H, Razeghi M. Optoelectronic Devices: III Nitrides. United Kingdom: Elsevier; 2005. 59 p. DOI: 10.1016/B978-008044426-0/50000-6

[34] Sardar K, Deepak FL, Govindaraj A, Seikh MM, Rao CN. InN nanocrystals, nanowires, and nanotubes. Small. 2005;**1**:91-94. DOI: 10.1002/smll.200400011

[35] Xiong Y, Xie Y, Li Z, Li X, Zhang R. Aqueous synthesis of group IIIA nitrides at low temperature. New Journal of Chemistry. 2004;**28**:214-217. DOI: 10.1039/b310373k

[36] Frank AC, Stowasser F, et al. Detonations of gallium azides: A simple route to hexagonal GaN nanocrystals. Journal of American Chemical Society. 1998;**120**:3512-3513. DOI: 10.1002/chin.199828028

[37] Sardar K, Dan M, et al. A simple single-source precursor route to the nanostructures of AlN, GaN and InN. Journal of Materials Chemistry. 2005;**15**:2175-2177. DOI: 10.1039/b502887f

[38] Chen C, Liang C. Syntheses of soluble GaN nanocrystals by a solution-phase reaction. Tamkang Journal of Science and Engineering. 2002;**5**:223-226

[39] Lamer VK, Dingar RH. Theory, production and mechanism of formation of monodispersed hydrosols. Journal of American Chemical Society. 1950;**72**:4847-4854. DOI: 10.1021/ja01167a001

[40] Mićić OI, Ahrenkiel SP, Bertram D, Nozik AJ. Synthesis, structure, and optical properties of colloidal GaN quantum dots. Applied Physics Letters. 1999;**75**:478-480. DOI: 10.1063/1.124414

[41] Janik JF, Wells RL. Gallium imide, $\{Ga(NH)_{3/2}\}_n$, a new polymeric precursor for gallium nitride powders. Chemistry of Materials. 1996;**8**:2708-2711. DOI: 10.1002/chin.199710018

[42] Sardar K, Rao CNR. New solvotherma routes for GaN nanocrystals. Advanced Materials. 2004;**16**:425-429. DOI: 10.1002/adma.200306050

[43] Pan GQ, Kordesch ME, Patten PGV. New pyrolysis route to GaN quantum dots. Chemistry of Materials. 2006;**18**:3915-3917. DOI: 10.1021/cm060368g

[44] Tranior JW, Rose K. Some properties of InN films prepared by reactive evaporation. Journal of Electronic Materials. 1974;**3**:821-828. DOI: 10.1007/BF02651400

[45] Westra KL, Lawson RPW, Brett MJJ. The effects of oxygen contamination on the properties of reactively sputtered indium nitride films. Journal of Vacuum Science & Technology A. 1998;**6**:1730-1732. DOI: 10.1116/1.575280

[46] Tansley TL, Foley CP. Optical band gap of indium nitride. Journal of Applied Physics. 1986;**59**:3241-3244. DOI: 10.1063/1.336906

[47] Morkoc H. Nitride Semiconductors and Devices. Heidelberg: Springer; 1999. 32 p. DOI: 10.1007/978-3-642-58562-3

[48] Cesar M, Ke YQ, Ji W, Guo H, Mi ZT. Band gap of $In_xGa_{1-x}N$: A first principles analysis. Applied Physics Letters. 2011;**98**:202107-202109. DOI: 10.1063/1.3592573

[49] Chen Z, Li YN, Cao CB, et al. Large-scale cubic InN nanocrystals by a combined solution- and vapor-phase method under silica confinement. Journal of American Chemical Society. 2012;**134**:780-783. DOI: 10.1021/ja209072v

[50] Xiao XY, Fischer AJ, Wang GT, et al. Quantum-size-controlled photoelectrochemical fabrication of epitaxial InGaN quantum dots. Nano Letters. 2014;**14**:5616-5620. DOI: 10.1021/nl502151k

[51] Xiao XY, Lu P, et al. Influence of pH on the quantum-size-controlled photoelectrochemical etching of epitaxial InGaN quantum dots. The Journal of Physical Chemistry C. 2015;**119**:28194-28198. DOI: 10.1021/acs.jpcc.5b09555

Chapter 5

CdTe Quantum Dot Fluorescence Thermometry of Rolling Bearing

Ke Yan and Bei Yan

Additional information is available at the end of the chapter

http://dx.doi.org/10.5772/intechopen.70866

Abstract

Temperature is one of the most important parameters affecting the service life and performance of a rolling element bearing component. In this paper, a nonintrusive method is developed to monitor the temperature variation of the inner raceway during bearing operation utilizing CdTe quantum dots as the temperature sensors. The CdTe quantum dots were synthesized and were used in constructing a sensor film by means of layer-by-layer electrostatic self-assembly method on an ultrathin glass slice. The peak wavelength shift of the fluorescence spectrum of the sensor film shows a linear and reversible relationship with temperature, and it is used to sense the temperature of the inner raceway. The resolution of the CdTe optothermal sensor is determined to be 0.14 nm/°C. The temperature measurement of rolling element bearing was conducted on a bearing test rig incorporated with an optical fiber fluorescence spectrum detecting system. To verify the accuracy of the temperature obtained by quantum dots sensor film, a thermocouple was used to test the temperature of the inner raceway right before and after the operation. Results show that the temperature obtained by the CdTe quantum dots film sensor is consistent with that by the thermocouple, with an error typically below 10% or smaller.

Keywords: high speed rolling bearing, inner ring temperature monitoring, quantum dots

1. Introduction

Rolling bearings are basic mechanical components widely used in machinery for low friction, high rigidity, and reliability. They are required to operate at high speed for long period of time under uneven conditions with minimum maintenance. The operating status of bearings directly affects the performance of rotating machinery. Bearing failure can make machine breakdown, lead to cost increase and even human death [1]. Hence, the development of strategies for monitoring bearing health conditions while in operation has been of significant importance.

The contact friction between the inner component leads to large heat generation and elevated temperature, which could cause thinner lubricant film, higher asperity contact, and reduction of material properties. Thus, the temperature is considered to be one of the most important parameters affecting the service life and performance of a rolling element bearing component. However, because of the complex structure and extreme operating conditions, instrument for real-time, nonintrusive monitoring of bearing temperatures has been limited. This is particularly true for the rolling element of a bearing, whose temperature is often indirectly obtained from the measured temperature of outer raceway. Indirect measurements are known to be error-prone. Thus far, direct measurement of the temperature of the inner bearing components such as the inner raceway and cage has eluded researchers [2–6]. Joshi [2] has developed a battery-powered telemeter and a remotely powered telemeter to measure the cage temperature in a tapered roller bearing. Also, Jia et al. [6] used a remotely powered wireless temperature sensor to monitor the cage temperature in real-time. However, the battery-powered telemeter has an extremely short functional life, and both the wireless ones are easily affected by the electromagnetic environment and are not suitable for high speed situations.

Recently, luminescent semiconductor nanocrystals, quantum dots, have attracted extensive attentions due to its unique optical properties and have been applied in light-emitting diodes, solar cells, and bio-labeling [7–9]. These semiconductor nanoparticles offer several advantages including narrow fluorescence emission, tunable wavelength, relatively high quantum yield, outstanding photo stability as well as flexible photo excitation. It also has been reported that the behavior of the luminescent properties of quantum dots with temperature has suitable characteristics for application as temperature probes [10–15]. The luminescence properties, such as the excited state lifetime, emission intensity, and peak wavelength, have been proven to be good indicators of temperature. The wide range of temperature in which luminescent properties change makes them very suitable for temperature sensing applications.

This paper presents a study on the use of CdTe quantum dots as thermal sensor to measure the temperature of inner raceway of rolling bearing while in operation. The quantum dots sensor film is fabricated by means of layer-by-layer electrostatic self-assembly method on an ultrathin glass slice. The peak wavelength shows a linear and reversible relationship as temperature changes. The factors that would affect the acquired fluorescence signal have been studied. Results show that this method is feasible and effective for the temperature measurement of rolling bearing, especially in very high speed conditions.

2. Sensor preparation and calibration

2.1. Sensor preparation

Colloidal solutions of CdTe quantum dots stabilized by TGA were prepared according to the method given by the previously reported paper [16]. Typically, 0.2 mmol $Cd(CH_3COO)_2 \cdot 2H_2O$

was dissolved into 50 ml deionized water in a three-neck flask and 18 μl TGA was added under stirring, then the pH was adjusted to 10.5–11 with 1 M NaOH solution. After that, 0.04 mmol K_2TeO_3, which was dissolved in 50 ml deionized water was added into the above solution. Then, 80 mg of $NaBH_4$ was added into the precursor solution. After the reactions proceeded for about 5 min, the flask was attached to a condenser and refluxed at 100°C under open-air condition. By controlling refluxing time, CdTe QDs with desired size and color can be obtained.

To implement the temperature measurement, QDs sensors were fabricated by the layer-by-layer electrostatic self-assembly (LBL ESA) technique [17, 18]. Quartz slides of 200 μm thickness were placed into Piranha Solution for 30 min for cleaning before the deposition of the QD coatings. Then the Quartz slides were immersed into 50 ml of a solution of 1%wt PDDA with the pH adjusted to 8.0 for the absorption of polycation for 15 min. Next, the substrates were cleaned in deionized water and dried by N_2. Then, the slides were removed into the CdTe QDs solution synthesized by ourselves for 10 min to absorb QDs followed by cleaned and dried. Repeating the above steps, a sensitive coating denoted by [PDDA/CdTe]n was formed, where n was chosen to be 15 for this paper. Afterward, the sensor films were cured at 150°C in a vacuum chamber. Thermal treatment yields a more repetitive and stable response when suffered to temperature change. When the fabrication was completed, the sensors were kept in darkness until the temperature response of emission spectrum was studied.

2.2. Sensor calibration

In order to utilize QDs as sensors for the temperature measurement of rolling bearings, the temperature-dependent emission properties of the sensor was first characterized using the experimental setup showing in **Figure 1**. The QDs sensor is placed on a heater cell with a thermocouple to monitor its temperature. Since the QDs have a wide absorption spectrum, a mercury lamp at 365 nm is used as the excitation light source. The light generated by the lamp is reflected by a dichroic mirror and directed to the QDs sensor through a focus lens, which is also used to prevent the excitation signal from masking the fluorescence of the QDs sensor. The fluorescence is collected by the same lens and leaded to an Andor Shamrock SR-303i spectrograph. Finally, an Andor iDus DU420A-BV CCD camera together with a computer is used to analyze the optical response of the QD with respect to changes in the temperature values. By adjusting the heater cell, its temperature varies and changes in wavelength and intensity of the luminescence emission of the quantum dots are registered.

To study the properties of the QDs sensor, the heater cell in **Figure 1** was adjusted to run several temperature cycles from room temperature to 70°C and back. **Figure 2(a)** shows the emission spectrum varies as the temperature increases and decreases. As it can be seen from the picture, the photoluminescence intensity decreases with the increase of temperature, while the peak wavelength red shifts and FWHM increases as the temperature increases. It means that there are three features that could be used to detect the variation of temperature. However, as the photoluminescence intensity is affected by the power of the excitation source as well as the distance between the focus lens and the QD sensor, it is not suitable for the temperature monitoring of rolling bearings, where slight or heavy vibration usually occurs. Besides, the average temperature sensitivity of the FWHM is generally small compared with peak

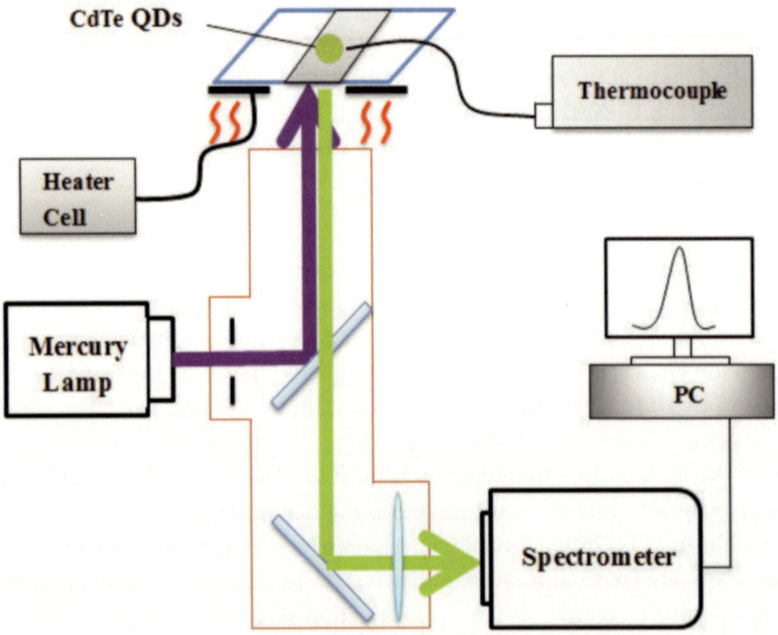

Figure 1. Schematic representation of the experimental setup.

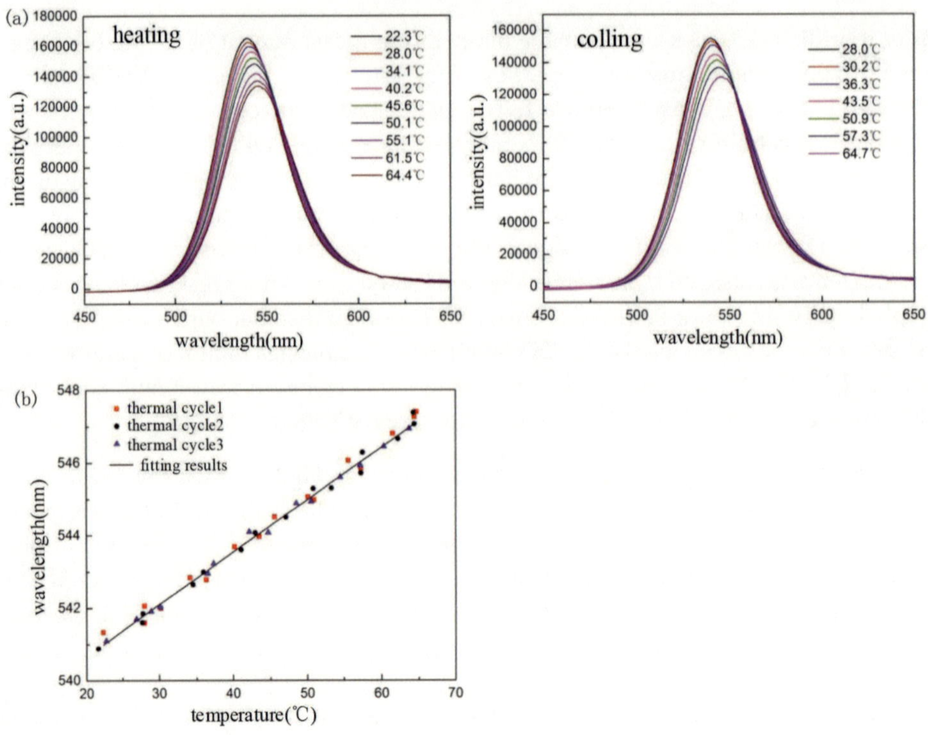

Figure 2. (a) Behavior of emission spectrum of QDs with different temperatures in heating and cooling. (b) Dependence of the emission peak wavelength with respect to the temperature in three thermal cycles.

wavelength. Therefore, here we choose the peak wavelength as the parameter for temperature measurement of rolling bearings. The temperature dependence of peak wavelength in three thermal cycles is depicted in **Figure 2(b)**. It is shown that the response of emission spectrum peak wavelength is linear and reversible as the temperature changes. The R square value with respect to the linear approximation is about 0.994 for both cases. And the sensitivity shown by the sensor is around 0.14 nm/°C. The wavelength shift can be explained by the fact that heat expands the crystalline of the quantum dots material and causes a change in the band gap [13], which is only decided by the properties of the quantum dots.

3. Rolling bearing temperature measurement

For the temperature measurement of rolling bearings, the influence of rotating to the fluorescence signal acquisition was studied first. The same setup shown in **Figure 1** was used with some modification. The QD sensor was mounted to a disk, which was drove by a motor, with its rotating speed detected by a photoelectric tachometer. The excitation light generated by the mercury lamp continuously is illuminated on the disk. What is different from the calibration state is that the QD sensor is excited at intervals when the disk is rotated with a certain speed. And the total amount of fluorescence detected by the CCD changes in time as the QD sensor moves into, though, and out of the focus lens's field of view, as is shown in **Figure 3**.

Generally, a specific exposure time is needed to collect fluorescence when the spectrograph is set to acquire the emission spectrum. Assuming that the exposure time is T, the angular velocity of the disk is ω, and the central angle of the QD sensor to the disk is α, the total time the QD sensor excited within the exposure time is: $t = T\alpha/2\pi$. This means that there is no difference whether the QD sensor is stable or in rotation, but the exposure time multiplies a factor of $\alpha/2\pi$. And it is proved by the results shown in **Figure 4**. We studied the

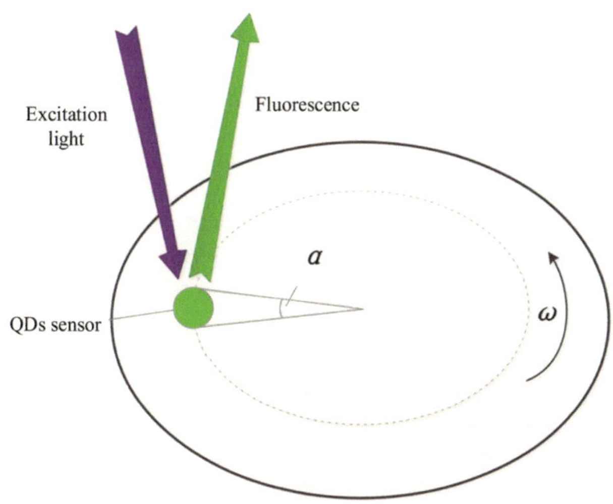

Figure 3. The rotation causes the QDs sensor to move with respect to the fluorescence collection input aperture.

effect under different speeds: 650, 1525, 2515, and 3012 r/min, with the exposure time of the spectrograph set at 500 ms. There are 20 spectrum lines captured every 1 min for different speeds. As is shown in the picture, the emission intensity acquired at different speed keeps nearly unchanged from 650 to 3012 r/min, which means that the rotating speed has no effect on the fluorescence signal. Therefore, we could take temperature measurement of rolling bearings by a common fluorescence spectrum measurement system, and the utilization of QD as temperature senor for rolling bearing thermometry could be applied in very high speed conditions.

3.1. Experimental setup

A new optical fiber fluorescence spectrum detecting system was established to measure the temperature of the inner raceway of the ball bearing of a bearing test rig, which is depicted in **Figure 5**. The bearing test rig was built on a rigid platform and one of its bearing was chosen as testing target. The cover of the bearing block was taken away for convenient measuring. The QDs sensor was mounted to the inner raceway by an epoxy binding agent. An optic fiber with a fluorescent probe was used to conduct the excitation light and collect the fluorescence. By setting the bearing test rig operating at different constant speeds, the fluorescence spectrum of the CdTe film sensor was acquired by the spectrograph (QEpro6500) every 1 min from the moment the setup started until running 20 min, which is thus used to obtain information

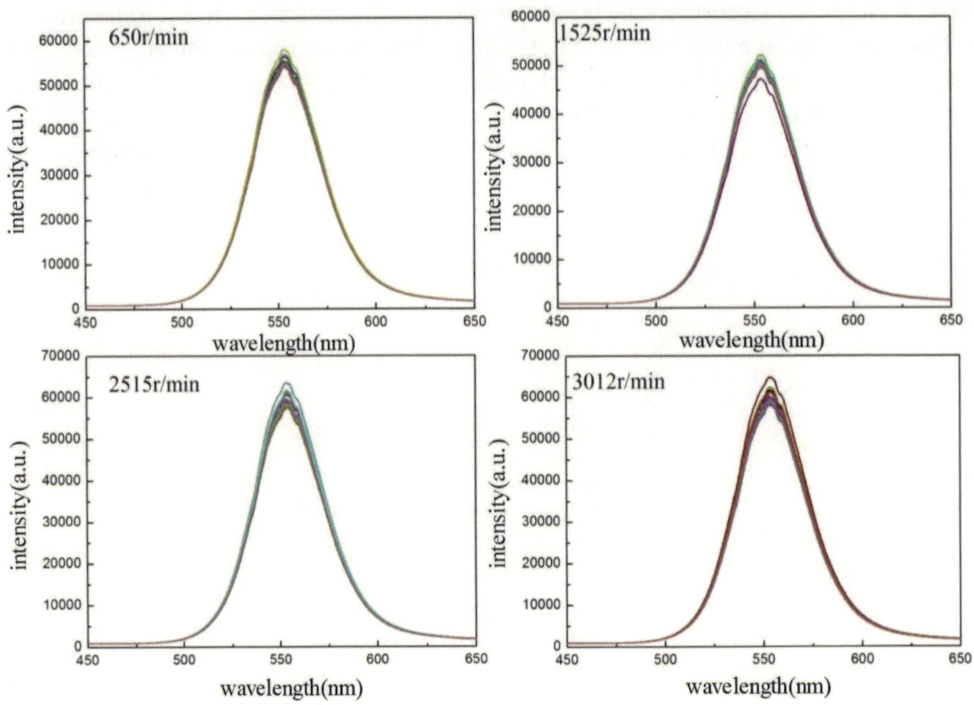

Figure 4. The influence of rotating speed of fluorescence signal.

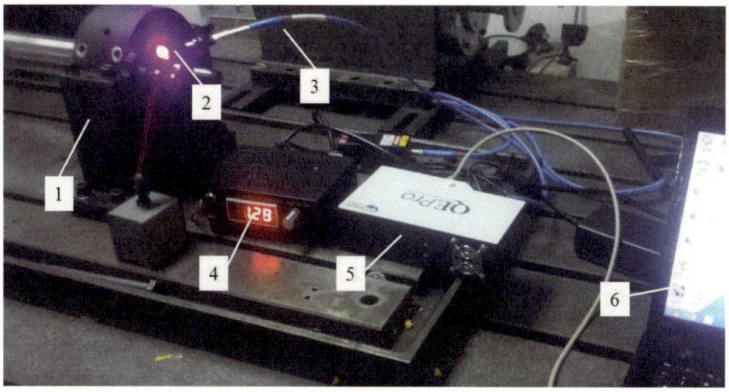

Figure 5. Rolling bearing temperature measurement system. (1) Testing bearing; (2) QDs sensor; (3) optic fiber; (4) 405 nm laser; (5) QEPro spectrograph; (6) computer. Inset: No excitation light irradiates on QDs sensor.

on the temperature variation of inner raceway. To verify the accuracy of the temperature obtained by quantum dots sensor film, a thermocouple was used to test the temperature of the inner raceway right before and after the operation.

3.2. Results and discussion

Temperature measurements were conducted at shaft speed of 1000, 1600, 2200, and 2800 r/min, respectively without any load. **Figure 6** depicts 20 fluorescence spectrum lines acquired every 1 min from the moment the setup started until running 20 min. As we can see from the picture, for different constant shaft speeds, the variation tendency of the fluorescence spectrum is similar to each other: the photoluminescence intensity decreases and the peak wavelength red shifts as time goes on, indicating that the temperature of the bearing inner raceway rises up in the testing 20 min. Comparing the fluorescence spectrum obtained at different shaft speeds, the red shift of the peak wavelength increases as the shaft speed increases, which means that the heat generation and the temperature rise vary with speed. The variation of bearing inner raceway temperature measured by QDs sensor with time at four different conditions of shaft speed is shown in **Figure 7**. It clearly shows that there are more heat generation and large temperature rise at higher speed.

The accuracy of the temperature obtained by the QDs sensor was studied by comparing with the temperature tested by a thermocouple. The thermocouple was used to test the temperature of the inner raceway at the point near QDs sensor right before and after the operation. Results show that the temperature acquired by these two methods has good consistency. The temperature of the inner raceway at 2800 r/min was about 26.1 and 50.2°C before and after the operation of the test rig by thermocouple, while the first and last obtained spectrum line indicate that the temperature was 28.7 and 51.4°C, respectively. The error of the temperature rise between these two methods is 5.8%. Because of the different measuring time and other affects, the temperature measured by two methods shows little difference, but the temperature rise error all blow 10% for different shaft speed.

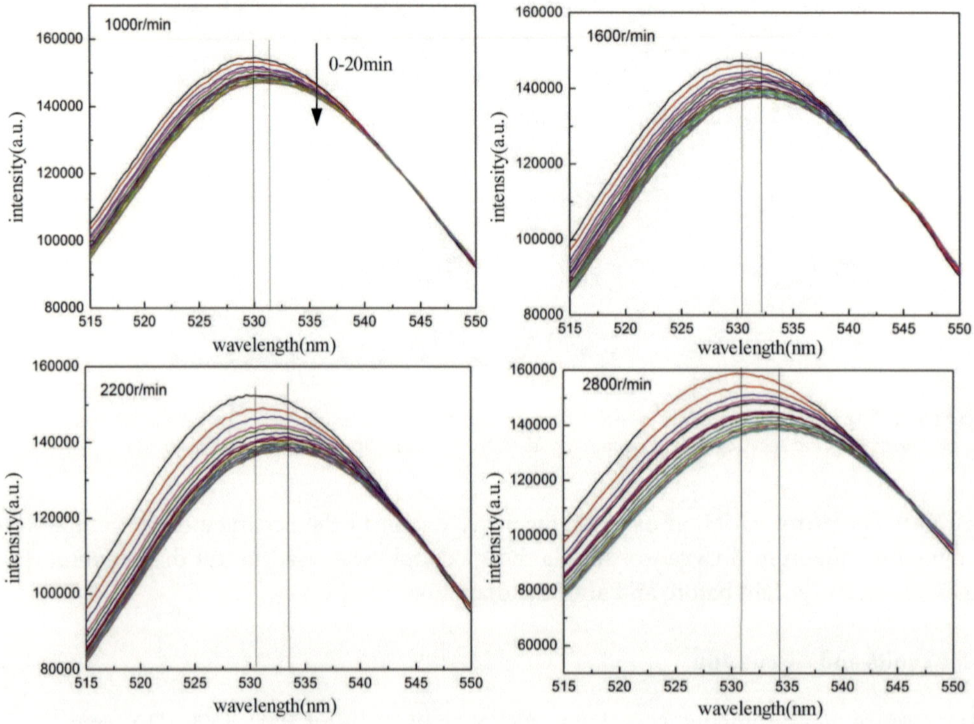

Figure 6. Fluorescence spectrum of different shaft speed in 20 min.

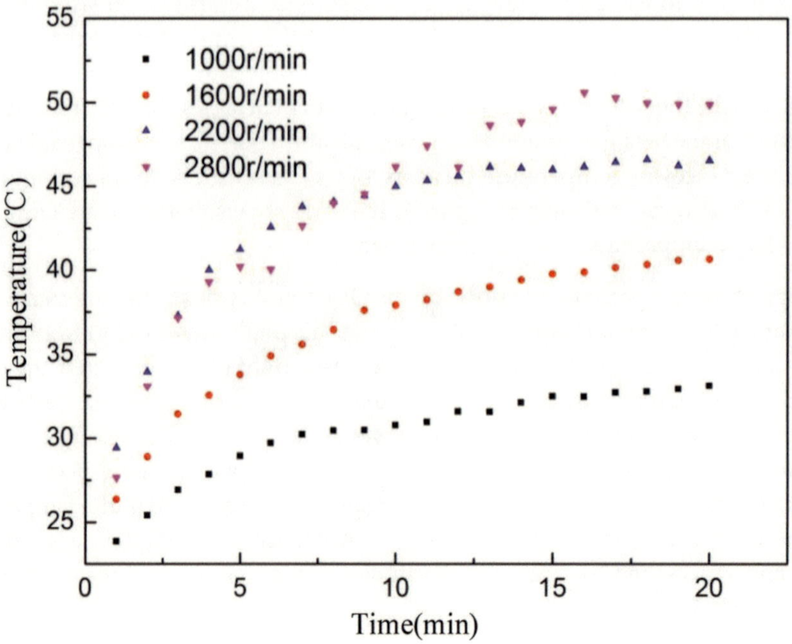

Figure 7. The temperature of inner raceway as a function of time at different speed.

http://dx.doi.org/10.5772/intechopen.70866

4. Conclusions

In this paper, we have proposed a quantum dot fluorescence-based thermometry method for rolling element bearings. Temperature sensor has been fabricated by the deposition of quantum dot films on quartz slide by means of layer-by-layer technique. It has been shown that the emission peak wavelength of the QD sensor has a very linear relationship with the temperature, making it applicable of noncontact temperature measurement of rotating surface. We have managed to take temperature measurement of rolling bearings by a common fluorescence spectrum measurement system. The rotating speed shows no effect on the acquired fluorescence signal, which makes the QD fluorescence-based thermometry method, suitable for very high rotating speed temperature measurement. The practical experiment proves that the CdTe quantum dot fluorescence thermometry could be a feasible and accurate temperature measurement method of bearing inner raceway in operation.

Acknowledgements

The research work is financially supported by the Youth Project of National Natural Science Foundation of China (Grant No. 51405375) and the China Postdoctoral Science Foundation.

Author details

Ke Yan[1]* and Bei Yan[2]

*Address all correspondence to: yanke@mail.xjtu.edu.cn

1 Key Laboratory of Education Ministry for Modern Design and Rotor-Bearing System, Xi'an Jiaotong University, Xi'an, China

2 State Key Laboratory for Manufacturing Systems Engineering, Xi'an Jiaotong University, Xi'an, China

References

[1] Harris TA, Kotzalas MN. Rolling Bearing Analysis. New York: Wiley; 2001

[2] Sadeghi F, Marble S, Joshi A. Bearing cage temperature measurement using radio telemetry. Proceedings of the Institution of Mechanical Engineers, Part J: Journal of Engineering Tribology. 2001;**215**(5):471-481

[3] Henao-Sepulveda JA, Toledo-Quiñones M, Jia Y. Contactless monitoring of ball bearing temperature. Proceedings of the Instrumentation and Measurement Technology Conference, IMTC 2005. Proceedings of the IEEE. Institute of Electrical and Electronics Engineers Inc,USA: 2005. pp. 1571-1573

[4] Nickel DA, Sadeghi F. In situ tribocomponent temperature measurement using a radio telemeter. Tribology Transactions. 1997;**40**(3):514-520

[5] Scott S, Sadeghi F, Peroulis D. An inherently-robust 300 C MEMS temperature sensor for wireless health monitoring of ball and rolling element bearings. Proceedings of IEEE Conference on Sensors. Institute of Electrical and Electronics Engineers Inc, USA; 2009. pp. 975-978

[6] Liu SC, Jia Y, Henao-Sepulveda J, Toledo-Quinones M. Wireless temperature sensor for bearing health monitoring. Proceedings of SPIE, Bellingham, WA. 2004;**5391**:368-376

[7] Wang XD, Wolfbeis OS, Meier RJ. Luminescent probes and sensors for temperature. Chemical Society Reviews. 2013;**42**(19):7834-7869

[8] Jorge P, Martins MA, Trindade T, Santos JL, Farahi F. Optical fiber sensing using quantum dots. Sensors. 2007;**7**(12):3489-3534

[9] Costa-Fernández JM, Pereiro R, Sanz-Medel A. The use of luminescent quantum dots for optical sensing. TrAC Trends in Analytical Chemistry. 2006;**25**(3):207-218

[10] Cutolo A, Bravo J, Goicoechea J, Corres JM, Arregui FJ, Matias IR, Culshaw B, López-Higuera JM. Fiber optic temperature sensor depositing quantum dots inside hollow core fibers using the layer by layer technique. Proceedings of SPIE - The International Society for Optical Engineering. 2007;**6619**:661919-661919-661914

[11] de Bastida G, Arregui FJ, Goicoechea J, Matias IR. Quantum dots-based optical fiber temperature sensors fabricated by layer-by-layer. Sensors Journal, IEEE. 2006;**6**(6):1378-1379

[12] Jorge PAS, Mayeh M, Benrashid R, Caldas P, Santos JL, Farahi F. Quantum dots as self-referenced optical fibre temperature probes for luminescent chemical sensors. Measurement Science and Technology. 2006;**17**(5):1032-1038

[13] Pugh-Thomas D, Walsh BM, Gupta MC. CdSe(ZnS) nanocomposite luminescent high temperature sensor. Nanotechnology. 2011;**22**(18):185503

[14] Walker GW, Sundar VC, Rudzinski CM, Wun AW, Bawendi MG, Nocera DG. Quantum-dot optical temperature probes. Applied Physics Letters. 2003;**83**(17):3555

[15] Wang H-l, Yang A-J, Sui C-H. Luminescent high temperature sensor based on the CdSe/ZnS quantum dot thin film. Optoelectronics Letters. 2013;**9**(6):421-424

[16] Wu S, Dou J, Zhang J, Zhang S. A simple and economical one-pot method to synthesize high-quality water soluble CdTe QDs. Journal of Materials Chemistry. 2012;**22**(29):14573

[17] Decher G. Fuzzy nanoassemblies: Toward layered polymeric multicomposites. Science. 1997;**277**(5330):1232-1237

[18] Crisp MT, Kotov NA. Preparation of nanoparticle coatings on surfaces of complex geometry. Nano Letters. 2003;**3**(2):173-177

Chapter 6

Quantum Dots-Based Nano-Coatings for Inhibition of Microbial Biofilms: A Mini Review

Eepsita Priyadarshini, Kamla Rawat and
Himadri Bihari Bohidar

Additional information is available at the end of the chapter

http://dx.doi.org/10.5772/intechopen.70785

Abstract

Infection of implants by microbial biofilm is chiefly caused by *Staphylococci, Pseudomonas* and *Candida* species. The growth of microbes by forming biofilms offers them protection from antibiotics, drugs and host defense mechanisms. The eradication of biofilms from implants and medical devices is difficult because of the protection by the biofilm forming pathogenic microbes. Hence, researches are focused on development of antibiofilm materials, which are basically constituted of antimicrobial substances or antimicrobial coatings. Nanomaterial-based coatings offer a promising solution in this regard. Quantum dots (QDs) are the group of semiconductor nanoparticles with high photoluminescent properties compared to conventional organic fluorophores. Thus, drug-conjugated QDs can be a promising alternative for biofilm treatment, and these can serve as excellent alternatives for the mitigation of recalcitrant biomaterial-associated infections caused by resistant strains. Furthermore, their use as antibiofilm coating would avoid the dispersion of antimicrobial agents in the surrounding cells and tissues, thereby minimizing the risks of developing microbial resistivity.

Keywords: quantum dots, microbial biofilms, fluorescence, infections, antibiofilm materials

1. Introduction

Quantum dots (QDs) represent a class of colloidal semiconductor nanocrystals having fluorescent properties that absorb photons at a particular (lower) wavelength and emit at a higher wavelength. These QDs are basically composed of a core and corona layer. The photoluminescence emission wavelength of QDs is directly proportional to its size. The core of the QDs may contain one or more heavy elements such as cadmium, selenium, zinc or tellurium. QDs possess significant superiority over the conventional fluorophores in terms of physicochemical and fluorescent properties. The distinguishable fluorescent properties, smaller size, photostability, resistivity to metabolic degradation and capability of conjugation to ligands/biomolecules make QDs a superior choice for biological applications compared to conventional fluorophores.

In the last three decades, several microbes (fungi, yeast and bacteria) have emerged as major human pathogens and have been responsible for causing life threatening diseases especially in immunecompromised individuals and patients with serious medical issues [1]. The widespread and prolonged use of antifungal agents and drugs for treating the infection caused by the pathogens has resulted in increasing incidences of multidrug resistance (MDR). Additionally, several mutant strains have developed that show high resistance to the antifungal drugs being used [1]. For example, *Candida albicans*, a dimorphic opportunistic pathogen, occurs as a normal commensal in humans but becomes pathogenic in immunecompromised individuals. The azole resistive clinical isolates of *C. albicans* result in cross-resistance to several unrelated drugs and this arises because of the phenomenon of multidrug resistance (MDR) [2, 3]. Similarly, *Pseudomonas aeruginosa* and *Staphylococcus aureus* are the two most pathogenic bacteria known to cause severe infection and biofilm formation [4]. Several mechanisms are responsible for development of MDR, some of which involve an overexpression of drug efflux pumps encoding genes such as *CDR1* and *CDR2* belonging to ATP-binding cassette [2, 5, 6]; overexpression of the drug and *MDR1* belonging to the major facilitator superfamily transporters [3, 6] and overexpression of mutations in *ERG11* and encoding the target enzyme of azoles, lanosterol 14α-demethylase [7]. Hence, microbial infection has become a major problem with concerns focusing on those that have become resistant to antibiotics. Around 2 million people are affected annually with antibiotic-resistant bacteria of which approximately 23,000 people die as per the studies of U.S. Center for Disease Control and Prevention [8].

Microbial communities adhere to a solid surface especially in surface/water interference forming biofilms [9]. Microbes attach to the surface by means of extracellular polymeric substances (EPS), and this acts as their survival means against harsh environmental conditions. Biofilm formation is however associated with surface deterioration and corrosion. In addition, pathogenic microbes form biofilms on medical devices and implants, and this has become a great concern in the arena of healthcare. Biofilm also enhances microbial activity and provides protection against harsh environmental conditions such as drugs, antibiotics and common sanitizers. Because of the emerging conditions of MDR, there is a demand for developing new drugs, antimicrobial agents and modifiers capable of inhibiting microbial growth and biofilm formation. With the necessity of developing antimicrobial agents with diverse functionality and ability to kill both strains of bacteria, nanomaterials have been widely investigated in this regard. Silver nanoparticles [10], copper oxide nanoparticles [11–13], metal oxide nanoparticles [12, 13] and even carbon nanomaterials [14] have been reported for their excellent antimicrobial efficiency. Among these, silver nanoparticles have been extensively used as antimicrobial and antibiofilm agents due to their broad spectrum antimicrobial activity, multiple cellular targets and minimum host toxicity. However, high concentration of silver is toxic to humans and its persistent use causes argyrosis and argia [15, 16]. Hence, the demand is for exploring novel nanomaterials with effective antimicrobial and antibiofilm properties along with biocompatibility. Therefore, the requirement must be targeted towards exploring novel biocompatible nanomaterials with effective antibiofilm and optical properties. QDs can be suitable alternatives because of their intriguing optical, fluorescence, high quantum yield, photostability and easy conjugation efficiency. QDs easily attach to microbial surface because of their small size and their dispersion stability is basically governed by colloidal theory [17]. These are excellent candidates in biomedical applications such as imaging, diagnosis and sensing and drug discovery. Developing QDs-based nanocomposites as coating materials on implants and catheters can thus combat pathogenic invasion and biofilm formation. QDs could be engineered with coating agents and conjugated with bioactive ligands or biorecognition elements for targeted treatment, biofilm visualization, and inhibition.

2. Biofilm formation, its mechanism and transmission

Biofilm can be defined as microbial cells enclosed in an exopolysaccharide matrix and adhered to a cell surface. Formation of biofilms by bacteria and fungus is a defense strategy for protection from environment. Microbes secrete extracellular polymeric substances (EPS) that act as a primary scaffold for attachment to solid substrate [18] and its basic constituents are proteins, polysaccharides, nucleic acids with some lipids and humic substances [19]. Three-dimensional study of the EPS layer suggested that it forms a gel-like network wherein microbes are embedded and it also maintains the attachment of bacteria to the solid substrate [20]. Stability to the 3D structure of EPS is rendered by the hydrophobic interactions as well as van der Waals attraction between amino acids/peptides and cations such as Ca^{2+} and Mg^{2+} [21]. Biofilm formation and its structure depend on the environmental conditions to which the bacteria are exposed. When cells are in a nutrient stress condition, an increase in EPS secretion occurs, which promotes hydrophobic interactions to allow attachment to solid substrate [22]. It has been suggested that the presence of a high concentration of EPS negatively affects the diffusion of lipophilic compounds (such as sanitizers, antibiotics and hydrocarbons), across the microbial cell surface [23, 24].

Among bacteria, *P. aeruginosa* is an opportunistic pathogen that causes a number of infections in humans. It develops resistance to antibiotics by forming biofilm matrices that comprise polysaccharide-EPS. It has been noticed that it forms and regulates biofilm via quorum sensing mechanism and therefore most of the researches have focused on disrupting the quorum sensing pathway [25]. Similarly, biofilm formation in *Staphylococcus epidermidis* has also been analyzed using different methods such as microtiter plate, congo red agar plate test and via molecular detection of the ica locus [26–28]. It was found that production of a slimy substance assisted in forming biofilm and was associated with virulence also. The production and formation of biofilm depend on the media constituents; however, the exact mechanism behind the formation of a mature biofilm is still being investigated. However, on the basis of in vitro experimental models, biofilm formation can be segregated into four different stages: (i) attachment of microbial cells to surface, (ii) formation of multi-layer structure via the accumulation and aggregation of cells, (iii) maturation of biofilm and (iv) detachment of cells from biofilm into planktonic state and initiation of a new biofilm cycle [29, 30]. The initial step of attachment is normally driven by hydrophobic, electrostatic and Lifshitze-van der Waals forces, and hence is nonspecific in

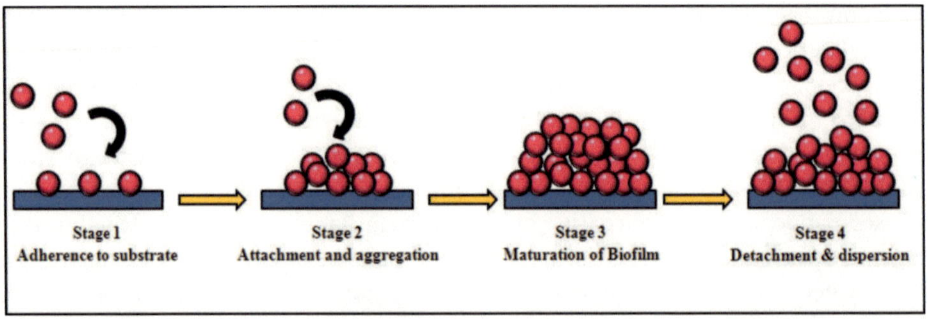

Figure 1. Schematic illustration of the different stages involved in biofilm formation and detachment.

nature. Additionally, certain specific proteins also assist in binding of the microbes to the surfaces [30]. The second step of accumulation is mediated via microbial surface components that recognize adhesive matrix molecules and occurs via an active process. This step involves the establishment of biofilm on the microbial surface. This process is followed by the maturation step. In this step, the characteristic features of the biofilm are formed on basis of specific microbial type. In the final step, where a new phase of invasion is initiated involves the detachment and dispersion of the microbes [31, 32]. **Figure 1** shows the schematic illustration of the different stages in the cycle of biofilm formation and detachment.

P. aeruginosa, *Vibrio cholerae* and some *mycobacterial* species are the common human pathogens that form biofilms and hence have the possibility of infecting humans. There are several mechanisms via which pathogenic microbes in the biofilm can initiate an infection. The seeding dispersal of a large number of pathogenic cells is one of the possible mechanisms that can initiate an infection, as the microbes are not sessile in a biofilm and hence can easily detach and initiate an infection. Secondly, the virulent phenotypes present in a biofilm can expand their colony and initiate infection. This is highly possible as biofilm has a huge heterogeneity in its phenotypical constitution [33, 34]. In addition to these, several other mechanisms have been hypothesized that could possibly allow the survival of a pathogenic organism and its transmission. For example, the detachment of pathogenic microbes from the biofilm, quorum sensing [35], co-aggregation and auto-aggregation [35, 36], modification in biosynthesis of EPS and metabolic pathways and genetic mutations [37] are important issues. However, the complete understanding of the mechanism of biofilm formation and virulence requires complete analysis of the pathogen's life cycle, environmental parameters and the different phenotypes.

3. Role of QDs in inhibiting biofilm formation

Biomedical implants are a necessity in modern health care; biofilm formation on these implants and devices is a major cause of their failure. Mostly *S. epidermidis* and *S. aureus* are observed in contaminated biomedical implants and devices [38]. Biofilms formed on implants and medical devices are difficult to remove as they are protected by exopolymeric matrix secreted by the pathogenic microbe [39]. Although a number of metal and their nanosize forms (silver, copper, gold etc.) have been used as antimicrobial agents, there efficiency is diminishing due to MDR. Investigations on the antibacterial and antifungal property of QDs have been conducted, which suggests that they can serve as excellent candidates for biomedical applications because of their solubility and biocompatibility.

Aqueous solubility and compatibility make graphene quantum dots (GQDs) useful in biomedicine. GQDs are reported to be biocompatible at cellular levels investigated via WST-1 assay, LDH production, ROS generation and *in vitro* and *in vivo* distribution [40]. GQDs also possess antibacterial property against *Escherichia coli* and *S. aureus*, and GQDs with low dose of H_2O_2-based band-aids have also been prepared based on the peroxidase-like property of these particles. The designed band-aids showed a good anti-disinfectant property. They analyzed the effect of formed GQDs on biofilm formation and destruction and observed a reduction in biofilm formation by *S. aureus* at 100 μg/mL and 100 mM of GQDs and H_2O_2 concentration, respectively. Furthermore, they also observed that GQDs alone also showed antibiofilm properties [41]. The studies thus suggested that appropriately designed GQDs had the ability to breakdown existing biofilms and simultaneously prevented the formation of new ones. Habiba et al. suggested the antimicrobial property of silver-graphene quantum dots against *P. aeruginosa* and *S. aureus*. They observed a synergistic effect between silver nanoparticles and GQDs with 25 and 50 g/ml of silver-graphene quantum dots inhibiting *S. aureus* and *P. aeruginosa* growth, respectively. Thus, the potential applicability of Ag-GQDs as fabrication and antibacterial coating agents was clearly established [42].

Furthermore, the use of semiconductor QDs will allow visualization of biofilm inhibition due to their fluorescent properties. The current methods being used for biofilm analysis are SEM, AFM, MRI and Raman spectroscopy that require lengthy and costly procedures apart from sample modulation, which sometimes provide partial details of the samples concerned [43, 44]. Other than this, conventional fluorescent dyes conjugated with carbohydrate recognition elements are used for biofilm analysis via confocal laser microscopy [45]. However, the use of a synthetic complex is sometimes toxic to cells thereby preventing in situ analysis. Therefore, QDs can be an exceptional solution for this. Moreover, amphiphilic carbon dots (CDs) have been shown to penetrate the EPS layer of *P. aeruginosa*, allowing direct visualization of its architecture, growth and how external agents affect its inhibition. The hydrocarbon side chains of CDs dock to the EPS network resulting in making the EPS scaffold highly fluorescent [46]. In yet another study, QDs with two varied surface chemistry [–COOH and polyethylene glycol (PEG) modified] were analyzed for their mobility and distribution in *P. aeruginosa* PAO1 biofilms. It was inferred that the QDs did not penetrate the bacterial cell but did colocalize with EPS matrix of the biofilm. While surface functionalization and QDs flow rate did not show any distinctive difference, analysis of center of density suggested that QDs with –COOH surface groups diffused easily compared to PEGlyated QDs. Biofilms treated with PEGlyated QDs had rough polysaccharide layers and cell distribution compared to –COOH functionalized QDs. It was thus concluded that treatment with nanomaterials can result in varying the structural parameters of biofilm [47]. The fluorescent property of QDs would thus allow recognition of biofilm formation at different growth stages and environmental conditions. Additionally, spectroscopic analysis can also be performed, which would allow better understanding of the phenomenon of binding of QDs to EPS. Conjugated QDs have also been used for biofilm imaging analysis. In a study, CdTe

QDs tagged with Concanavalin A for labeling the saccharide molecules on the surface of *C. albicans* was studied. It relied on the ability of Concanavalin A to specifically bind to α-D mannose and glucose residues of saccharides. They observed that almost 93% of cells were labeled with the modified CdTe particles and were highly specific in activity [48]. Similarly, CdSe/ZnS QDs surface capped by 3-mercaptopropionic acid (MPA) and the amino acids (leucine or phenylalanine) were also used for labeling the biofilm produced by *Shewanella*. Amphiphilic core/shell CdSe/ZnS QDs were used for labeling the hydrophobic microdomains of biofilm produced by *Shewanella oneidensis*, a Gram-negative bacteria. It was inferred that CdSe/ZnS@dihydrolipoic acid-Leu or CdSe/ZnS@dihydrolipoic acid-Phe QDs showed increased hydrophobicity in comparison to CdSe-core QDs capped with 3-mercaptopropionic acid (MPA). Thus, the functional group on QD surface and the ligand density played an integral role in interaction with biofilm matrix. While the hydrophilic MPA-capped QDs were homogeneously associated, DHLA-Leu and DHLA-Phe QDs were specifically confined assisting in identifying the hydrophobic microdomains of biofilm. Hence, appropriate conjugation of surface functional groups can significantly dictate their interaction with biofilm [49]. Quite recently, selenium nanoparticles have been reported for their tremendous potential in biofilm inhibition in *C. albicans*. For the study, selenium nanoparticles were synthesized via laser ablation method and were used to analyze biofilm inhibition. They observed a very good attachment of selenium nanoparticles to the *Candida* surface, which was due to electrostatic attraction between the positively charged surface of *Candida* and negatively charged Se nanoparticles. The particles affect the cellular morphology of the fungus by substitution of sulfur groups of amino acids by the Se particles. This consequently altered the protein structure and damaged Candida morphology. Size and crystallinity of particles had a significant effect on biofilm inhibition [50].

Figure 2 presents the mode of action of quantum dots. The application of QDs as antibiofilm agents can inhibit microbial biofilm at two stages. It can act at the initial stage, where its presence would hinder further attachment of microbial cells to the solid substrate thereby preventing the progression to mature biofilm stage and EPS secretion. Secondly, QDs can act on the matured biofilm, where its penetration into the cells would result in killing of the microbes and subsequent dispersion of the formed biofilm.

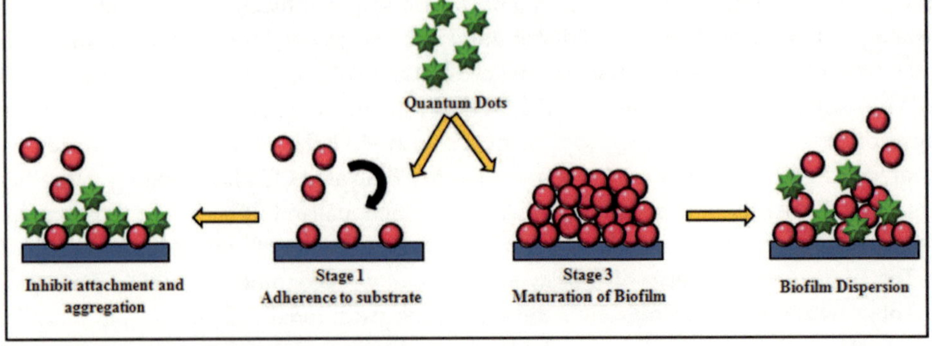

Figure 2. Schematic illustration showing the possible mode of action of antimicrobial quantum dots on biofilm.

http://dx.doi.org/10.5772/intechopen.70785

With this, we envision that QD-based antibiofilm coatings can be promising probes in investigating biofilm imaging, treatment and their eradication. Furthermore, their broad spectrum activity and minimal host toxicity are additional advantages in this regard. Hence, the use of semiconductor QDs would not only allow detecting the inhibition process but also favor their visible monitoring.

4. Conclusion and future perspective

There is a steady increase in the use of QDs. Despite the several advantages offered by QDs, with some improvements, these can emerge as excellent probes for biological applications. Focus should be towards improved protocols for functionalizing the surface of QDs simultaneously making sure that its properties remain unaltered and secondly, appropriately modifying the surface of QDs so that they do not aggregate in a protein-rich solution or cystol. These methods along with the said advantages would assist in utilizing QDs for biological and biomedical applications. Furthermore, the QDs can be tagged with antimicrobial drugs or drugs can be encapsulated inside the QD core thereby increasing the potency of drugs even at low concentration. Synergistic effect of silver nanoparticles with antibiotics such as penicillin G, amoxicillin, erythromycin, clindamycin and vancomycin is known. Therefore, studies on the synergism between QDs and drug molecules have to be analyzed in detail. This would also assist in providing insights into the molecular mechanism of action of QDs and any kind of cellular changes occurring in the pathogen upon its interaction with pathogenic microbes. Additionally, QDs labeling would allow a high throughput analysis of biofilm inhibition and disruptions that will have significant effect in healthcare sector to identify and combat biofilm formation and pathogenic infections.

Acknowledgements

This research was supported by DST-PURSE-II funding. KR acknowledges the receipt of a DST-Inspire Faculty award from Department of Science and Technology, Government of India.

Author details

Eepsita Priyadarshini[1], Kamla Rawat[2,3] and Himadri Bihari Bohidar[1,2*]

*Address all correspondence to: bohi0700@mail.jnu.ac.in

1 School of Physical Sciences, Jawaharlal Nehru University, New Delhi, India

2 Special Centre for Nano Sciences, Jawaharlal Nehru University, New Delhi, India

3 Inter University Accelerator Centre, New Delhi, India

References

[1] Barelle CJ, Priest CL, MacCallum DM, Gow NA, Odds FC, Brown AJ. Niche-specific regulation of central metabolic pathways in a fungal pathogen. Cellular Microbiology. 2006 Jun 1;**8**(6):961-971

[2] Franz R, Kelly SL, Lamb DC, Kelly DE, Ruhnke M, Morschhäuser J. Multiple molecular mechanisms contribute to a stepwise development of fluconazole resistance in clinical *Candida albicans* strains. Antimicrobial Agents and Chemotherapy. 1998 Dec 1;**42**(12):3065-3072

[3] White TC, Marr KA, Bowden RA. Clinical, cellular, and molecular factors that contribute to antifungal drug resistance. Clinical Microbiology Reviews. 1998 Apr 1;**11**(2):382-402

[4] Bowler PG, Duerden BI, Armstrong DG. Wound microbiology and associated approaches to wound management. Clinical Microbiology Reviews. 2001 Apr 1;**14**(2):244-269

[5] Kohli A, Smriti NF, Mukhopadhyay K, Rattan A, Prasad R. In vitro low-level resistance to azoles in *Candida albicans* is associated with changes in membrane lipid fluidity and asymmetry. Antimicrobial Agents and Chemotherapy. 2002 Apr 1;**46**(4):1046-1052

[6] Stoldt VR, Sonneborn A, Leuker CE, Ernst JF. Efg1p, an essential regulator of morphogenesis of the human pathogen *Candida albicans*, is a member of a conserved class of bHLH proteins regulating morphogenetic processes in fungi. The EMBO Journal. 1997 Apr 15;**16**(8):1982-1991

[7] Prasad T, Saini P, Gaur NA, Vishwakarma RA, Khan LA, Haq QM, Prasad R. Functional analysis of CaIPT1, a sphingolipid biosynthetic gene involved in multidrug resistance and morphogenesis of *Candida albicans*. Antimicrobial Agents and Chemotherapy. 2005 Aug 1;**49**(8):3442-3452

[8] US Department of Health and Human Services, Centers for Disease Control and Prevention, Antibiotic Resistance Threats in the United States, 2013. US Centers for Disease Control and Prevention. 2013. Available from: http://www.cdc.gov/drugresistance/threat-report-2013/pdf/ar-threats-2013-508.pdf [Accessed: Jan 24, 2015]

[9] Costerton JW, Lewandowski Z, Caldwell DE, Korber DR, Lappin-Scott HM. Microbial biofilms. Annual Reviews in Microbiology. 1995 Oct;**49**(1):711-745

[10] Pal S, Tak YK, Song JM. Does the antibacterial activity of silver nanoparticles depend on the shape of the nanoparticle? A study of the Gram-negative bacterium *Escherichia coli*. Applied and Environmental Microbiology. 2007 Mar 15;**73**(6):1712-1720

[11] Ruparelia JP, Chatterjee AK, Duttagupta SP, Mukherji S. Strain specificity in antimicrobial activity of silver and copper nanoparticles. Acta Biomaterialia. 2008 May 31;**4**(3):707-716

[12] Azam A, Ahmed AS, Oves M, Khan MS, Memic A. Size-dependent antimicrobial properties of CuO nanoparticles against Gram-positive and-negative bacterial strains. International Journal of Nanomedicine. 2012;**7**:3527

[13] Azam A, Ahmed AS, Oves M, Khan MS, Habib SS, Memic A. Antimicrobial activity of metal oxide nanoparticles against Gram-positive and Gram-negative bacteria: A comparative study. International Journal of Nanomedicine. 2012;**7**:6003

[14] Ristic BZ, Milenkovic MM, Dakic IR, Todorovic-Markovic BM, Milosavljevic MS, Budimir MD, Paunovic VG, Dramicanin MD, Markovic ZM, Trajkovic VS. Photodynamic antibacterial effect of graphene quantum dots. Biomaterials. 2014 May 31;**35**(15):4428-4435

[15] White JM, Powell AM, Brady K, Russell-Jones R. Severe generalized argyria secondary to ingestion of colloidal silver protein. Clinical and Experimental Dermatology. 2003 May 1;**28**(3):254-256

[16] Kwon HB, Lee JH, Lee SH, Lee AY, Choi JS, Ahn YS. A case of argyria following colloidal silver ingestion. Annals of Dermatology. 2009 Aug 1;**21**(3):308-310

[17] Israelachvili J. Intermolecular and Surface Forces. 2nd ed. San Diego, CA: Academic Press Inc; 1992

[18] Flemming HC, Neu TR, Wozniak DJ. The EPS matrix: The house of biofilm cells. Journal of Bacteriology. 2007 Nov 15;**189**(22):7945-7947

[19] Sutherland IW. Biofilm exopolysaccharides: A strong and sticky framework. Microbiology. 2001 Jan 1;**147**(1):3-9

[20] Flemming HC, Wingender J, Mayer C, Korstgens V, Borchard W. Cohesiveness in biofilm matrix polymers. In: Symposia-Society for General Microbiology 2000 Jan 1. Cambridge: Cambridge University Press; 1999. p. 87-106

[21] Kitazawa H, Toba T, Itoh T, Kumano N, Adachi S, Yamaguchi T. Antitumoral activity of slime-forming, encapsulated *Lactococcus lactis* subsp. *cremoris* isolated from Scandinavian ropy sour milk, viili. Animal Science and Technology (Japan). 1991; ISSN: 0021-5309

[22] Sheng X, Ting YP, Pehkonen SO. The influence of ionic strength, nutrients and pH on bacterial adhesion to metals. Journal of Colloid and Interface Science. 2008 May 15;**321**(2):256-264

[23] Ryu JH, Beuchat LR. Biofilm formation by *Escherichia coli* O157: H7 on stainless steel: Effect of exopolysaccharide and curli production on its resistance to chlorine. Applied and environmental microbiology. 2005 Jan 1;**71**(1):247-254

[24] Kang YS, Park W. Protection against diesel oil toxicity by sodium chloride-induced exopolysaccharides in Acinetobacter sp. strain DR1. Journal of Bioscience and Bioengineering. 2010 Feb 28;**109**(2):118-123

[25] Koutsoudis MD, Tsaltas D, Minogue TD, von Bodman SB. Quorum-sensing regulation governs bacterial adhesion, biofilm development, and host colonization in Pantoea stewartii subspecies stewartii. Proceedings of the National Academy of Sciences. 2006 Apr 11;**103**(15):5983-5988

[26] Christensen GD, Simpson WA, Younger JJ, Baddour LM, Barrett FF, Melton DM, Beachey EH. Adherence of coagulase-negative staphylococci to plastic tissue culture plates: A

quantitative model for the adherence of staphylococci to medical devices. Journal of Clinical Microbiology. 1985 Dec 1;**22**(6):996-1006

[27] Freeman DJ, Falkiner FR, Keane CT. New method for detecting slime production by coagulase negative staphylococci. Journal of Clinical Pathology. 1989 Aug 1;**42**(8):872-874

[28] Arciola CR, Campoccia D, Gamberini S, Cervellati M, Donati E, Montanaro L. Detection of slime production by means of an optimised Congo red agar plate test based on a colourimetric scale in *Staphylococcus epidermidis* clinical isolates genotyped for ica locus. Biomaterials. 2002 Nov 30;**23**(21):4233-4239

[29] Costerton JW, Montanaro L, Arciola CR. Biofilm in implant infections: Its production and regulation. The International Journal of Artificial Organs. 2005 Nov;**28**(11):1062-1068

[30] Mack D, Becker P, Chatterjee I, Dobinsky S, Knobloch JK, Peters G, Rohde H, Herrmann M. Mechanisms of biofilm formation in *Staphylococcus epidermidis* and *Staphylococcus aureus*: Functional molecules, regulatory circuits, and adaptive responses. International Journal of Medical Microbiology. 2004 Sep 24;**294**(2):203-212

[31] Patti JM, Allen BL, McGavin MJ, Hook M. MSCRAMM-mediated adherence of microorganisms to host tissues. Annual reviews in. Microbiology. 1994 Oct;**48**(1):585-617

[32] Speziale P, Pietrocola G, Rindi S, Provenzano M, Provenza G, Di Poto A, Visai L, Arciola CR. Structural and functional role of *Staphylococcus aureus* surface components recognizing adhesive matrix molecules of the host. Future Microbiology. 2009 Dec 8;**4**(10):1337-1352

[33] Hall-Stoodley L, Stoodley P. Biofilm formation and dispersal and the transmission of human pathogens. Trends in Microbiology. 2005 Jan 31;**13**(1):7-10

[34] Purevdorj B, Costerton JW, Stoodley P. Influence of hydrodynamics and cell signaling on the structure and behavior of *Pseudomonas aeruginosa* biofilms. Applied and Environmental Microbiology. 2002 Sep 1;**68**(9):4457-4464

[35] Hammer BK, Bassler BL. Quorum sensing controls biofilm formation in *Vibrio cholerae*. Molecular Microbiology. 2003 Oct 1;**50**(1):101-104

[36] von Götz F, Häussler S, Jordan D, Saravanamuthu SS, Wehmhöner D, Strüßmann A, Lauber J, Attree I, Buer J, Tümmler B, Steinmetz I. Expression analysis of a highly adherent and cytotoxic small colony variant of *Pseudomonas aeruginosa* isolated from a lung of a patient with cystic fibrosis. Journal of Bacteriology. 2004 Jun 15;**186**(12):3837-3847

[37] Yildiz FH, Schoolnik GK. *Vibrio cholerae* O1 el tor: Identification of a gene cluster required for the rugose colony type, exopolysaccharide production, chlorine resistance, and biofilm formation. Proceedings of the National Academy of Sciences. 1999 Mar 30;**96**(7):4028-4033

[38] Khalil H, Williams RJ, Stenbeck G, Henderson B, Meghji S, Nair SP. Invasion of bone cells by *Staphylococcus epidermidis*. Microbes and Infection. 2007 Apr 30;**9**(4):460-465

[39] Subbiahdoss G, Sharifi S, Grijpma DW, Laurent S, van der Mei HC, Mahmoudi M, Busscher HJ. Magnetic targeting of surface-modified superparamagnetic iron oxide nanoparticles

yields antibacterial efficacy against biofilms of gentamicin-resistant staphylococci. Acta Biomaterialia 2012 Jul 31;**8**(6):2047-2055

[40] Chong Y, Ma Y, Shen H, Tu X, Zhou X, Xu J, Dai J, Fan S, Zhang Z. The in vitro and in vivo toxicity of graphene quantum dots. Biomaterials. 2014 Jun 30;**35**(19):5041-5048

[41] Sun H, Gao N, Dong K, Ren J, Graphene QX. Quantum dots-band-aids used for wound disinfection. ACS Nano. 2014 Jun 2;**8**(6):6202-6210

[42] Habiba K, Bracho-Rincon DP, Gonzalez-Feliciano JA, Villalobos-Santos JC, Makarov VI, Ortiz D, Avalos JA, Gonzalez CI, Weiner BR, Morell G. Synergistic antibacterial activity of PEGylated silver-graphene quantum dots nanocomposites. Applied Materials Today. 2015 Dec 31;**1**(2):80-87

[43] Denkhaus E, Meisen S, Telgheder U, Wingender J. Chemical and physical methods for characterisation of biofilms. Microchimica Acta. 2007 Apr 1;**158**(1-2):1-27

[44] Wolf G, Crespo JG, Reis MA. Optical and spectroscopic methods for biofilm examination and monitoring. Reviews in Environmental Science and Biotechnology. 2002 Sep 1;**1**(3):227-251

[45] Neu TR, Swerhone GD, Lawrence JR. Assessment of lectin-binding analysis for in situ detection of glycoconjugates in biofilm systems. Microbiology. 2001 Feb 1;**147**(2):299-313

[46] Ritenberg M, Nandi S, Kolusheva S, Dandela R, Meijler MM, Jelinek R. Imaging *Pseudomonas aeruginosa* biofilm extracellular polymer scaffolds with amphiphilic carbon dots. ACS Chemical Biology. 2016 Feb 26;**11**(5):1265-1270

[47] Morrow JB, Holbrook RD. Association of quantum dot nanoparticles with biofilm. Journal of Environmental Quality. 2010 Nov 1;**39**(6):1934-1941

[48] Tenório DP, Andrade CG, Cabral Filho PE, Sabino CP, Kato IT, Carvalho LB, Alves S, Ribeiro MS, Fontes A, Santos BS. CdTe quantum dots conjugated to concanavalin A as potential fluorescent molecular probes for saccharides detection in *Candida albicans*. Journal of Photochemistry and Photobiology B: Biology. 2015 Jan 31;**142**:237-243

[49] Aldeek F, Mustin C, Balan L, Roques-Carmes T, Fontaine-Aupart MP, Schneider R. Surface-engineered quantum dots for the labeling of hydrophobic microdomains in bacterial biofilms. Biomaterials. 2011 Aug 31;**32**(23):5459-5470

[50] Guisbiers G, Lara HH, Mendoza-Cruz R, Naranjo G, Vincent BA, Peralta XG, Nash KL. Inhibition of *Candida albicans* biofilm by pure selenium nanoparticles synthesized by pulsed laser ablation in liquids. Nanomedicine: Nanotechnology, Biology and Medicine. 2017 Apr 30;**13**(3):1095-1103

Chapter 7

Biomolecule-Conjugated Quantum Dot Nanosensors as Probes for Cellular Dynamic Events in Living Cells

Jung Y. Huang

Additional information is available at the end of the chapter

http://dx.doi.org/10.5772/intechopen.72858

Abstract

A single-molecule tracking/imaging technique with semiconductor quantum dot (QD) nanosensors conjugated with appropriate peptides or antibodies is appealing for probing cellular dynamic events in living cells. We developed a 2D analysis of single-molecule trajectories using normalized variance versus mean square displacement (MSD) to provide high-quality statistics sampled by nanosensors while preserving single-molecule sensitivity. This plot can be more informative than MSD alone to reflect the diffusive dynamics of a protein in its cellular environment. We illustrate the performance of this technique with selected examples, which are designed to expose the functionalities and importance in live cells. Our findings suggest that biomolecule-conjugated QD nanosensors can be used to reveal interactions, stoichiometries, and conformations of proteins, and provide an understanding of the mode of the interaction, stable states, and dynamical pathways of biomolecules in live cells.

Keywords: quantum dot, single-particle tracking, fluorescence imaging, stochastic thermodynamics, single-molecule trajectory, living cell, plasma membrane, epidermal growth factor receptor, lipid domain, actin filaments, cell-penetrating peptides

1. Introduction

Studying the movement of individual biomolecules in live cells and their interactions with the surrounding microenvironment would greatly improve our understanding of how biomolecules behave in their native cellular environment [1, 2]. Deciphering those functions and relevant regulation mechanisms is also important for developing new therapeutic strategies for diseases [2]. The major factors affecting protein mobility include local viscosity, protein–protein interaction, molecular crowding, and dimensionality of accessible space [3]. However, such factors are difficult to reconstitute in vitro using purified constituents. Therefore, there is a compelling demand for a tool to directly access the properties of the molecular assemblies and kinetics of interaction in live cells. Single-molecule imaging and tracking based on fluorescence microscopy have been developed to meet this challenge [4, 5].

The primary factor controlling the motion of a protein in a living cell is often not the friction in the cellular medium but the interactions with its molecular partners, which often result in a transient stall or transport of molecules [3]. The binding energies between the protein of interest and its interacting partners are also of interest because regulatory processes can be mediated by changes in these binding energies. Biological media are spatially inhomogeneous, which is poorly conveyed by measuring just a few, sparse single-molecule trajectories. Thus, to fully realize the potential of a single-molecule imaging and tracking technique, an efficient and reliable analytical method is required to help extract useful information from the large amounts of trajectory data. This type of analysis usually involves the computing of the mean square displacement (MSD) along the trajectories of the molecules [6, 7].

The key component of the single-molecule imaging and tracking technique is a set of bright fluorophores with different emission wavelengths. Semiconductor quantum dots (QDs) have unique optical properties, such as high emission efficiency, wavelength tunability, and long-term stability, which make them appealing as *in vivo* and *in vitro* fluorophores [8, 9]. The ability to make water-soluble QDs that can be targeted to specific biomolecules has led to a variety of applications in cellular sensing and imaging. For example, Zhang et al. developed a QDs-labeled silica nanoprobe for the detection of apoptotic cells in response to therapy [10]. Different classes of fluorescent nanoprobes were also developed for the imaging of cellular metal ions [11], which serve as essential cofactors in energy metabolism, signal transduction, and nucleic acid processing. Recently, Jainb et al. developed a synthetic technique of QD immunoconstructs by coupling antibodies (Ab) to QD [12]. The resulting QD-Ab conjugates can maintain a high and stable quantum yield for *in vivo* environments and acts as an ideal nanosensor to specific antigene. Therefore, a molecular-level of understanding of the cellular functions in the context of their native environments becomes possible.

Labeled biomolecules in their native environments can be considered a mesosystem with a length scale ranging from a few nanometers to <1 μm. Our understanding of thermodynamically equilibrated mesosystems roots solidly in equilibrium statistical mechanics. For small deviations from the equilibrium, researchers can invoke the linear response theory to relate the transport properties caused by the external fields to the equilibrium correlation functions. Beyond this linear response regime, no universally exact results are currently available [13].

Under non-equilibrium conditions, the temperature of a mesosystem in solution remains well-defined, yielding a value that is the same as that of the embedding solution [13]. For a complex biomolecular system comprised of N relatively rigid domains, the configuration can be described by a 3N–dimensional vector of Cartesian coordinates. The interactions among these rigid units introduce cooperative couplings between the units that yield a separation of time scales [14]. The resulting time-scale separation occurs between the observable slow degrees of freedom of the system and the fast ones that are made up by both the system and thermal bath. The collection of the slow degrees of freedom offers a natural approach to define the states of a system. The state changes with time, either due to the external driving or from ever-present thermal fluctuations that trace out a trajectory. The thermodynamic quantities defined along the trajectories follow a distribution with some universal constraints [14].

Stochastic thermodynamics is a relatively new subject, which focuses on the description of the individual trajectories [14]. This framework can serve as a solid foundation for single-molecule technique but has not been sufficiently clarified in the literature. In this chapter, we first briefly review some basic concepts of stochastic thermodynamics that are specifically relevant to the analysis of the trajectories from a single-molecule optical imaging and tracking technique. Repetitive measurements of the cellular locations $\tilde{x}(t)$ of nanosensors provide an entire history of that observation. Provided that the repetitive measurements could scan all of the system degrees of freedom, the time evolution of the system may be encoded into the trace of this nanosensor. Based on this understanding, we illustrated the procedures of implementing such concepts in the analysis of trajectories for receptor proteins on the plasma membranes of living cells and obtained useful information from this important system.

Note that the separation of time scales can also render the dimensionality of a mesosystem much lower than the 3N–dimensional coordinate space. Thus, the trajectories through 3N–dimensional space are effectively restrained to an intrinsic manifold of much lower dimensionality [15, 16]. To show the usefulness of this concept in single molecule tracking, we took advantage of the brightness and photostability of QDs to investigate the translocation behavior of the human immunodeficiency virus 1 (HIV-1) transactivator of transcription peptide (TatP)-conjugated quantum dot (TatP-QD) nanosensors in complex cellular terrains [17]. As TatP-QDs translocate across the plasma membranes of living cells, the particles can be viewed as nanoscale pens [18] to record the influence of the hierarchical structure of the cellular environment on TatP-QD trajectories. Analysis of the resulting three-dimensional (3D) trajectories disclosed the interaction between the TatP-QDs and bioactive groups on the plasma membrane [19, 20]. An understanding of the cellular uptake of TatP is also essential for the development of TatP-based delivery strategies for therapeutic applications.

This chapter aims to expose the connections between the framework of stochastic thermodynamics, single-molecule optical tracking, and trajectory analysis. Applications were mentioned to illustrate the type of information that can be deduced from these studies. The chapter is not a complete review on relevant subjects. Many important research studies have not been mentioned or referred. The author simply hopes this article will encourage interested readers to design new experiments that would fill in the holes of this article.

2. Formalism of single-molecule trajectory analysis

2.1. A brief overview of stochastic thermodynamics

For a mesosystem in contact with a heat bath, the probability of finding it in a specific microstate is given by the Boltzmann factor (i.e., $\exp[-U(x,p)/k_BT\ ]$ [13], where U denotes the total internal energy with x and p being the generalized coordinates and momentums of particles enclosed in the system). Stochastic thermodynamics provides the framework for extending the notions of work, heat and entropy production from classical thermodynamics to individual trajectories of non-equilibrium processes. It brings out the fluctuation-dissipation theorems (FDT) to constrain the probability distributions for work, heat, and entropy production along each trajectory [14]. Some milestone developments in this discipline include:

1) Both the steady-state and transient FDT were valid for a large class of systems, including chaotic dynamics [21], driven Langevin dynamics [22], and driven diffusive dynamics [23].

2) The Jarzynski relation (JR) was derived [24, 25], which relates the free energy difference between two equilibrium states to the average work done to drive the system from one state to the other along a non-equilibrium process. For non-equilibrium systems driven by time-dependent forces, a refinement of the JR became extremely useful to determine the free energy landscapes of biomolecules [26, 27].

3) The exchanged heat and applied work could also be rigorously defined along individual trajectories of the driven Brownian motion. The entropy produced in a medium could thereby be related to the stochastic action, which also serves as the weight of trajectories [28].

2.2. Trajectory analysis of single-molecule stochastic processes

2.2.1. Two-dimensional plot of normalized variance and mean square displacement of single-molecule trajectories

In a living cell, a biomolecule subjected to random influences can explore its possible outcomes and evolve to yield dispersion over state space. This evolution contains contributions from both deterministic and stochastic forces. The time-scale separation mentioned above implies that the dynamics will become Markovian and follow a generalized Langevin equation [29]

$$\partial_t \vec{x}_k(t) = -\nabla_k U\left(\vec{x}_k(t)\right)/\gamma \ + \sqrt{2D}dW_t = F_k(t)/\gamma \ + f_k(t)/\gamma, \tag{1}$$

where subindex k represents the k-th particle at the position $\vec{x_k}$. The frictional parameter γ is relevant to the diffusion coefficient D with $\gamma.D = k_B T$. The total deterministic force acting on the diffusive particle is expressed as $F_k(t)$. The stochastic force $f_k(t)$ follows the Weiner process with zero mean and a delta correlation of $< f_k(t+\tau)\cdot f_k(t) >= 2\gamma k_B T\delta(\tau)$.

By rewriting Eq. (1) as $dx_k = F_k/\gamma \ \cdot dt + \sqrt{2D}dW_t$ and invoking the stochastic chain rule, we derived an equation for the square displacement [30]

$$(dx_k)^2 = 2F_k/\gamma \cdot x_k dt + 2Ddt + 2\sqrt{2D}x_k dW_t. \tag{2}$$

From Eq. (2), we further derived the local MSD $\overline{R_k^2}$ as

$$\overline{R_k^2} = \overline{dx_k^2} = 2/\gamma \ \cdot \overline{F_k x_k}\, dt + 2Ddt, \tag{3}$$

and similarly, the variance of the square displacement $Var\left(dx_k^2\right) = \overline{\left(dx_k^2\right)^2} - \left[\overline{dx_k^2}\right]^2$ as

$$Var\left(dx_k^2\right) = 8D\left(\overline{x_k}\right)^2 dt - (4/\gamma)\cdot Var[F_k x_k]\,(dt)^2, \tag{4}$$

where $Var[.]$ denotes the variance operation. A normalized variance of square displacement was defined as $V\left(R_k^2\right) = Var\left[R_k^2\right]/\left(\overline{R_k^2}\right)^2$, which yielded [30]

$$V(R_k^{\,2}) = \frac{(2\gamma/dt\ )(\overline{x_k})^2 - Var[F_k x_k]}{\gamma D + \overline{F_k x_k}}. \tag{5}$$

We applied this function to display the relative influence on the trajectories by deterministic forces $F_k(t)$ and by the stochastic force f_k. As particles diffuse under a force field, $V(R^2)$ increases with force strength [30]. As a particle diffuses near a short-range force field, it can be stalled briefly by the force field, resulting in large variances in the diffusion step size. In contrast, $V\left(\overline{R_\tau^{\,2}}\right)$ will fall below the free diffusion limit ($V(R^2) = 2$) when a probing particle moves in an environment where its surrounding medium can be polarized by the particle either electrically or orientationally [30]. This dressing effect could lead to smaller variances, and therefore smaller $V(R^2)$. We proposed a histogram of $\overline{R_\tau^{\,2}}$ and normalized variance $V(R^2)$ in a contour plot. Here, the MSD values were used to quantify the diffusion of a probing particle, and $V\left(\overline{R_\tau^{\,2}}\right)$ revealed the nature (e.g., deterministic or stochastic) of the interactive forces involved [31].

In the following, we presented some simulated results to illustrate the features of this ad hoc data-driven methodology in the framework of stochastic thermodynamics. Using Eq. (1), we first simulated 2D motions of Brownian particles in a force field, which had a potential energy surface $U\left(\vec{x}\right) = k\left(\vec{x} - \vec{x}_0\right)^T \cdot \left(\vec{x} - \vec{x}_0\right)/2\ + \sum_i \alpha_i \exp\left[-\left(\vec{x} - \vec{x}_{i0}\right)^T \cdot \left(\vec{x} - \vec{x}_{i0}\right)/w_i^{\,2}\right]$ (see **Figure 1a**). The potential energy surface comprised two isotropic Gaussian wells and a long-range harmonic potential to prevent the particle from drifting off to infinity. The central positions of the harmonic potential and two Gaussian wells are located at (3, 3), (2, 0) and (0, 2), respectively. The strength parameters of the harmonic and Gaussian potentials were chosen to yield deterministic forces, which were a factor of 0.14 and 3.8 to that of the stochastic force

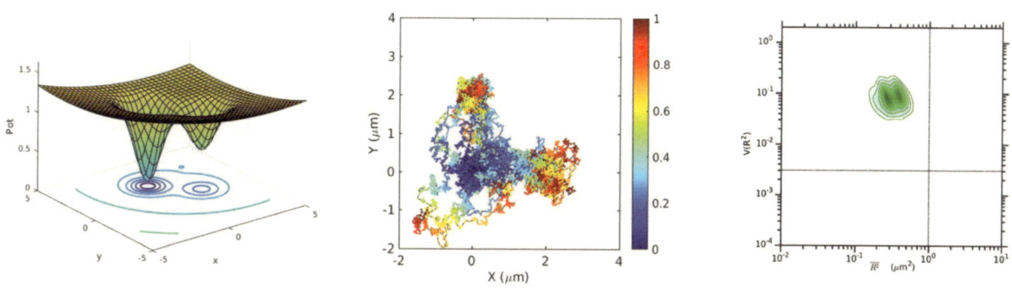

(a) (b) (c)

Figure 1 (a) Plot of the potential energy surface with $U\left(\vec{x}\right) = 0.05\left[0.1\left(\vec{x} - \vec{x}_0\right)^T \cdot \left(\vec{x} - \vec{x}_0\right) - 11\left(e^{-\left(\vec{x}-\vec{x}_{10}\right)^T \cdot \left(\vec{x}-\vec{x}_{10}\right)} + 2e^{-\left(\vec{x}-\vec{x}_{20}\right)^T \cdot \left(\vec{x}-\vec{x}_{20}\right)}\right)\right]$, where $\vec{x}_0 = (3, 3)$, $\vec{x}_{10} = (2, 0)$, and $\vec{x}_{20} = (0, 2)$. The energy surface was shifted upwardly by 1 to make the contour plot visible. (b) Twenty-five randomly selected trajectories were displayed with appearing times coded by different colors. (c) the statistics of 500 trajectories were summarized in the $V(R^2)$-$\overline{R_\tau^{\,2}}$ plot, exhibiting a peak at $V(R^2) \simeq 0.1$, $\overline{R_\tau^{\,2}} \simeq 0.27$ and 0.4.

($\sqrt{2D}/\gamma = 0.35$). We updated the locations of the Brownian particles every 0.01 s using the Euler–Maruyama solver, starting at initial position (0, 0). We generated a total of 500 trajectories, each of a 10 s duration. **Figure 1b** displays 25 randomly selected trajectories with appearing times coded by different colors.

As the particles move toward the center of the harmonic potential, they are attracted to the two Gaussian wells. Well 2, centered at (2, 0), had the same width but was deeper than well 1 by a factor of 2. Thus, at the end of the simulation, the particles near (2, 0) were about twice that of those near well 1. As displayed in **Figure 1c**, the diffusion yielded a dual-peak structure at $V(R^2) \simeq 0.1$, $\overline{R_\tau^2} \simeq 0.27$ and 0.4 in the $V(R^2)$-$\overline{R_\tau^2}$ plot, indicating that as a particle repeatedly visits or stays in a spatial region, the characteristic $V\left(\overline{R_\tau^2}\right)$ and $\overline{R_\tau^2}$ of the location will be imposed on the trajectories.

Next, we reduced the width of well 2 by a factor 3 while keeping its depth at the same value (see **Figure 2a**. At the end of the simulation, the particles located near (0, 2) became one third that of those near Well 1 (see red spots in **Figure 2b**). **Figure 2c** displays a peak at $V(R^2) \simeq 0.25$ and $\overline{R_\tau^2} \simeq 0.25$. Although the population at well 2 was lower, its influence on the trajectories with a higher $V(R^2)$ value was visible. For a brief summary of this simulation, we would like to point out an attractive feature of the $V(R^2)$-$\overline{R_\tau^2}$ plot. When a molecule repeatedly visits or stays in a spatial region, the characteristic $V\left(\overline{R_\tau^2}\right)$ and $\overline{R_\tau^2}$ of the location will be imposed on the trajectories, which then results in the formation of a peak at the corresponding position on the plot.

We used the hidden Markov model (HMM) to further reveal the dynamics by identifying the underlying state changes and their corresponding occupation probability π_i and transition rates $r_{i\to j}$. Note that in the ergodic limit, the system will reach an equilibrium with a distribution of

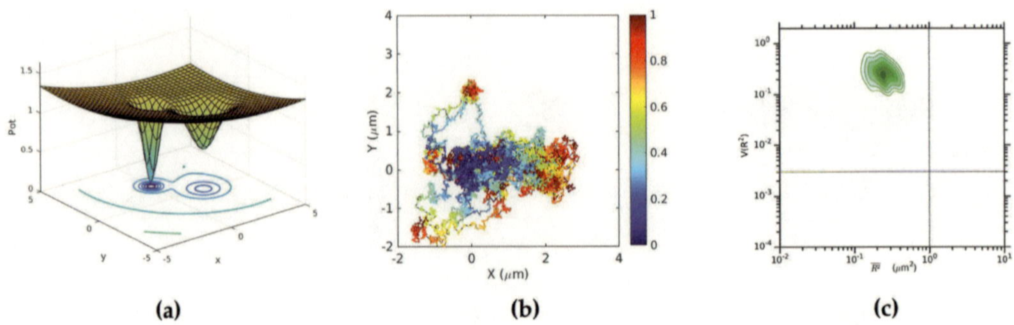

Figure 2 (a) Plot of the potential energy surface with $U(\vec{x}) = 0.05\left[0.1\left(\vec{x} - \vec{x}_0\right)^T \cdot \left(\vec{x} - \vec{x}_0\right) - 11\left(e^{-\left(\vec{x}-\vec{x}_{10}\right)^T \cdot \left(\vec{x}-\vec{x}_{10}\right)} + 2e^{-3\left(\vec{x}-\vec{x}_{20}\right)^T \cdot \left(\vec{x}-\vec{x}_{20}\right)}\right)\right]$, where $\vec{x}_0 = (3,3)$, $\vec{x}_{10} = (2,0)$, and $\vec{x}_{20} = (0,2)$. The energy surface was shifted upwardly by 1 to make the contour plot visible. (b) twenty-five randomly selected trajectories from the simulation were displayed. (c) the statistics of 500 trajectories were summarized in the $V(R^2)$-$\overline{R_\tau^2}$ plot, exhibiting a peak at $V(R^2) \simeq 0.25$, $\overline{R_\tau^2} \simeq 0.25$.

$p_{eq}(x) = e^{-\beta U(x)}/Z_{eq}$, where $U(x)$ is the potential of mean force (PMF) with x denoting the generalized coordinates, and Z_{eq} representing the equilibrium partition function [32]. We can measure the static dispersion of the system using entropy $S\left[p_{eq}(x)\right] = -\int dx p_{eq}(x) \ln\left[p_{eq}(x)\right]$ as a caliber [14, 33]. For non-equilibrium processes, the entropy produced along a trajectory with time resolution Δt becomes $S(\Delta t) \equiv \sum_{i,j}^{N} \pi_i r_{i\to j}(\Delta t) \ln r_{i\to j}(\Delta t)$, which required averaging over the stochastic trajectories to display the degree of dynamic dispersion. For coarser time resolutions (i.e., larger Δt), the transition rates converged to their equilibrium values and the information about the dynamics is lost.

2.2.2. Spectral-embedding analysis of single-molecule trajectories

Conformational trajectories of a biomolecular system, comprising N relatively rigid domains, can be displayed in a 3N–dimensional phase space. As noted above, cooperative couplings between these rigid units often yield a separation of time scales, which causes the system's slow degrees of freedom to be separated from the fast ones made up by the system and thermal bath. An intrinsic manifold of much lower dimensionality is thus embedded in the high-dimensional configuration trajectories. Unfortunately, the projection of dynamical configurations $f: R^{3N} \to M(R^m)$ into a reduced dimensional space, which is specified by m collective variables $\tilde{\phi} = [\phi_1, \phi_2, \ldots, \phi_m] \in M$, is highly nonlinear and unavailable from analytical theory. The first issue encountered in depicting the complex dynamics in a low-dimensional space is how to identify a set of appropriate slow variables $\tilde{\phi}$. In recent years, a number of machine learning approaches have been developed to infer such mappings by discovering low-dimensional manifolds within high-dimensional trajectories [15].

Recently, Wang and Ferguson successfully applied the generalized Takens Delay Embedding Theorem [34] to retrieve a low-dimensional representation of the free energy landscape from univariate time series of single-molecule physical observable. The authors also determined that the univariate time series could be expanded into a high-dimensional space in which the dynamics were equivalent to those of the molecular motions in real space. Single-molecule optical techniques based on a variety of nanosensors can provide the time series of experimentally accessible observables. By measuring the impact of cellular environments on the trajectory ensemble of those nanosensors, it is possible to reveal the influence of the cellular environments. **Figure 3** presents a conceptual drawing to illustrate the translocation process of biomolecule-conjugated quantum dot nanosensors across the cellular plasma membrane.

We assumed the trajectory ensemble $\vec{x}(t)$ of the nanosensors to be generated by a stochastic process governed by Eq. (1). Here, $\vec{x}(t)$ was implicitly dependent on the generalized coordinates of the fast degrees of freedom $\vec{\xi}_i(t)$ because the probing particles could move in n different realizations of the local environment with interaction potentials $U_i\left(\vec{x}\left(\vec{\xi}\right)\right); i = 1, .., n$. In the following, we will describe a projection of $\vec{x}(t)$ on slow degrees of freedom to disclose the influences of $U_i\left(\vec{x}\left(\vec{\xi}\right)\right); i = 1, .., n$.

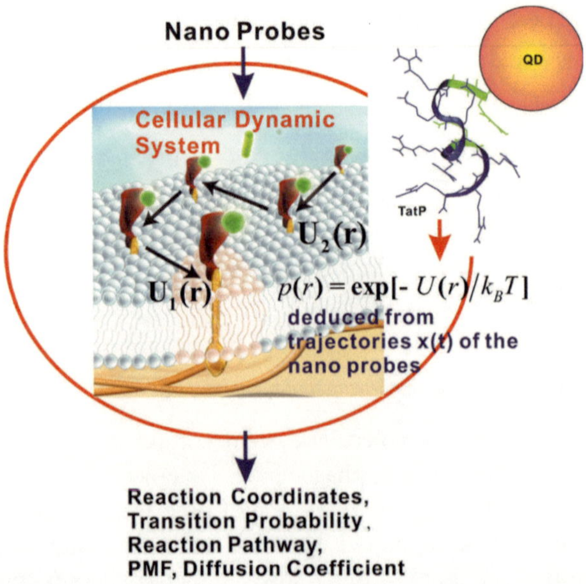

Figure 3 A schematic showing the translocation process of biomolecule-conjugated QDs that depict cellular dynamic processes by recording the impact of cellular environments on the trajectory ensemble of the nanosensors.

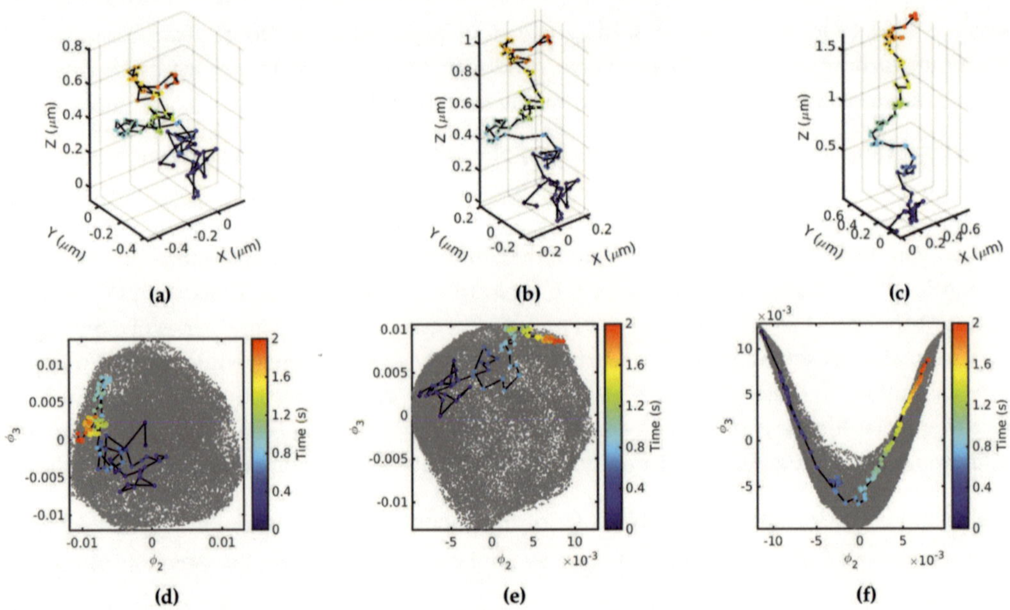

Figure 4 Simulated trajectories of particle diffusing (a,d) isotropically with $|F_k/f_k| = 0$ $D = 0.05\mu m^2/s$) or under a unidirectional force field with (b,e) $|F_k/f_k| = 0.01$ and (c,f) $|F_k/f_k| = 0.03$ were displayed with appearing times color-coded. For each case, a total of 300 simulated trajectories (gray in d, e, and f) were shown on a low-dimensional manifold of two principal spectral-embedding eigenvectors. The trajectories shown in (a, b, and c) were reproduced on the manifold with appearing times color-coded.

We first simulated 3D Brownian motion with Eq. (1) under the three conditions: isotropic diffusion with $|F_k/f_k| = 0$, $D = 0.05\,\mu m^2/s$ and under z-unidirectional force fields with $|F_k/f_k| = 0.01$ and 0.03. Starting at the origin we generated an ensemble of 300 trajectories for each case. Every trajectory contained 100 diffusion steps with a time resolution of 0.02 s. We display three typical trajectories in **Figure 4a–c**, which clearly exhibit increased spread along z due to the action of the unidirectional force field.

Spectral-embedding analysis can be implemented in the diffusion-map framework to enable an efficient construction of good slow observables and thereby can expose the low-dimensional manifolds underlying the high-dimensional datasets [35]. A graph-based method provided a discretized approximation of the manifold for efficiently constructing eigen-decomposition of the datasets [36]. We assembled the time-delayed vector $\tilde{x}_k(t) = \left[\vec{x}_k^{\,(i)}\right]_{i=1,s} = \left[\vec{x}_k(0), \vec{x}_k(\tau), \ldots, \vec{x}_k((s-1)\tau)\right]$ and projected the time series onto a low-dimensional manifold by exploiting the spectral-embedding technique [35]. The first eigenvector was trivial because its eigenvalue gave only the data density in a cluster. We then focused on the next two eigenvectors, ϕ_2 and ϕ_3, which offered the most critical information on the interactions between the nanosensors and their environments. **Figure 4d–f** display the ϕ_2-ϕ_3 plot of the simulated trajectory data, shown in gray. For comparison, the trajectories presented in **Figure 4a–c** are also reproduced here with appearing times color-coded. A V-shaped distribution in the ϕ_2-ϕ_3 plot developed gradually with force field strength, suggesting the V-shaped feature was a useful indicator of directed movement under a force field [17].

In **Figure 5a**, the isotropic diffusion produced statistics with a clear peak at $V\left(R^2\right) \simeq 0.0023$ and $\overline{R_\tau{}^2} \simeq 0.86$ in the $V(R^2)$-$\overline{R_\tau{}^2}$ plot. As field strength F_k/f_k increased to 0.03, the peak shifted upwardly to $V\left(R^2\right) \simeq 0.014$ and $\overline{R_\tau{}^2} \simeq 0.59$, indicating a larger diffusion stepsize variation under the force field. Thus, both of the $V\left(R^2\right)$-$\overline{R_\tau{}^2}$ analysis and spectral-embedding technique can reveal the influences of the cellular environment on nanosensors but from different viewpoints.

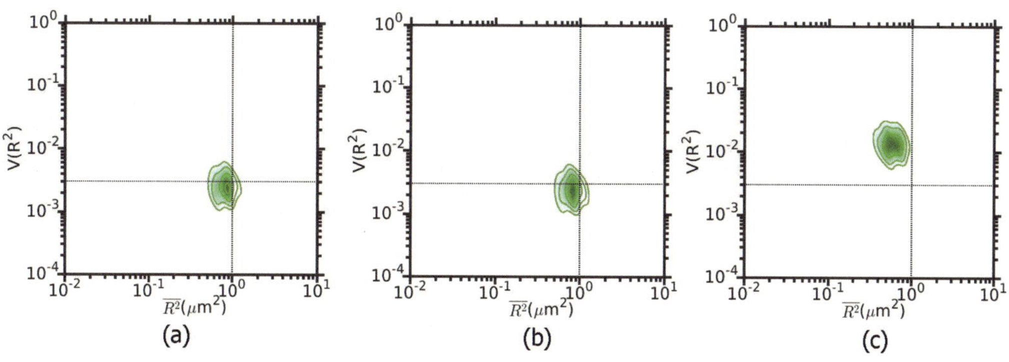

Figure 5 2D contour plot of $\left[V\left(\overline{R_\tau{}^2}\right)\right]$-$\overline{[R_\tau(t)]^2}$ histogram from simulated trajectories of particle diffusing (a) isotropically with $|F_k/f_k| = 0$, $D = 0.05\,\mu m^2/s$) or under a unidirectional force field with (b) $|F_k/f_k| = 0.01$ and (c)$|F_k/f_k| = 0.03$.

3. Apparatus and experimental procedures

3.1. Optical setup

The schematic of our single-particle fluorescence microscopy apparatus with light-sheet excitation is shown in **Figure 6a**. The output beam from a solid-state laser with different wavelengths was shaped into a light sheet of 3 μm thickness at the beam waist, yielding a diffraction-limited beam propagation with a Rayleigh range of 41 μm in the x direction. By using a galvanometer scanner, the light sheet can be positioned at a sample in a range of 34 μm along the y and z directions with an accuracy of 0.5 μm [17].

A 60 × 1.45 numerical aperture oil immersion objective lens (APON 60XOTIRFM, Olympus) was used to ensure both high spatial resolution and high photon collection efficiency. However, this objective lens had a limited depth of field (~500 nm). For accurately resolving the depth of a fluorophore, we exploited the astigmatism created by a cylindrical lens (CL2). The combination of the CL2 (f = 100 cm and an imaging lens (f = 20 cm) with a separation of 5.5 cm generates an effective focal length of 17.5 cm on the sagittal focal plane and 20 cm on the tangential focal plane, which encodes the fluorophore depth as an elliptically shaped point spread function (PSF; see **Figure 6a**). We also inserted an electrically tunable lens (ETL) at the pupil plane of a 4f optical system (formed by the relay lens and the imaging lens) to enable fast acquisition of images at different depths. The fluorescence images of fluorophores on living cells were recorded with a scientific complementary metal-oxide semiconductor (sCMOS) camera (ORCA-Flash 4.0 V2, Hamamatsu). Fluorescence images of quantum dot nanosensors on a living HeLa cell acquired with this apparatus are displayed without CL2 in **Figure 6b** and with CL2 in **Figure 6c**, respectively. Elliptically shaped spots were observable, which were localized within 1 μm of the imaging plane at a lateral accuracy of 27 nm and an axial accuracy of 52 nm [17].

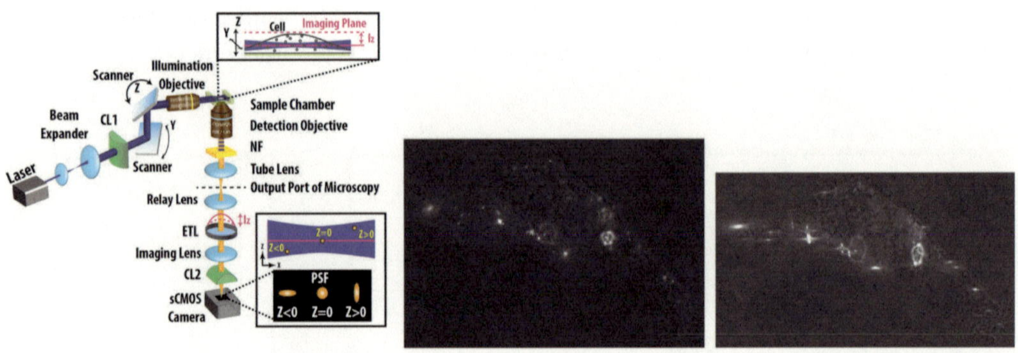

(a) (b) (c)

Figure 6 (a) Schematic of the light-sheet microscope used to record 3D trajectories of probing nanoparticles in a living cell. The excitation beam was shaped to form a 3 μm-thick light sheet, giving a Rayleigh range of 41 μm. A two-dimensional (2D) scanner was inserted to move the light sheet by 34 μm along the y and z directions at the sample position. The imaging arm was perpendicular to the excitation direction, and an imaging plane in the sample was relayed and imaged to a sCMOS camera. The position of the imaging plane was adjusted using an ETL to yield a set of depth-resolved images. An astigmatism was introduced using a CL2 to encode information on the fluorophore depth into an elliptically distorted point spread function. Fluorescence images of Qdot585 for a living HeLa cell acquired with this apparatus without CL2 (b) or with CL2 (c) are shown. (a) has been reproduced from ref. [17].

3.2. Linking localized coordinates of nanosensors for generating 3D trajectories

Single-particle trajectories were recorded for as long as 100 s, with a frame time of 25 ms. The localized coordinates of the nanosensors were extracted from a set of images acquired by synchronously scanning the light sheet and imaging focal plane. Connecting the acquired location coordinates to generate 3D trajectories was challenging. We first carried out multiple particle tracking by solving a linear-assignment problem [37] to identify the assignment matrix between the measured location coordinates and their predicted positions. A Kalman filter was also implemented to provide an optimal estimate of Brownian motion in the presence of Gaussian noise [38]. To verify the functionality of our linking method for 3D–trajectory generation, we simulated a group of particles diffusing in a spatial region with a different number of densities and diffusion coefficients. The resulting 3D trajectories were coarse-grained to yield time series of location coordinates with the same data-taking procedure as that used in our light-sheet microscope. The simulation results are shown in **Figure 7**, which indicated a particle density lower than 0.01 $/\mu m^2$, $D < 0.1\,\mu m^2/s$, and a linking accuracy that could be higher than 98%.

3.3. Cell culture and reagents

HeLa and A431 cells were cultured in Dulbecco's Modified Eagle's medium (DMEM) without phenol red supplemented with 10% (v/v) fetal bovine serum. MCF12A cells were cultured in a 1:1 mixture of DMEM and Ham's F12 medium containing 20 ng/mL Human EGF, 0.01 mg/mL bovine insulin, 500 ng/mL hydrocortisone, and 5%(v/v) horse serum [39]. Before single-molecule live-cell imaging was performed, the cells were plated in an eight-well chamber slide. When a 70% confluence was reached, HeLa and A431 cells were deprived of serum for 24 h and MCF12A cells for 3 h.

To label EGFR, anti-EGFR antibody (10 nM; Thermo Scientific) was conjugated with Qdot525 (from Invitrogen, Carlsbad, CA, USA). Cells were incubated with the EGFR-Ab-Qdot525 for 15 min and washed three times with phosphate buffered saline (PBS). Fluorescent EGF (EGF-Qdot585) was synthesized by conjugating biotin-EGF (from Invitrogen) to Qdot585-streptavidin in PBS. To activate EGFRs, cells were incubated in the presence of 40 ng/mL EGF-Qdot585 [31, 39].

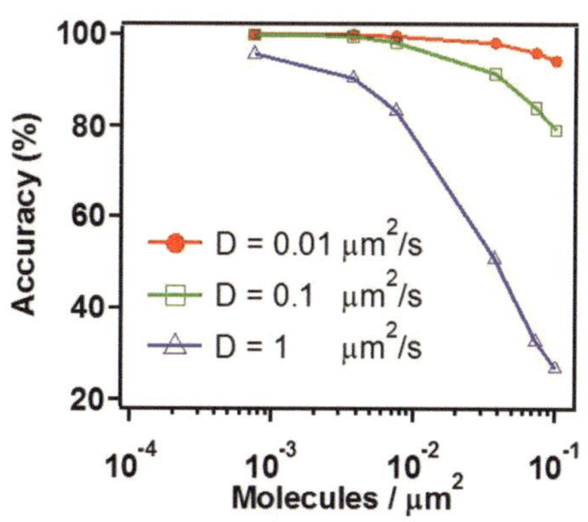

Figure 7 Linking accuracy of localized coordinates as a function of number density of Brownian particles. Trajectories of a group of particles diffusing in a spatial region with different particle densities and diffusion coefficients were simulated with Eq. (1) and then coarse-grained with the same sampling scheme as that used in our scanning light-sheet microscope.

To sequester cholesterol molecules on the plasma membranes, cells were treated with 10 μg/ml nystatin for 1 h before staining with either the antibody or EGF. To disrupt the actin filaments, the cells were pretreated with 10 μm cytochalasin D (Cyto D) for 1 h.

The N terminals of the Tat peptides (from Invitrogen) were biotinylated. Conjugated TatP-QDs were prepared by incubating 20-nm diameter streptavidin-coated Qdot585 in PBS with excess biotinylated TatP (5 μm TatP:50 nM Qdot585) at room temperature for 30 min. Although streptavidin is a tetramer and each subunit can bind biotin with equal affinity, the covalent attachment of streptavidin to the surface of a quantum dot makes two of the four binding sites inaccessible to the biotinylated TatP. As each QD has approximately 5 to 10 streptavidin molecules on its surface, we estimated that an average of 14 Tat peptides was conjugated to each QD.

4. Experimental results

The plasma membrane of a living cell is not merely a sea of lipids and proteins, but is more complex with individual components organized into spatially distinct compartments to yield strategic advantages for protein function and signaling [40]. Organizing proteins and lipids into nanodomains could also shield these assemblies from other proteins to tailor specific interactions, thereby mediating signal transduction to relay cellular messages from the external environment to the nucleus [41, 42].

The first event of cellular signaling occurs at various types of receptor proteins in the plasma membrane. To faithfully sense a signal that varies in space and time, live cells face an optimization problem of placing a set of distributed and mobile receptors by balancing two opposing objectives [43]: 1) the need to locally assemble the nanosensors to reduce the estimation noise; and 2) the need to spread these nanosensors to reduce spatial sensing errors. Receptor signaling dysregulation is attributed to the pathogenesis of several diseases [44, 45]. Therefore, understanding the interactions, molecular processes and relevant structures of such signaling assemblies is imperative. One such receptor protein is the epidermal growth factor receptor (EGFR), which can drive cell growth and survival [44]. There is tremendous interest in unraveling how those transducing proteins diffuse and interact on plasma membranes of living cells. However, it remains a challenge to study these cellular events at the single-molecule level in a live cell as living cells are highly heterogeneous and stochastically dynamic.

Based on our current knowledge of molecular diffusion in the plasma membrane, there are two types of interactions between a receptor and its local environment [46, 47]. First, the protein can induce a local ordering of the surrounding lipid molecules via a lipid-protein interaction. In addition, the cortical actin framework can induce membrane compartments [30]. To study the diffusing behaviors of EGFRs and the interaction with their cellular environment, we tagged EGFRs with antibody-conjugated quantum dots (Qdot525-Ab) and exploited fluorescent EGF, which was synthesized by conjugating EGF with quantum dots (QD585-EGF), to activate the EGFRs. By using this scheme, we could study the diffusive dynamics of paired EGFRs by selecting a pair of liganded and unliganded EGFR or a pair of liganded EGFRs, and follow their relative motions [31]. To appreciate the potential of biomolecule-conjugated QD nanosensors in a live cell study, we will briefly review some previous results of applying single-molecule tracking techniques to EGFR studies [31, 39].

4.1. Ligand binding induced receptor protein translocation in plasma membranes of living cells

Typical single-molecule tracks of unliganded Qdot525-Ab-EGFR and liganded Qdot585-EGF-EGFR on live cells exhibited confined diffusion interspaced by directed movement [31]. We binned the measured MSD in a histogram to deduce the probability density function of the diffusion coefficient. **Figure 8a** shows the data taken at a frame rate of $\tau = 25$ ms. For unliganded EGFR at rest, two sets of diffusers were observed with the diffusion coefficient of the fast species peaking at 9 $\mu m^2/s$ and the slower one at 0.3 $\mu m^2/s$. EGF activation suppressed the population of $0.3\,\mu m^2/s < D < 6\ \mu m^2/s$, whereas it increased the populations of D $D \leq 0.1\ \mu m^2/s$ and $D = 9\ \mu m^2/s$. It is clearly shown that receptor ligation can affect the diffusion of EGFR and leads to a population change in the two diffusion states.

We used Cytochalasin D to disrupt the cellular actin frameworks. As presented in **Figure 8b**, the major population of unliganded Qdot525-Ab-EGFR (open symbols) on EGF-activated cells shifted from the slow state ($D < 0.1\ \mu m^2/s$) to the fast state ($0.1\,\mu m^2/s < D < 2\ \mu m^2/s$) after Cytochalasin D treatment. This influence is even more pronounced on liganded Qdot585-EGF-EGFR (filled symbols). As pointed out previously, the primary effector controlling the motion of a protein in a living cell is often not due to the friction in the cellular medium but the interactions with its molecular partners and microenvironment. To illustrate this subject further, we used the $V(R^2) - \overline{R^2}$ technique to analyze the statistics of single-molecule trajectories, specifically focusing on the slow diffusion of EGFR.

For live HeLa cells at rest, the $V(R^2) - \overline{R^2}$ plot of measured tracks of unliganded Qdot525-Ab-EGFR are presented in **Figure 9a**. In this plot, peak 2 was the most populated and stable state

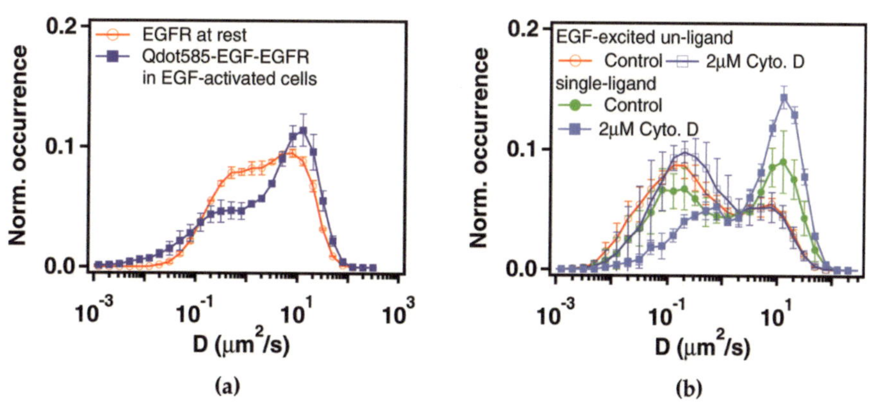

Figure 8 (a) Histogram of the diffusion coefficient of unliganded EGFR (open red circles) in the cells at rest and the liganded EGFR (solid blue squares) in EGF-activated cells. (b) Histogram of the diffusion coefficient of unliganded EGFR (open symbols) and the singly liganded species (filled symbols) on activated HeLa cells without (circles) or with (squares) Cytochalasin D pretreatment.

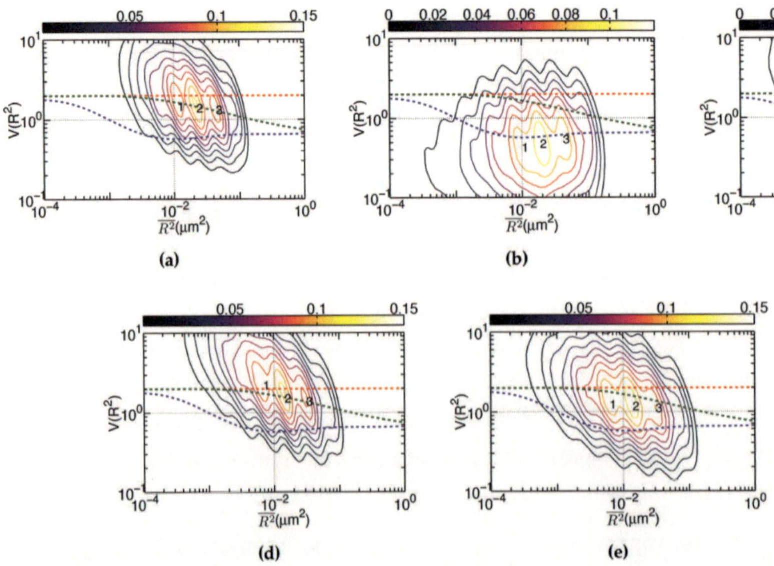

Figure 9 (a) The plot of $V(R^2) - \overline{R^2}$ for unliganded Qdot525-Ab-EGFR on live HeLa cells at rest. Simulated curves of the peak positions for receptor molecules under free Brownian motion (red dash line), diffusive motion with the confinement from actin corrals alone (green dash line), or both the actin corrals and lipid raft domains (blue dash line) are included for comparison. The same plot is shown for liganded Qdot585-EGF-EGFR in (b) EGF-activated cells, and (c) activated cells pretreated with nystatin. (d) Unliganded Qdot525-Ab-EGFR on live EGF-activated HeLa cells. (e) the same plot as (d) for unliganded Qdot525-Ab-EGFR on activated cells pretreated with nystatin. (a–c) have been reproduced from ref. [31].

among the three peaks detected, located at the $\left(\overline{R^2}, V(R^2)\right)$ coordinates of (0.01, 1.45), (0.02, 1.39), and (0.04, 1.33), respectively. Three simulated curves were plotted for proteins under free Brownian motion (red dash line), confinement by actin corrals alone (green dash line) or by both actin corrals and lipid domains (blue dash curves) and presented in **Figure 9a** for comparison. The peak positions of the three states clearly fell on the curve of the actin confinement, indicating that these unliganded receptor molecules were not free diffusers, but instead confined by the actin corrals alone.

With the EGFR at rest as the control, we proceeded to examine the diffusion of liganded Qdot585-EGF-EGFR. **Figure 9b** shows the $V(R^2) - \overline{R^2}$ plot of the liganded EGFR on activated HeLa cells. Three peaks were found at (0.01, 0.42), (0.02, 0.47), and (0.04, 0.54). These peak positions agreed better with the model that included the confinement effects of actin corrals and lipid raft domains. We used nystatin on the cells to sequester the membrane cholesterol and disrupt the lipid raft domains. **Figure 9c** displays the $V(R^2) - \overline{R^2}$ plot of the liganded EGFR on EGF-activated cells pretreated by nystatin. The three peaks of **Figure 9c**, at (0.01, 1.39), (0.02, 1.33), and (0.04, 1.28), fell again on the curve for the actin confinement model.

For unliganded Qdot525-Ab-EGFR on EGF-activated cells, the corresponding $V(R^2) - \overline{R^2}$ plot is presented in **Figure 9d**. Although peak 1 appeared to be produced by the free diffusion proteins, peaks 2 and 3 agree with the confinement model of the actin corrals alone. Qdot525-Ab-EGFR on the nystatin-pretreated cells (shown in **Figure 9e**) revealed similar peaks at (0.006, 1.24), (0.014, 1.18), and (0.031,1.12), suggesting that lipid raft domains on activated cells play a minor role in restraining the motion of unliganded Qdot525-Ab-EGFR.

We proposed the following picture to explain our experimental results: Unliganded EGFRs at rest may locate outside the cholesterol-enriched lipid domains. EGF binding causes the receptors to move into the cholesterol-enriched lipid domains. Pretreatment of cells with nystatin, which can disrupt these lipid domains, results in local environmental changes of the ligand-bound EGFR. This interpretation is further supported by the observations shown in **Figure 9d** and **e** in which nystatin pretreatment did not alter the peak positions of unliganded EGFR on EGF-activated cells.

The $V(R^2)$ peak values of unliganded EGFRs in native cells at rest were near the free diffusion limit (**Figure 9a**). However, those unliganded species could have higher $V(R^2)$ peak values in highly EGF-activated cells due to much stronger confinements from EGF-promoted actin polymerization [30]. EGF ligation reduced the $V(R^2)$ values of Qdot585-EGF-EGFR to below the free diffusion limit (**Figure 9b**) due to the dressing effect by the lipid domain. These experimental findings were verified in three cell lines; including two cancer cell lines (HeLa and A431) and one non-tumorigenic breast epithelial cell line (MCF12A). These cell lines possess a wide range of EGFR expression levels and concentrations of membrane cholesterol. Therefore, the experimental results may represent a general behavior of unliganded and activated receptors in live cells.

4.2. Correlated motion of receptor proteins in plasma membrane of live cells

Receptor dimerization plays a critical role in initializing a signal cascade [48]. Do nearby receptor proteins move correlatively prior to dimer formation? Imagine when a receptor protein moves in the plasma membrane of a live cell, it may induce order in its surrounding lipid molecules through the protein-lipid interaction. A receptor protein and the induced lipid ordering can be viewed as a lipid-dressed protein. As two nearby proteins move in the plasma membrane, they may interact with each other through the ordered lipid molecules.

We can simulate the diffusive behaviors of two dressed proteins in proximity using coupled Langevin equations [49]. To display the mutual correlation between the two trajectories quantitatively, we expressed the position vectors as $\vec{x}_k(t) = A_k(t)e^{i\theta_k(t)}$, and defined the degree of mutual correlation as

$$C(\tau) = \mathrm{Re}\left[\frac{\sum \vec{x}_1^{\,*}(t)\cdot\vec{x}_2(t+\tau)}{\sqrt{\sum\left|\vec{x}_1(t)\right|^2}\sqrt{\sum\left|\vec{x}_2(t)\right|^2}}\right] = \frac{\sum_t A_1(t)A_2(t+\tau)\cos\left[\theta_2(t+\tau)-\theta_1(t)\right]}{\sqrt{\sum_t A_1^{\,2}}\sqrt{\sum_t A_2^{\,2}}}. \quad (6)$$

The summations were taken over a time mesh along the single-molecule tracks. By using this approach, we simulated the correlated motion of two Brownian-like particles with their spatial separation perturbed by a correlated thermal fluctuation [49]. We carried out the simulations from an initial condition that positioned one receptor at $\vec{x}_1 = (0,0)$ and places the other randomly within a 1-μm radius circle centered at (0,0). The coupled Langevin equations were then solved to generate a pair of trajectories, and the degree of mutual correlation was calculated using Eq. (6). **Figure 10a** and **b** display the histograms of simulated trajectories with mutual correlations of 0 and 0.5, respectively. **Figure 10c** illustrates the histogram of correlation in experimental single-molecule tracks of paired Qdot525-Ab-EGFR and Qdot585-EGF-EGFR on live HeLa cells. As shown, the tracks clearly exhibited a correlation peak at 0.5.

By using this method, we were able to select those highly correlated segments from the single-molecule tracks and analyzed the correlated motion [31]. We plotted the $V(R^2) - \overline{R^2}$ plot of the motion of the unliganded Qdot525-Ab-EGFR correlated with liganded Qdot585-EGF-EGFR in **Figure 11a**. As the unliganded Qdot525-Ab-EGFR moved correlatively with a nearby liganded Qdot585-EGF-EGFR, the diffusion motility $\overline{R^2}$ of state 1 decreased drastically to near 10^{-3}, accompanied by a reduction of $V(R^2)$ to 0.1. It is interesting to note that the $V(R^2) - \overline{R^2}$ plot of the reverse case (i.e., Qdot585-EGF-EGFR relative to Qdot525-Ab-EGFR) differed in $\overline{R^2}$ (**Figure 11b**). The resident time of the liganded EGFR in state 2 became longer, and the $V(R^2)$ of both states 1 and 2 increased to 1, indicating that the liganded and unliganded EGFR resided in different lipid environments.

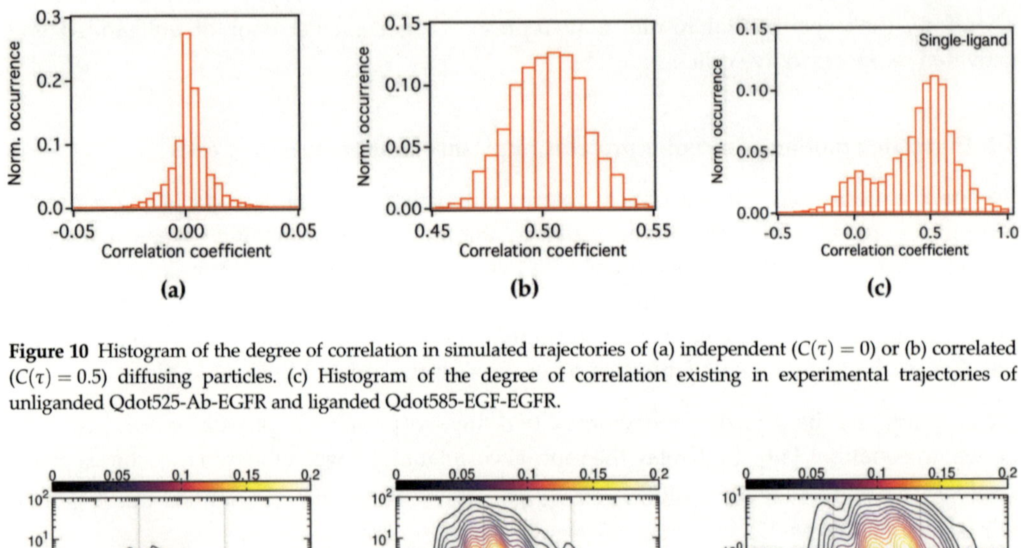

Figure 10 Histogram of the degree of correlation in simulated trajectories of (a) independent ($C(\tau) = 0$) or (b) correlated ($C(\tau) = 0.5$) diffusing particles. (c) Histogram of the degree of correlation existing in experimental trajectories of unliganded Qdot525-Ab-EGFR and liganded Qdot585-EGF-EGFR.

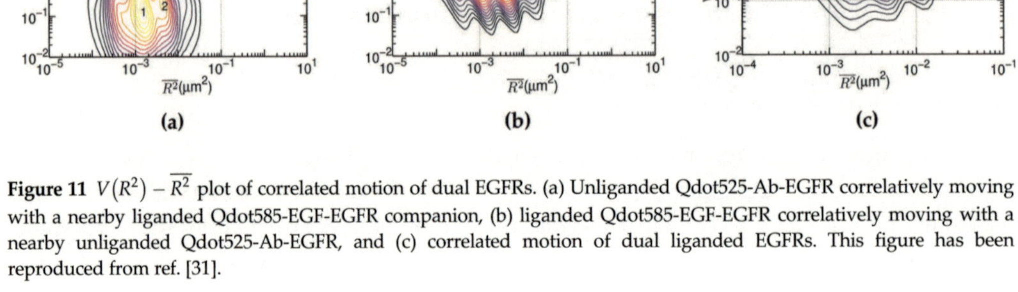

Figure 11 $V(R^2) - \overline{R^2}$ plot of correlated motion of dual EGFRs. (a) Unliganded Qdot525-Ab-EGFR correlatively moving with a nearby liganded Qdot585-EGF-EGFR companion, (b) liganded Qdot585-EGF-EGFR correlatively moving with a nearby unliganded Qdot525-Ab-EGFR, and (c) correlated motion of dual liganded EGFRs. This figure has been reproduced from ref. [31].

The $V(R^2) - \overline{R^2}$ plot of the correlated motion between two liganded EGFRs is presented in **Figure 11C**. Compared to the plot shown in **Figure 11b**, the $V(R^2)$ values of the two major states were decreased when the unliganded companion of **Figure 11b** was replaced by the liganded EGFR, perhaps because the lipid raft domains surrounding the two receptor molecules merged and yielded a larger dressing force during the highly correlated motion.

4.3. Cholesterol-mediated interaction between liganded EGF receptors

Ligand binding promotes receptor dimerization and leads to a downstream signaling cascade. Researchers have increasingly determined that lipid domains rich in raft sphingolipids (GM1) and cholesterol can facilitate signaling receptors to form a dimer [50–52]. The recent identification of cholesterol-dependent nanoassemblies with biophysical techniques also suggests that a cholesterol-mediated interaction exists between lipid domains to affect the organization, stability, and function of membrane receptor proteins [52–54].

We selected and analyzed those highly correlated segments from single-molecule trajectories to reveal the influence of cholesterol-mediated interactions [39]. **Figure 12** displays the $V\left(\overline{R_\tau^2}\right)$-$\overline{R_\tau^2}$ plots of paired Qdot585-EGF-EGFRs in three different cell lines. The contour plots are more scattered, indicating that these data are indeed highly sensitive to receptor interaction.

The $V\left(\overline{R_\tau^2}\right)$ value of the correlated Qdot585-EGF-EGFRs is considerably lower in A431 cells, which may be attributed to effective receptor-lipid and receptor-receptor interactions in these cells [33, 34]. To inspect the nature of the interactions and their relevance to receptor-induced lipid ordering, we again took advantages of the drug effects of nystatin and MβCD. **Figure 12b and c** display the $V\left(\overline{R_\tau^2}\right)$-$\overline{R_\tau^2}$ plots for the three cell lines pretreated with nystatin and MβCD. Correlated Qdot585-EGF-EGFRs appeared to have a weaker interaction in the nystatin-treated A431 cells as evidenced by an increased $V\left(\overline{R_\tau^2}\right)$ value. This observation may be explained by a

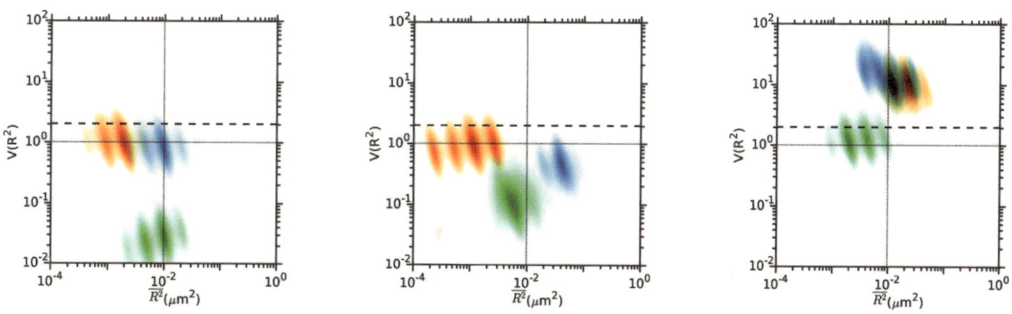

(a) **(b)** **(c)**

Figure 12 $V\left(\overline{R_\tau^2}\right)$-$\overline{R_\tau^2}$ plots of correlated Qdot585-EGF-EGFRs diffusing in the plasma membrane of (a) native cells, (b) nystatin-pretreated cells, and (c) MβCD-pretreated cells. Trajectory segments with a degree of correlation exceeding 0.8 were selected for analysis. Data are shown in red for HeLa cells, green for A431 cells, and blue for MCF12A cells. This figure has been reproduced from ref. [39].

less stable lipid domain due to a lower amount of cholesterol, which resulted in a larger variance in the diffusing step size of the correlated receptors. In contrast, the interaction became stronger in nystatin-treated MCF-12A cells, suggesting the effect of the cholesterol-mediated interaction was opposite to that of the receptor-lipid interaction. The $V\left(\overline{R_{\tau}^{2}}\right)$ of correlated Qdot585-EGF-EGFR in A431 increased by two orders of magnitude from 10^{-2} for native cells to 1 for MβCD treated cells. Of note, the $V\left(\overline{R_{\tau}^{2}}\right)$ value could be increased to higher than 10 in MβCD treated HeLa and MCF-12A cells, which suggests that a deterministic interaction dominated because the screening effect from the membrane cholesterol was reduced after membrane cholesterol was depleted. These results identified a vital role for membrane cholesterol in mediating the interaction between liganded receptors in the three cell lines.

4.4. Nonraft lipids and sphingolipids in live plasma membranes segregate into separated nanodomains

Previous data analysis implicitly assumed the coexistence of different lipid phases in plasma membranes. Indeed, lipid–lipid interactions were capable of inducing liquid ordered (Lo)-liquid disordered (Ld) phase coexistence in model lipid membranes [55, 56]. It was conjectured that plasma membrane composition is poised for selective and functional raft clustering at physiological temperatures [57]. However, such lipid nanodomains have remained largely unresolved in the plasma membrane of living cells. Researchers recently used a fluorescence correlation technique to successfully distinguish between free and anomalous molecular diffusion in a 30-nm focal spot of a stimulated emission depletion (STED) nanoscope [58]. The observed differences were attributed to transient cholesterol-assisted and cytoskeleton-dependent binding of sphingolipids to other membrane constituents. However, the optical force acting on the highly excited lipid molecules by the STED spot may not be negligible.

In our current study, we investigated lipid nanodomains in live plasma membranes at a much lower excitation level with light-sheet microscopy. We probed the nonraft lipids in living HeLa cells with carbocyanine dyes 1,1′-didodecyl- 3,3,3′,3′-tetramethylindocarbocyanine perchlorate (DiI-C12), which serves as an excellent lipophilic fluorescent probe with a strong partition tendency into the Ld phase [59]. The preference originated from the fact that highly packed lipids in Lo phase usually exclude exogenous molecules. BODIPY FL C_5-ganglioside GM_1 (BODIPY C_5-GM1) from ThermoFisher Scientific was used as a direct indication of lipid rafts. By synchronously adjusting the light-sheet position and the focal plane of the high NA objective lens, we were alble to acquire a 3D image of DiI-C12 and BODIPY C_5-GM1 in a living HeLa cell. **Figure 13a** displays a typical image of the scan with DiI-C12 shown in blue and BODIPY C_5-GM1 in orange. The 3D point spread function (PSF) of the light-sheet microscope was deduced (shown in **Figure 13b**) using a fluorescent bead with a diameter of 100 nm. We exploited this PSF to deconvolve the image [60] of DiI-C12 and BODIPY C_5-GM1. The resulting deconvolved image is shown in **Figure 13a**.

Following the formalism developed by Veatch et al. [61], we calculated both the auto-correlation function $A(r) = \langle\rho(R)\rho(R-r)\rangle/\langle\rho(r)\rangle^2$ and the cross-correlation function $C(r) = \langle\rho_1(R)\rho_2(R-r)\rangle/(\langle\rho_1(r)\rangle\langle\rho_2(r)\rangle)$ of DiI-C12 and BODIPY C_5-GM1. The results are presented in **Figure 14,** which indicates that both species may form clusters with an average diameter of less than 150 nm.

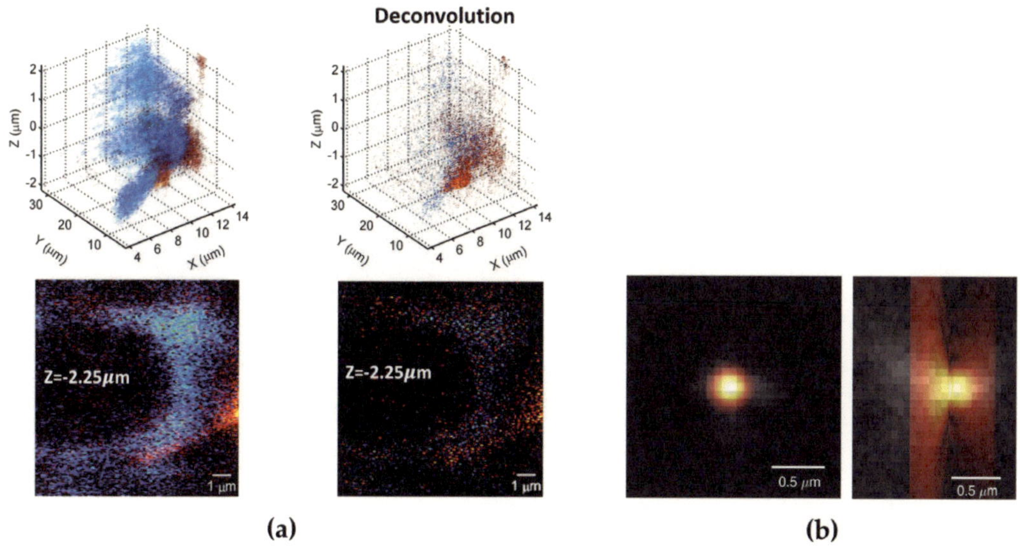

Figure 13 (a) 3D fluorescence distribution (left) and its deconvoluted image (right) from DiI-C12 (blue) and BODIPY C_5-GM1 (orange) fluorophores in a living HeLa cell. The cross section (z = -2.25 μm) of the distribution was shown below. (b)the point spread function of the light-sheet microscope on the xy- (left) and xz-plane (right) emitted from a fluorescent bead with a diameter of 100 nm.

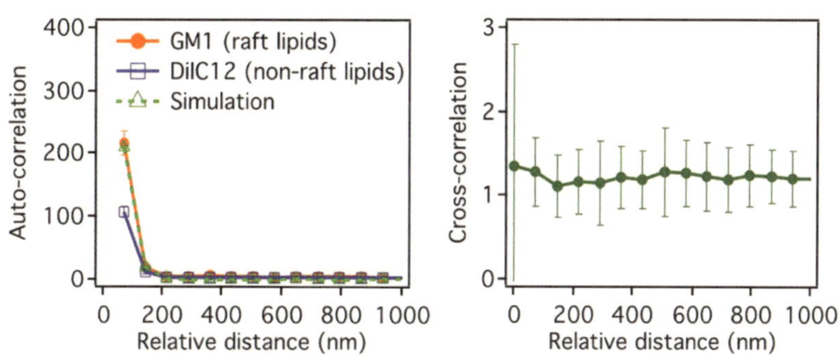

Figure 14 The auto-(left) and cross-(right) correlation function of DiI-C12 and BODIPY C_5-GM1 calculated from the deconvoluted image of **Figure 13a**.

To retrieve information about the lipid clustering process from the measured pair correlation, we simulated lipid clustering dynamics. First, we randomly distributed M clusters in an imaging region with each cluster containing N molecules in a circle with diameter R. **Figure 15a** illustrates an example of two randomly distributed clusters (M = 2) at a spatial resolution of 2 nm. We then binned the molecules in a cluster to the pixel size of the camera used (see **Figure 15b**). By using these cluster images, we calculated the auto- and cross-correlation functions with three different lipid clustering models: 1) a random clustering model, 2) an aggregation model, and 3) a segregation model. The random clustering model assumed that two lipid species were independently and randomly distributed in an area. In contrast, the lipid molecules in the aggregation and segregation models were stochastically distributed in each cluster with a Gaussian distribution. There is a major difference between

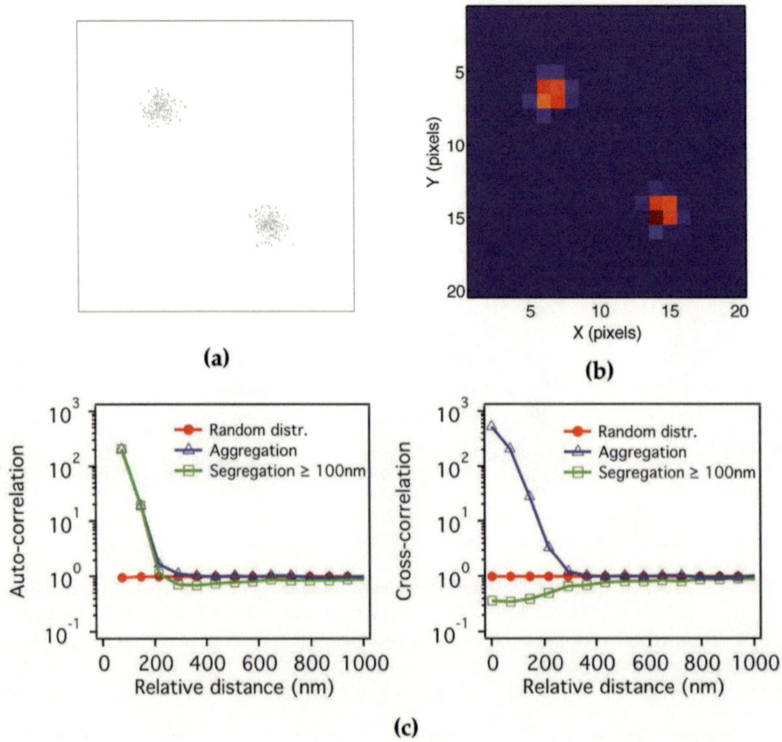

Figure 15 (a) Two randomly distributed clusters (M = 2) at a spatial resolution of 2 nm was prepared in a simulation area, (b) the two clusters were binned to the pixel size of the camera used. (c) the auto- (left) and cross- (right) correlation function using the parameters: R = 72 nm, N = 100, and M = 100 (2.8 clusters/μm^2) were simulated with the random clustering model (filled circles), aggregation model (open triangles), and segregation model (open squares).

the latter two models. In the aggregation model, any two clusters will attract each other to form an overlapping cluster with the same center of mass (but with different radii). In the segregation model, two clusters will experience a repulsive force to yield the minimum separation distance Δ. To simulate the segregation process, we first used a random number generator to produce the center position X_1 of one cluster and then used the multiplicative congruential generation algorithm to position the other cluster to lie in the interval of $[0,\ X_1 - \Delta)$ or $(X_1 + \Delta,\ 1]$. In this way, the second cluster was located randomly but was excluded from the neighborhood of X_1 with a minimum separation Δ. For a fair comparison, we kept the number of clusters, cluster size, and number of molecules in each cluster at the same values. **Figure 15** shows the simulation results for the following parameters: the waist size of the cluster was 72 nm, the number of molecules in a cluster was 100, and the number of clusters in an image was 100 (i.e., 2.8 clusters/μm^2), which yielded the short-distance autocorrelation and agreed well with the experimental result shown in **Figure 14**.

By comparing **Figure 15c** with **Figure 14**, we concluded that our data could not fit to the model of random clustering. From the comparison of cross-correlation shown in **Figures 15c** and **14**, we could remove the aggregation model, and we concluded that our data was better described by the segregation model with a segregation distance <100 nm.

It was discovered that the raft lipid species GM1 can be tightened by pentameric cholera toxin-β (CTxB), which initiates a minimum raft coalescence to form the GM1 nanodomains [62]. The plasma membranes in our study were in their native state without perturbations from either intense laser spot or cross linking reagent. Thus, our evidence for segregation of nonraft lipids and GM1 into separate nanodomains supports the idea that such a phase coexistence in a native plasma membrane not only exists but also is a general behavior of living cells.

4.5. Probing translocation of HIV-1 tat peptides in living cells with tat-conjugated quantum dot nanosensors

Viral infection can initiate at entry points on plasma membranes via lipid domains. Drug delivery may benefit from our understanding of this entry process because upon arriving at target tissues, drug molecules must also cross the plasma membrane to reach the sites of action. It is of particular interest to make drug molecules that cross cellular membranes directly to avoid the complications of vesicle-mediated internalization pathways. Recently, cell-penetrating peptides (CPPs), which are short sequences (8 to 30) of amino acids (aa) with a net positive charge in water [63], were found to exhibit such a membrane-crossing capability. An 11 aa segment in the trans activator of a transcription protein of the human immunodeficiency virus is a prototypical example of a CPP that can effectively penetrate a cell [64, 65]. The interactions involved in the approach to developing a TatP-coated nanoscale probe may determine whether the uptake of the probe succeeds or fails. To illustrate the potential of biomolecule-conjugated QDs as a cellular dynamic probe, in this section we briefly discuss the results of the translocation of TatP-conjugated QDs across the plasma membranes of live cells using the single-molecule tracking technique [17].

The first step for cellular internalization may involve some form of interaction between the Tat peptides and the surface of the cell. The strong anionic charge present on the glycosaminoglycan (GAG) chains of the proteoglycans (PGs) makes them favorable first-binding sites for the cationic Tat peptides [20, 66, 67]. To verify this scenario, we treated cells with Heparinase III enzyme (HSase) to cleave heparan sulfate (HS) groups from heparan sulfate proteoglycans (HSPGs). We observed a reduction in TatP-QD internalization of 74% at 30 min. Treatment with Cyto D, which can inhibit actin polymerization and thereby disrupt the cellular actin framework [68], resulted in a similar drop in TatP-QD internalization. The results indicate that both HS-mediated binding and the interaction with intracellular actin filaments are crucial for the rapid intake of TatP-QDs.

4.5.1. TatP-QDs approaching cell surface aggregate at selected regions of plasma membrane

For single-particle tracking, we prepared a cell culture medium containing 1 μm free TatPs and 1 nM TatP-QD nanosensors. The major species of free TatPs were used to restructure the environment of the membrane-peptide interaction, whereas TatP-QDs served as nanoscale dynamic pens to depict the landscape of the membrane-peptide interaction. We conducted single-particle trajectory analysis of the TatP-QDs with light-sheet microscopy to reveal the translocation dynamics. A unique affordance of our light-sheet microscope was the ability to track TatP-QDs in parallel, providing a global view of the dynamics of the approaching TatP-QDs. However, due to the limited image-taking speed of the camera used, we were only able to track TatP-QDs within a short distance from the cell surface.

Without external interaction, these Tat-QD nanosensors were expected to traverse the extracellular space through a random walk search, attach to the membrane, and then diffuse to find a suitable entrance site. **Figure 16a** displays three trajectories of TatP-QDs, color-coded to indicate the approaching times. The green surface depicts the cell surface rendered from the phase contrast images taken by scanning the imaging focal plane at different z positions in the cell. The determination of the cell profile was limited by the diffraction effect of the objective lens used, yielding a resolution of 200 nm in the lateral plane and 500 nm in the axial direction. As indicated in the top inset, the initial approaching trajectories of some of the Tat-QDs resembled directed movement under a force field, and the motion became more diffuse as the TatP-QDs come closer to the cell surface. A longer observation period accumulated more approaching events and revealed the trajectory aggregates at selected regions of the plasma membrane (**Figure 16b**).

The binding affinity of TatP for HSPGs was greater than that for anionic lipids by 2 to 3 orders of magnitude. Given that the anionic HSPG chains on the plasma membrane [20, 69, 70] may be favorable binding sites for cationic CPPs, we hypothesized that the trajectory aggregates were caused by HS groups in the HSPG chains. To verify this hypothesis, we treated the cells with HSase to cleave the HS groups from the HSPGs, which revealed considerably fewer and more randomly positioned spots in the extracellular space. Thus, the observed trajectory aggregation seems to be caused by the binding to HS groups on the membranes and suggests that HSPGs play a critical role in redirecting the TatP entry process toward spatially restricted sites on the plasma membrane.

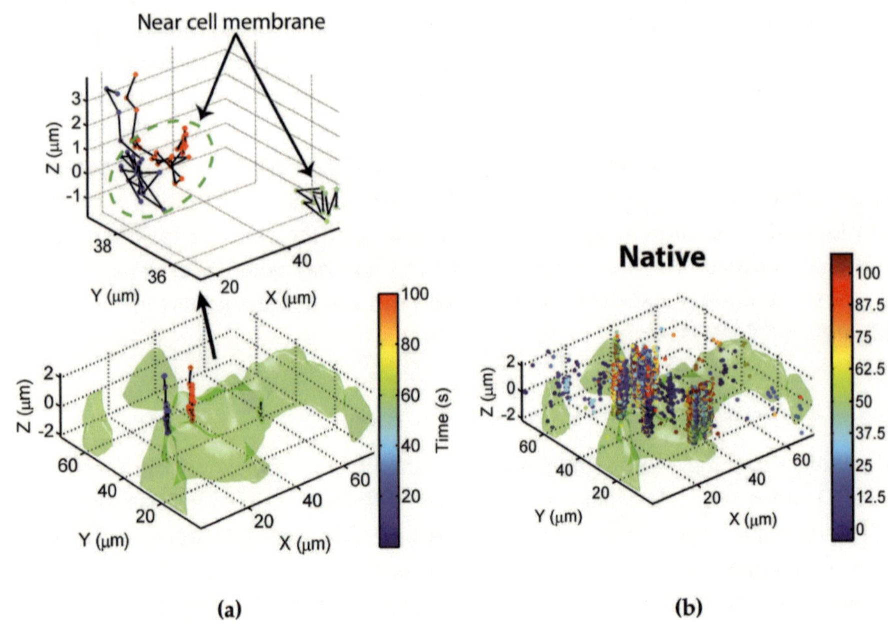

Figure 16 (a) Three trajectories of TatP-QDs near a living HeLa cell (green) were color-coded to indicate their appearing times. The green profile denotes the cellular surface rendered from optical sectioning phase contrast images; (b) when duration was increased to acquire information on more approaching events, trajectory aggregates were observed at selected regions on a native HeLa cell. This figure has been reproduced from ref. [17].

4.5.2. Spectral-embedding analysis of trajectory aggregates of TatP-QDs

As TatP-QDs translocate across the plasma membrane of a living cell, the probing particles can record the influences of the cellular environment on their trajectories. We considered the trajectory $\vec{r}\,(t)$ to be produced by a stochastic process with influences on the probing particle by its local environment, which may have different realizations with the corresponding interaction potentials $U_i\left(\vec{r} - \vec{\xi}\right)$; $i = 1, .., n$. Thus, the trajectory coordinates of the nanosensors could implicitly record the configurations of the local environment. Recently, Wang and Ferguson generated a reconstruction of single-molecule free-energy surfaces from time-series data of a physical observable by using the generalized Takens Delay Embedding Theorem [34]. Here we focused on retrieving the eigenmodes of the trajectory coordinates of TatP-QDs to identify the mechanism underlying the trajectory aggregation of TatP-QDs.

We used the spectral-embedding technique [35] to extract a low-dimensional manifold from a set of trajectories: $\vec{r}_i(t); i = 1, .., n$. A graph-based method provided a useful discretized approximation of the manifold [36] and enabled an efficient construction of the eigen-decomposition. The first eigenvector we retrieved was trivial with the corresponding eigenvalue giving only the data density in a cluster. We then focused on the next two eigenvectors, ϕ_2 and ϕ_3, which offered the most critical information on the interactions between TatP-QDs and their cellular environments.

Figure 17 presents two trajectory aggregates of the TatP-QDs: one (left) is near a living cell, and the other (right) is directly on top of the cell surface. All of the coordinates of the trajectory aggregates are shown in gray. The location coordinates of the TatP-QDs associated with the left

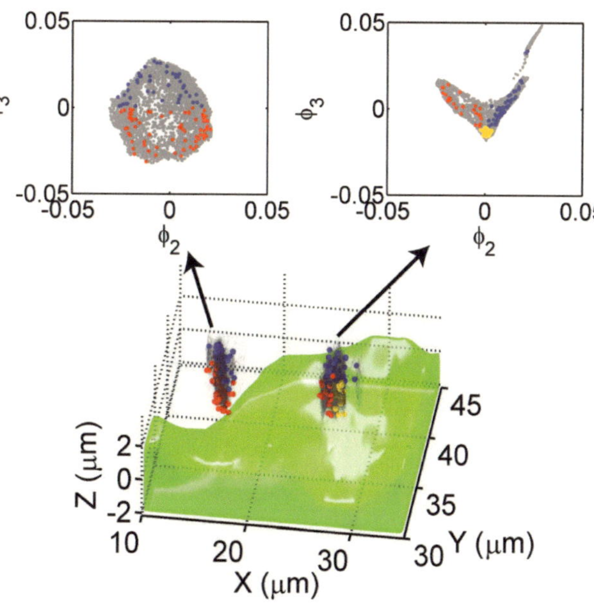

Figure 17 Two trajectory clusters (gray) of TatP-QDs near a living HeLa cell (green) are presented on a manifold of spectral-embedding eigenvectors (top inset). For each trajectory cluster, the $\left[V\left(\overline{R_\tau{}^2}\right)\right]\text{-}\overline{[R_\tau(t)]}^2$ coordinates of the trajectory segments within 2% variance of the peak shown in **Figure 18** are displayed in blue on the far side, red near the $z = 0$ plane, and yellow for those closest to the cell membrane. This figure has been reproduced from ref. [17].

cluster present a nearly circular distribution on the ϕ_2-ϕ_3 plane. The right cluster, however, displays a V-shaped distribution. Our results indicate that the spectrally decomposed structure of the trajectory aggregates provides the information on the interaction of the TatP-QDs with their cellular environments.

4.5.3. Influence of actin framework on translocation of TatP-QDs

The findings of a recent study indicated that on attachment to a membrane surface, Tat peptides can remodel the actin framework in an actin-encapsulated giant unilamellar vesicles (GUV) [69]. However, it remains unclear whether such multiplexed membrane and cytoskeletal interactions can also occur in a living cell. The trajectories of nanoscale probing particles may provide the answer. To extract relevant stochastic and geometrical structures from the data and gain insights into the mechanism that generated the data, we generated the $V\left(\overline{R_\tau{}^2}\right)$-$\overline{R_\tau{}^2}$ plots of the trajectories in **Figure 18**, which summarizes the single-particle diffusion statistics from 23,382 TatP-QD trajectories.

A single peak at the coordinates (0.15, 0.21), which is well below the free diffusion limit of $V\left(\overline{R_\tau{}^2}\right) = 2$, is shown in **Figure 18a**, suggesting that the TatP-QDs did not diffuse freely near a native HeLa cell. The peak split into two and shifted downward to $V\left(\overline{R_\tau{}^2}\right) = 0.06$ for Cyto D-treated cells (**Figure 18b**), indicating that TatP-QDs experience a stronger interaction with a strained cellular membrane. This finding is understandable because, without the support of an actin framework, the plasma membrane may develop a higher local curvature as a result of TatP-QD attachment. In native cells, the effect of the interaction between Tat peptides and the cell membrane may be counter balanced by that of the Tat and actin filaments, resulting in a higher $V\left(\overline{R_\tau{}^2}\right)$.

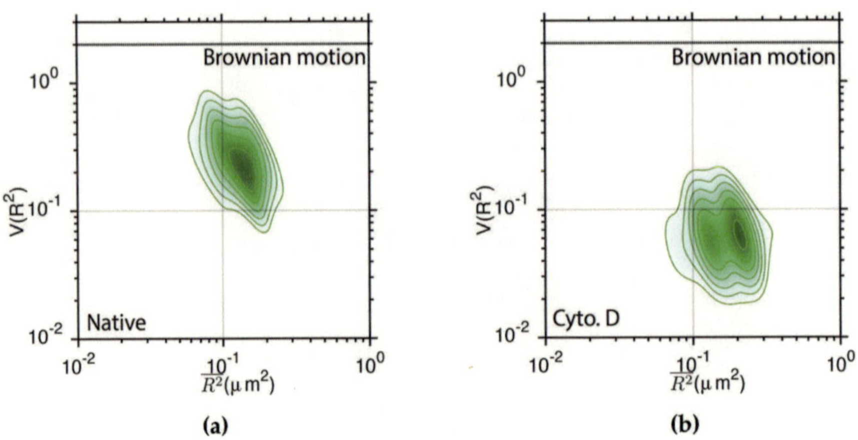

Figure 18 2D contour plot of $\left[V\left(\overline{R_\tau{}^2}\right)\right]$-$\overline{[R_\tau(t)]^2}$ histogram for TatP-QDs moving in the neighborhood of (a) living HeLa cell and (b) Cyto D-pretreated cell. This figure has been reproduced from ref. [17].

We also analyzed each trajectory aggregate by selecting segments that fell within 2% variance of the $\left[V\left(\overline{R_\tau^2}\right)\right]$-$\overline{[R_\tau(t)]^2}$ peak. Labeling the resulting $\left[V\left(\overline{R_\tau^2}\right)\right]$-$\overline{[R_\tau(t)]^2}$ coordinates on the trajectories offered insight into the environmental influences on the TatP-QDs. For example in the left trajectory cluster of **Figure 17**, these special points are shown in blue on the far side and red near the z = 0 plane. The blue dots were uniformly distributed at the rim of the circle on the ϕ_2-ϕ_3 plane, and the distribution of the red dots, which were close to the cell membrane, appeared to be denser on the $\phi_2 > 0$ side. In the trajectory aggregate directly on top of the cell, the blue dots were located at the right leg ($\phi_2 > 0$) and the red points dots were concentrated at the left leg ($\phi_2 < 0$) of a V-shaped distribution. Yellow dots, which represent trajectory segments closest to the cell membrane, aggregated at the tip of the V-shaped distribution, suggesting the formation of hot spots of interaction on the cell membrane, which may be supported by specifically oriented actin filaments.

4.5.4. *Classification of TatP-QD trajectories*

We also applied spectral embedding [17, 70] to classify 23,382 TatP-QD trajectories measured on 30 cells. In **Figure 19**, the resulting circular or V-shaped distributions on the ϕ_2-ϕ_3 plane are displayed in green. For classification, the norm of the residuals, defined as the sum of the squared deviation from the circular distribution of free diffusion, was used as the metric. The coordinates (blue) within 2% variance of the $\left[V\left(\overline{R_\tau^2}\right)\right]$-$\overline{[R_\tau(t)]^2}$ peak with the corresponding contours were also included for comparison. As shown in **Figure 20**, the class of circular

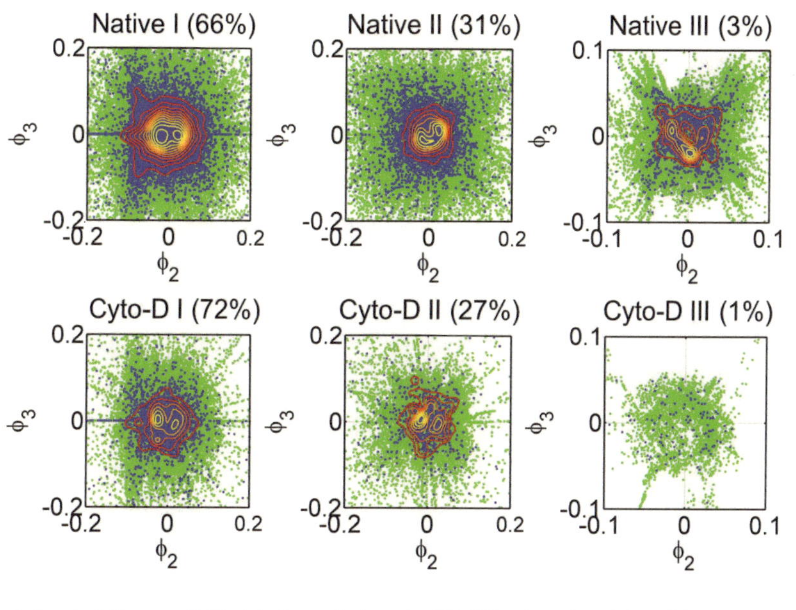

Figure 19 Spectral embedding manifold plots (green in insets) of 23,382 trajectories of TatP-QDs measured on 30 living HeLa cells (up row) and 5112 trajectories measured on Cyto D-treated cells (bottom row). The $\left[V\left(\overline{R_\tau^2}\right)\right]$-$\overline{[R_\tau(t)]^2}$ coordinates of the trajectory segments within 2% variance of the peaks shown in **Figure 18** are displayed in blue, with associated contour curves revealing the peak profiles. This figure has been reproduced from ref. [17].

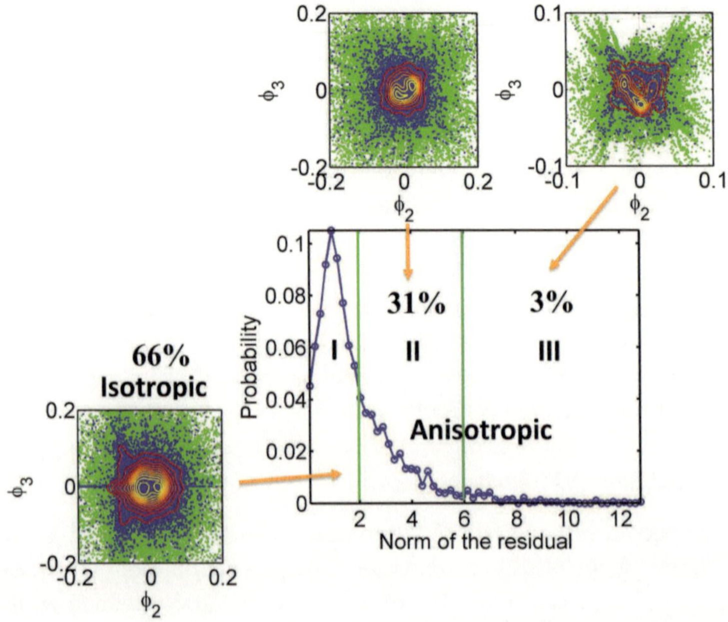

Figure 20 Classification (center) of spectral embedding manifold plots (green in insets) of 23,382 trajectories of TatP-QDs measured on 30 living HeLa cells. The $\left[V\left(R_\tau^2\right)\right]-\overline{[R_\tau(t)]^2}$ coordinates of the trajectory segments within 2% variance of the peaks shown in **Figure 18** are displayed in blue, with associated contour curves revealing the peak profiles. This figure has been reproduced from ref. [17].

distribution with a norm of the residuals of 0.91 ± 0.34 contained about 66% of the data from the native cells. The moderate (2 to 6) and highly anisotropic (>6) trajectory data occupied 31 and 3%, respectively. We acquired 5112 trajectories for the Cyto D-treated cells. The proportions of the moderate and highly anisotropic trajectories decreased to 27 and 1%, respectively. Treatment with Cyto D reduced the cellular uptake of the TatP-QDs to 25% of that of the native cells, indicating that both the isotropic and moderate anisotropic classes played a minor role in the initial cellular uptake. The trajectories belonging to the highly anisotropic class resulted in 75% uptake. These findings indicate the formation of funnel passages for the TatP-QDs due to the combined effect of HS-binding and actin remodeling.

5. Conclusion

Probing the distribution and mobility of proteins in live cellular environments is crucial for understanding cellular functions and regulatory mechanisms, which also serve as the basis for developing therapeutic strategies. Factors that affect protein mobility are difficult to reconstitute in vitro using purified constituents. Single-molecule imaging and tracking provide direct access to probe the properties of molecular assemblies and the kinetics of the interaction in live cellular environments. However, biological media are spatially inhomogeneous, which is poorly conveyed by measuring just a few, sparse single-molecule trajectories. Finding a way to efficiently and reliably extract useful information from a large amount of trajectory data is an obstacle of this technique.

A biomolecule subjected to random influences can explore its possible outcomes and evolves to yield a dispersion over its state space. The evolution may contain both contributions from deterministic and stochastic forces. To provide high-quality statistics sampled by appropriate probing biomolecules while preserving single-molecule sensitivity, we developed a 2D analysis of single-molecule trajectories with $V\left(\overline{R_\tau^{\ 2}}\right)$ and $\overline{R_\tau^{\ 2}}$ (normalized variance-vs-MSD). Here the MSDs were used to quantify the diffusion of a protein in its cellular environment, and the normalized variance discloses the nature of these interactions. Thus, the plot can be more sensitive than MSD alone to reflect the diffusive dynamics of a protein in cellular environments. We applied this 2D analysis technique to the dimerization processes of EGFRs in live cells under varying cellular conditions. Based on this study, we found that unliganded species appear to remain outside the cholesterol-enriched lipid domains. After ligand binding, EGFR molecules may relocate to lipid raft domains. This experimental finding was verified using three cell lines with a wide range of EGFR expression levels and membrane cholesterol concentrations, suggesting that these results may represent a general behavior of unliganded and activated receptors in live cells. Reigada et al. recently applied near-field scanning optical microscopy to fixed monocytes and found that raftophilic proteins did not physically intermix at the nanoscale with CTxB-GM1 nanodomains but converged within a characteristic distance [62]. Our result of unliganded EGFR agrees with this finding but on live cells without CTxB tightening.

Selectively tagging EGFR species with semiconductor quantum dots allowed us to monitor the correlated motions of unliganded and liganded species. Paired liganded receptors, which diffused in proximity on the plasma membrane interacted with each other and caused the receptors to move correlatively. The correlated motions can be caused by the correlated fluctuations in the lipid environment, which occur when the two receptors are closely separated. The correlated motion can be changed by manipulating either the distribution or total quantity of the membrane cholesterol, suggesting that the membrane cholesterol plays a vital role in mediating the interactions between the liganded receptors. Our quantitative 2D analysis method can capture the dynamic receptor interactions at the single-molecule level, providing details that are often obscured in other methods.

We further used the HIV-1 Tat peptide-conjugated QD as a nanosensor to illustrate the translocation dynamics of the Tat peptides in living cells. By using spectral-embedding analysis, we extracted an intrinsic low-dimensional manifold, which was formed by the isotropic diffusion and a fraction of the directed movement, from the measured trajectories. Our result suggest that HSPGs play a significant role in redirecting the TatP-QD entry process toward spatially restricted sites on the plasma membrane. We further applied 2D analysis of $V\left(\overline{R_\tau^{\ 2}}\right)$-$\overline{R_\tau^{\ 2}}$ to determine the underlying cause of the trajectory aggregation. We found that the membrane deformation induced by the Tat-peptide attachment increased with the disruption of the actin framework, which resulted in higher interactions on the TatP-QDs. In native cells, the Tat peptides could remodel the actin framework to reduce their interaction with the local membrane environment.

Semiconductor quantum dots conjugated with appropriate peptides or antibodies are appealing for probing cellular dynamic events in living cells. The nanosensors have the advantages of high emission efficiency, wavelength tunability, and long-term stability, which have led to a variety of applications in cellular sensing and imaging. Biomolecule-conjugated QD nanosensors are also useful for studying the interactions, stoichiometries, and conformational changes of proteins in living cells, which provides an understanding of the mode of the interaction and free-energy surfaces, and can reveal the stable states and dynamic pathways of biomolecules in live cells. The application examples presented in this chapter clearly support the use of biomolecule-conjugated QDs as probes for the cellular dynamics in living cells.

Acknowledgements

This research was funded by the Ministry of Science and Technology of the Republic of China (grant number MOST 106-2112-M-009-019-MY3). Parts of this chapter are taken from the authors' former work with permission of the Creative Commons Attribution license.

Conflicts of interest

The author declares that he has no competing interests.

Nomenclature

Ab	antibody
CPP	cell-penetrating peptide
CTxB	cholera toxin-β
cyto D	cytochalasin D
ETL	electrically tunable lens
FDT	fluctuation-dissipation theorems
GAG	glycosaminoglycan
GUV	giant unilamellar vesicles
HS	heparan sulfate
HSPG	heparan sulfate proteoglycan
MβCD	methyl β-cyclodextrin
MSD	mean square displacement
PG	proteoglycan
PMF	potential of mean force
PSF	point spread function
QD	quantum dot

http://dx.doi.org/10.5772/intechopen.72858

sCMOS complementary metal-oxide semiconductor

TatP transactivator of transcription (Tat) peptide

TatP-QD TatP-conjugated quantum dot

Author details

Jung Y. Huang

Address all correspondence to: jyhuang@faculty.nctu.edu.tw

Department of Photonics and The T.K.B. Research Center for Photonics, Chiao Tung University, Hsinchu, Taiwan

References

[1] Lippincott-Schwartz J, Snapp E, Kenworthy A. Studying protein dynamics in living cells. Nature Reviews Molecular Cell Biology. 2001;**2**:444-456

[2] Hung MC, Link W. Protein localization in disease and therapy. Journal of Cell Science. 2011;**124**:3381-3392

[3] Masson JB, Dionne P, Salvatico C, Renner M, Specht CG, Triller A, Dahan T. Mapping the energy and diffusion landscapes of membrane proteins at the cell surface using high-density single-molecule imaging and Bayesian inference: Application to the multiscale dynamics of glycine receptors in the neuronal membrane. Biophysical Journal. 2014;**106**:74-83

[4] Kusumi A, Tsunoyama TA, Hirosawa KM, Kasai RS, Fujiwara TK. Tracking single molecules at work in living cells. Biophysical Journal. 2014;**106**:74-83

[5] Liu Z, Lavis LD, Betzig E. Imaging live-cell dynamics and structure at the single-molecule level. Molecular Cell. 2015;**58**:644-659

[6] Serag MF, Abadi M, Habuchi S. Single-molecule diffusion and conformational dynamics by spatial integration of temporal fluctuations. Nature Communications. 2014;**5**:5123

[7] Michalet X. Mean square displacement analysis of single-particle trajectories with localization error: Brownian motion in an isotropic medium. Physical Review E. 2010;**82**:041914

[8] Medintz IL, Uyeda HT, Goldman ER, Mattoussi H. Quantum dot bioconjugates for imaging, labelling and sensing. Nature Materials. 2005;**4**:435-446

[9] Jaiswal JK, Simon SM. Imaging live cells using quantum dots. Cold Spring Harbor Protocols. 2015. p. doi:10.1101/pdb.top086322

[10] Zhang JJ, Zheng TT, Cheng FF, Zhang JR, Zhu JJ. Toward the early evaluation of therapeutic effects: An electrochemical platform for ultrasensitive detection of apoptotic cells. Analytical Chemistry. 2011;**83**:7902-7909

[11] Zhang JJ, Cheng FF, Li JJ, Zhu JJ, Lu Y. Fluorescent nanoprobes for sensing and imaging of metal ions: Recent advances and future perspectives. Nano Today. 2016;**11**:309-329

[12] Hana HS, Niemeyerb E, Huang Y, Kamoun WS, Martin JD, Bhaumik J, Chen Y, Roberge S, Cui J, Martin MR, Fukumura D, Jain RK, Bawendi MG, Duda DG. Quantum dot/antibody conjugates for in vivo cytometric imaging in mice. Proceedings of the National Academy of Sciences of the United States of America. 2015;**112**:1350-1355

[13] Wereszczynski J, McCammon JA. Statistical mechanics and molecular dynamics in evaluating thermodynamic properties of biomolecular recognition. Quarterly Reviews of Biophysics. 2012;**45**:1-25

[14] Seifert U. Stochastic thermodynamics, fluctuation theorems and molecular machines. Reports on Progress in Physics. 2012;**75**:126001-126058

[15] Singera A, Erban R, Kevrekidis IG, Coifman RR. Detecting intrinsic slow variables in stochastic dynamical systems by anisotropic diffusion maps. Proceedings of the National Academy of Sciences of the United States of America. 2009;**106**:16090-16095

[16] Coifman R, Kevrekidis I, Lafon S, Maggioni M, Nadler B. Diffusion maps, reduction coordinates and low dimensional representation of stochastic systems. SIAM Multiscale Model Simulation. 2008;**7**:842-864

[17] Lin CY, Huang JY, Lo LW. Depicting binding-mediated translocation of HIV-1 tat peptides in living cells with Nanoscale pens of tat-conjugated quantum dots. Sensors. 2017;**17**:315-314

[18] Welsher K, Yang H. Multi-resolution 3D visualization of the early stages of cellular uptake of peptide-coated nanoparticles. Nature Nanotechnology. 2014;**9**:198-203

[19] Belting M. Heparan sulfate proteoglycan as a plasma membrane carrier. Trends in Biochemical Sciences. 2003;**28**:145-151

[20] Yang J, Tsutsumi H, Furuta T, Sakurai M, Mihara H. Interaction of amphiphilic alpha-helical cell-penetrating peptides with heparan sulfate. Organic & Biomolecular Chemistry. 2014;**12**:4673-4681

[21] Gallavotti G, Cohen EGD. Dynamical ensembles in nonequilibrium statistical mechanics. Physical Review Letters. 1995;**74**:2694

[22] Kurchan J. Fluctuation theorem for stochastic dynamics. Journal of Physics A: Mathematical and General. 1998;**31**:3719

[23] Lebowitz JL, Spohn HA. Gallavotti-Cohen-type symmetry in the large deviation functional for stochastic dynamics. Journal of Statistical Physics. 1999;**95**:333

[24] Jarzynski C. Nonequilibrium equality for free energy differences. Physical Review Letters. 1997;**78**:2690

[25] Jarzynski C. Equilibrium free-energy differences from nonequilibrium measurements: A master-equation approach. Physical Review E. 1997;**56**:5018

[26] Crooks GE. Path-ensemble averages in systems driven far from equilibrium. Physical Review E. 2000;**61**:2361

[27] Hummer G, Szabo A. Free energy reconstruction from nonequilibrium single-molecule pulling experiments. Proceedings of the National Academy of Sciences of the United States of America. 2001;**98**:3658

[28] Hummer G, Szabo A. Time-reversal and entropy. Journal of Statistical Physics. 2003;**110**:269

[29] Zwanzig R. Nonequilibrium Statistical Mechanics. Oxford, UK: Oxford University Press; 2001

[30] Lin CY, Huang JY, Lo LW. Energetic modeling and single-molecule verification of dynamic regulation on receptor complexes by actin corrals and lipid raft domains. The Journal of Chemical Physics. 2014;**141**:215102

[31] Lin CYL, Huang JY, Lo LW. Unraveling the impact of lipid domains on the dimerization processes of single-molecule EGFRs of live cells. Biochimica et Biophysica Acta. 1848;**2015**: 886-893

[32] Tome T, de Oliveira MJ. Entropy production in irreversible systems described by a Fokker-Planck equation. Physical Review E. 2010;**82**:021120

[33] Jaynes E. The minimum entropy production principle. Annual Review of Physical Chemistry. 1980;**31**:579-601

[34] Wang J, Ferguson AL. Nonlinear reconstruction of single-molecule free-energy surfaces from univariate time series. Physical Review E. 2016;**93**:032412-032428

[35] Nadler B, Lafon S, Coifman RR, Kevrekidis IG. Diffusion maps, spectral clustering and reaction coordinates of dynamical systems. Applied and Computational Harmonic Analysis. 2006;**21**:113-132

[36] von Luxburg, U. A Tutorial on spectral clustering. Statistics and Computing. 2007;**174**:395–416

[37] Jaqaman K, Loerke D, Mettlen M, Kuwata H, Grinstein S, Schmid SL, Danuser G. Robust single-particle tracking in live-cell time-lapse sequences. Nature Methods. 2008;**5**:695-702

[38] Wu PH, Agarwal A, Hess H, Khargonekar PP, Tseng Y. Analysis of video-based microscopic particle trajectories using Kalman filtering. Biophysical Journal. 2010;**98**:2822-2830

[39] Lin CY, Huang JY, Lo LW. Exploring in vivo cholesterol-mediated interactions between activated EGF receptors in plasma membrane with single-molecule optical tracking. BMC Biophysics. 2016;**9**:1-11

[40] Kusumi A, Fujiwara TK, Morone N, Yoshida KJ, Chadda R, Xie M, Kasai RS, Suzuki KGN. Membrane mechanisms for signal transduction: The coupling of the meso-scale raft domains to membrane-skeleton-induced compartments and dynamic protein complexes. Seminars in cell and. Developmental Biology. 2012;**23**:126-144

[41] Balla T. Phosphoinositides: Tiny lipids with Giant impact on cell regulation. Physiological Reviews. 2013;**93**:1019-1137

[42] Jaqaman K, Grinstein S. Regulation from within: The cytoskeleton in transmembrane signaling. Trends in Cell Biology. 2012;**22**:515-526

[43] Iyengar G, Rao M. A cellular solution to an information-processing problem. Proceedings of the National Academy of Sciences of the United States of America. 2014;**111**:12402-12407

[44] Gschwind A, Fischer OM, Ullrich A. The discovery of receptor tyrosine kinases: Targets for cancer therapy. Nature Reviews. Cancer. 2004;**4**:361-370

[45] Mendelsohn J, Baselga J. The EGF receptor family as targets for cancer therapy. Oncogene. 2000;**19**:6550-6565

[46] Gómez-Llobregat J, Buceta J, Reigada R. Interplay of cytoskeletal activity and lipid phase stability in dynamic protein recruitment and clustering. Scientific Reports. 2013;**3**:2608, 1–8

[47] Kusumi A, Nakada C, Ritchie K, Murase K, Suzuki K, Murakoshi H, Kasai RS, Kondo J, Fujiwara T. Paradigm shift of the plasma membrane concept from the two-dimensional continuum fluid to the partitioned fluid: High-speed single-molecule tracking of membrane molecules. Annual Review of Biophysics and Biomolecular Structure. 2005;**34**:351-378

[48] Low-Nam ST, Lidke KA, Cutler PJ, Roovers RC, van Bergen en Henegouwen PMP, Wilson BS, Lidke DS. ErbB1 dimerization is promoted by domain co-confinement and stabilized by ligand binding. Nature Structural & Molecular Biology. 2007;**174**:395-416

[49] Huang JY, Lin CY. Exploring the stochastic dynamics of correlated movement of receptor proteins in plasma membranes in vivo. The Journal of Chemical Physics. 2015;**143**:225101-225107

[50] Hess ST, Gould TJ, Gudheti MV, Maas SA, Mills KD, Zimmerberg J. Dynamic clustered distribution of hemagglutinin resolved at 40 nm in living cell membranes discriminates between raft theories. Proceedings of the National Academy of Sciences of the United States of America. 2007;**104**:17370-17375

[51] Lingwood D, Simons K. Lipid rafts as a membrane-organizing principle. Science. 2010;**327**: 46-50

[52] Orr G, Hu D, Özgxelik S, Opresko LK, Wiley HS, Colson SD. Cholesterol dictates the freedom of EGF receptors and HER2 in the plane of the membrane. Biophysical Journal. 2005;**89**:1362-1373

[53] Ringerike T, Blystad FD, Levy FO, Madshus IH, Stang E. Cholesterol is important in control of EGF receptor kinase activity but EGF receptors are not concentrated in caveolae. Journal of Cell Science. 2002;**115**:1331-1340

[54] Linda JP, Casey L. Cholesterol levels modulate EGF receptor-mediated signaling by altering receptor function and trafficking. Biochemistry. 2002;**41**:10315-10322

[55] Shlomovitz R, Maibaum L, Schick M. A unified picture of rafts: Lipid phase behavior in a multicomponent membrane. Biophysical Journal. 2014;**106**:1979-1985

[56] Shlomovitz R, Schick M. Model of a raft in both leaves of an asymmetric lipid bilayer. Biophysical Journal. 2013;**105**:1406-1413

[57] Lingwood D, Ries J, Schwille P, Simons K. Plasma membranes are poised for activation of raft phase coalescence at physiological temperature. Proceedings of the National Academy of Sciences of the United States of America. 2008;**105**:10005-10010

[58] Mueller V, Ringemann C, Honigmann A, Schwarzmann G, Medda R, Leutenegger M, Polyakova S, Belov VN, Hell SW, Eggeling C. STED Nanoscopy reveals molecular details of cholesterol- and cytoskeleton-modulated lipid interactions in living cells. Biophysical Journal. 2011;**101**:1651-1660

[59] Baumgart T, Hunt G, Farkas E, Webb W, Feigenson G. Fluorescence probe partitioning between lo/Ld phases in lipid membranes. Biochimica et Biophysica Acta. 2007;**1768**:2182-2194

[60] Manzo C, van Zanten TS, Saha S, Torreno-Pina JA, Mayor S, Garcia-Parajo MF. PSF decomposition of nanoscopy images via Bayesian analysis unravels distinct molecular organization of the cell membrane. Scientific Reports. 2014;**4**:4353–4358

[61] Veatch SL, Machta BB, Shelby SA, Chiang EN, Holowka DA, Baird BA. Correlation functions quantify super-resolution images and estimate apparent clustering due to over-counting. PLoS One. 2012;**7**:e31457

[62] van Zanten TS, Gómez J, Manzo C, Cambi C, Buceta J, Reigada R, Garcia-Parajo MF. Direct mapping of nanoscale compositional connectivity on intact cell membranes. Proceedings of National Academy of Sciences USA. 2010;**107**:15437–15442

[63] Wang F, Wang Y, Zhang X, Zhang W, Guo S, Jin F. Recent progress of cell-penetrating peptides as new carriers for intracellular cargo delivery. Journal of Controlled Release. 2014;**174**:126-136

[64] Frankel AD, Pabo CO. Cellular uptake of the TAT protein from human immunodeficiency virus. Cell. 1988;**55**:1189-1193

[65] Vives E, Brodin P, Lebleu B. A truncated HIV-1 TAT protein basic domain rapidly translocates through the plasma membrane and accumulates in the cell nucleus. The Journal of Biological Chemistry. 1997;**272**:16010-16017

[66] Tyagi M, Rusnati M, Presta M, Giacca M. Internalization of HIV-1 tat requires cell surface heparan sulfate proteoglycans. The Journal of Biological Chemistry. 2001;**276**:3254-3261

[67] Nakase I, Tadokoro A, Kawabata N, Takeuchi T, Katoh H, Hiramoto K, Negishi M, Nomizu M, Sugiura Y, Futaki S. Interaction of arginine-rich peptides with membrane-associated proteoglycans is crucial for induction of actin organization and macropinocytosis. Biochemistry. 2007;**46**:492-501

[68] Goddette DW, Frieden C. Actin polymerization. The mechanism of action of cytochalasin D. The Journal of Biological Chemistry. 1986;**261**:15974-15980

[69] Mishra A, Lai GH, Schmidt NW, Sun VZ, Rodriguez AR, Tong R, Tang L, Cheng J, Deming TJ, Kamei DT, Wong GCL. Translocation of HIV TAT peptide and analogues induced by multiplexed membrane and cytoskeletal interactions. Proceedings of the National Academy of Sciences of the United States of America. 2011;**108**:16883-16888

[70] Nadler B, Lafon Se, Coifman RR, Kevrekidis IG. Diffusion maps, spectral clustering and reaction coordinates of dynamical systems. arXiv.org. 2008. p. arXiv:math/0503445

Chapter 8

Redox-Mediated Quantum Dots as Fluorescence Probe and Their Biological Application

Wei Ma

Additional information is available at the end of the chapter

http://dx.doi.org/10.5772/intechopen.70761

Abstract

Semiconductor quantum dots (QDs) as a new class of fluorescent labels have become valuable fluorescent platforms for biological applications due to their unique optical properties. In addition to their well-known size-dependent emission spectra, QDs are extremely sensitive to the presence of additional charges either on their surfaces or in the surrounding environment, which leads to a variety of optical properties and electronic consequences. By using thiols as bridges between QDs and redox-active ligands, the fluorescence effects of functionalized QD conjugates were investigated because QDs are prone to exchange electrons or energy with the attached ligands upon excitation, resulting in their fluorescence change. The recovery/enhancement or quenching of the QD conjugate fluorescence could be reversibly tuned with the transformation with the redox state of surface ligands. Moreover, quenching of the QD emission is highly dependent on the relative position of the oxidation levels of QDs and the redox-active ligand used. Importantly, the utility of these systems could enhance the compatibility of functionalized QDs in biological systems and can be used for monitoring the fluorescence change to trace in vitro and intracellular target analyte sensing. We believe that redox-mediated quantum dots as fluorescence probe are a significant step forward toward biosensing.

Keywords: quantum dots, redox-mediation, charge transfer, fluorescence, biosensor

1. Introduction

Semiconductor quantum dots (QDs) or nanocrystals with sizes smaller than the so-called Bohr exciton radius (a few nanometers), resulting in an effect called quantum confinement due to the appearance of discrete energy states in both the conduction and valence bands [1, 2]. Optoelectronics of colloidal QDs offer a compelling combination of solution processing and fluorescence tunability through quantum size effects [3, 4]. They, however, are affected by a variety of parameters including defects in the nanocrystal structure and the surface or with the surrounding medium [5]. QDs, in particular, have a large-area solution processing on their surfaces and are always capped with functional ligands, which provide surface passivation and promote compatibility with the surrounding medium [6, 7]. These ligands along with the surrounding matrix alter the overall optical and electronic properties of QDs as a result of efficient elimination of the surface native defects, often attributed to the saturation of dangling bonds, improved passivation, and higher packing densities [8–10]. So far, the processibility of colloidal QDs is also exploited in a diversity of applications by fine-tuning their surface ligand characteristics of the semiconductor nanoparticles [11–14]. For example, a water-soluble surface ligand is required for biological sensors [15], an electron conductive layer is important for photoelectric devices [16], and a polymerizable surface is needed to make fluorescence polymer composites [17]. Unlike most organic dyes, QDs are also highly sensitive to charge transfer, thus altering their fluorescence properties [18, 19]. Notably, coupling redox-active ligands to the QDs surface can promote transfer of external electrons (and holes) to QD [18, 19]. Due to an efficient Auger recombination, the presence of additional charges can lead to quenching of the QD fluorescence [20]. The quenching degree of QD fluorescence depends on the location of the added charge, with a complete quenching observed for charges existing in the QD core, due to the strong spatial overlap between charge(s) and exciton, whereas partial quenching is observed for charge(s) locating on the QD surface (due to weaker overlap with the exciton) [19, 20]. When electron transfer between QDs and the molecules bound to their surface occurs, the nanocrystal and its attached ligand molecule exist in highly reactive charged forms long enough to interact with the surrounding environment. The redox-active moiety-functionalized QDs may promote the transfer of external electrons and holes to either the QDs core conduction band or the QDs surface states [21]. Therefore, controlling charge transfer of redox-active surface ligands across functionalized QD conjugates has been attracting increasing interests for advanced diagnostics and in vivo imaging as well as ultrasensitive biosensing [22, 23]. Redox-active compounds including metal complexes, ions, and dyes have already been investigated for use in photo-induced electron-transfer QD sensing. Since the development of high-performance QDs and the advent of excellent coupling techniques to modify them with biological systems [23–25], there has been a urgent need to exploit the interactions of QDs with the redox-active ligand for sensing [26, 27]. A few preliminary researches have reported the redox-active ligand-functionalized QDs and their use to monitor specific biological events. Biofunctional QDs enjoy increasing interest in basic and applied science because of the many possible applications of these structures to fields including proteomics, microarray technology, and biosensors. It is expected that these redox-active ligand-functionalized nanocrystal will be able to perform specific functions, such as biorecognition in the context of an electrical measurement, better than either purely organic or inorganic systems.

2. Quinone/hydroquinone as redox-active surface ligands of QDs

Quinone/hydroquinone is ubiquitous in nature and constitutes an important class of naturally occurring redox molecules [28]. It is well-known that quinone/hydroquinone fulfills a universal and possibly unique function in electron transfer and energy conserving system [29]. Especially, a number of quinones/hydroquinones have the critical biological functions involving

http://dx.doi.org/10.5772/intechopen.70761

brain activity and neurotransmission (i.e., dopamine), blood clotting (i.e., vitamin K), protein post-translational modification (i.e., topaquinone), cellular signaling molecule metabolism (i.e., estrogens and catecholamines), and antioxidant metabolism (i.e., ubiquinone and tocopherol congeners) [30–32]. Redox moiety was introduced into the surface ligands to achieve the redox-switchable fluorescence properties that could be useful for signal multiplexing, since QDs are highly sensitive to the electron-transfer processes. Recently, the research of functionalized QDs fabrication of redox quinone/hydroquinone on the surface of nanocrystals enjoys increasing interest and performs the specific functions, such as biosensing, ultrasensitive detection, and biomimetic research.

3. Ubiquinone-quantum dot bioconjugates and their application

In particular, ubiquinones [coenzyme Q, (CoQ)] are composed of the redox-active ubiquinonyl ring with a tail of isoprenoid units in different homolog forms occurring in nature, which are the only lipid-soluble antioxidant and plays a very important role in the cell membrane physiology [33]. CoQ acts as a mobile electron carrier in the energy-transducing membranes of mitochondria, which can be reduced by NAD(P)H-dependent enzymes. The reduced form $CoQH_2$ is a potent radical scavenger and antioxidant that protects membranes and lipoproteins from peroxidations as a potent radical scavenger [34, 35]. The redox state can be determined not only by the extent of oxidation (oxidative stress), but also by that of reduction (enzymatic reaction). As well-known, fluorescence enhancement/quenching in QDs can be switched by electrochemically modulating electron transfer between attached molecules and QDs (**Figure 1**) [36]. For this purpose, three CoQ disulfide derivatives ($[CoQC_nS]_2$) possessing the basic ubiquinone structure of 2,3-dimethoxy-5-methyl-1,4-benzoquinone with different mercaptoalkyl side chain lengths at the 6-position (n = 1, 5, and 10) (**Figure 2**, left). The emission of functionalized QDs can be reversibly tuned in two directions, enhancement or

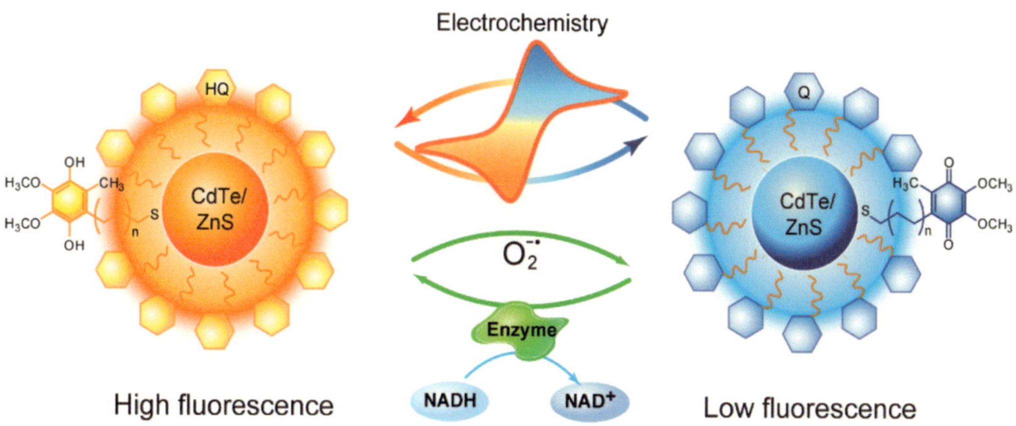

Figure 1. Schematic of fluorescence enhancement/quenching characteristics of $CoQH_2$ and CoQ-functionalized CdTe/ZnS QDs. Adapted with permission from [36]. Copyright 2011 Wiley-VCH Verlag GmbH & Co. KGaA.

quenching, depending on the different redox state of substrates bound to the surface of the QDs (**Figure 2**, right). Following photoexcitation of functionalized QD bioconjugates, the conductive band electron of QDs is transported to the lowest unoccupied molecular orbital of the oxidized ubiquinone acceptor and the electron is then went back to the valence band of QDs via nonradiative pathways. Thus, ubiquinones play the surface trap states acting as nonradiative de-excitation paths for photo-induced electron carriers, resulting in fluorescence quenching. It is worth noting that reduced ubiquinol ligands on the surface of QDs yield an obvious fluorescence enhancement. In this case, the photo-excited $CoQH_2$-QD bioconjugates decay to the ground state because the ubiquinols serve as poor electron donors. This switching results in recovering a high fluorescence compared to bare QDs. Furthermore, the reduced ubiquinols provide an efficient passivation of the surface trap states to overcome the potential surface defects, leading to a significantly enhanced fluorescence in $CoQH_2$-QD bioconjugates. According to energy band, bandgap of surface-capping ligand ubiquinol is larger than that of CdSe/ZnS QDs and hole trapping is also negligible. Upon photoexcitation, the resulting electrons and holes are confined in the surface regions of the ubiquinol-functionalized QDs, thus increasing the fluorescence. In addition, the fluorescence efficiency and stability of $CoQH_2$-QD bioconjugates against photo-oxidation has shown significant improvement due to the antioxidation effect of ubiquinol. Therefore, there is the remarkable fluorescence difference between CoQ and $CoQH_2$-capped QDs. Notably, the capping layer of reduced ubiquinol ligands enhances the QDs' fluorescence intensity significantly, while a modification using the oxidized ubiquinone ligands presents efficient quenching on fluorescence intensity of QDs under the identical conditions (**Figure 2**, right). We show fluorescence quenching efficiency to be dependent on alkyl chain spacer length of surface ligands, as more pronounced quenching was observed for C_2 spacer-modified QDs. Surface-attached CdTe/ZnS QDs exploiting coenzyme Q derivatives CoQ and $CoQH_2$ can be chemically attached to the surface of the QDs in an effort to mimic the electron transfer in the part of mitochondrial respiratory chain. Our system is extremely sensitive to NADH and superoxide radical ($O2^{\bullet -}$) species, and mimics a biological electron-transfer system in the part of the mitochondrial respiratory chain. In addition, in situ fluorescence spectra-electrochemical results further validate that the reduced state of ubiquinols significantly increase the fluorescence of QD bioconjugates, while the oxidized state of the ubiquinones decrease the fluorescence at varying degrees.

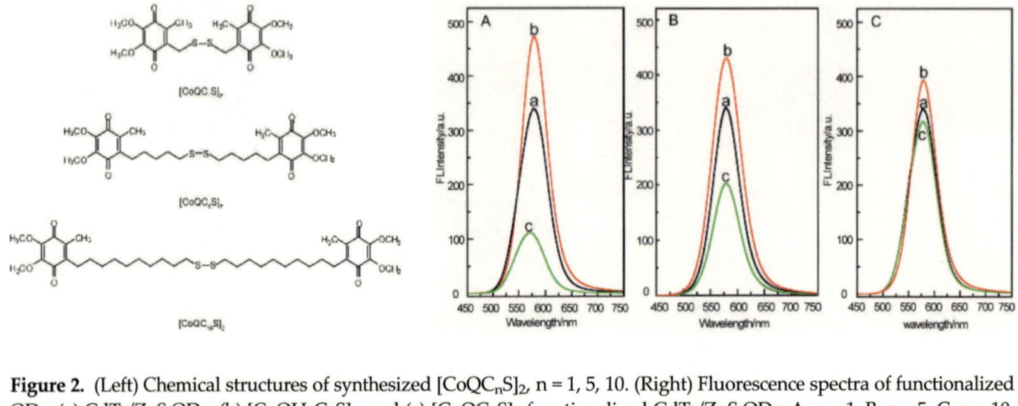

Figure 2. (Left) Chemical structures of synthesized $[CoQC_nS]_2$, n = 1, 5, 10. (Right) Fluorescence spectra of functionalized QDs. (a) CdTe/ZnS QDs, (b) $[CoQH_2C_nS]_2$, and (c) $[CoQC_nS]_2$-functionalized CdTe/ZnS QDs. A: n = 1, B: n = 5, C: n = 10. Adapted with permission from [36]. Copyright 2011 Wiley-VCH Verlag GmbH & Co. KGaA.

To further enhance the compatibility of ubiquinone-QD bioconjugates in biological system, the ligands Q_2NS, Q_5NS, and $Q_{10}NS$ were designed and synthesized by a facile click reaction between ubiquinone with terminal alkynes and alkylazide-disulfides via copper(I) tris (benzyltriazolylmethyl) amine catalyzed 1,2,3-triazole formation [37] (**Figure 3a**). In this system, the quinoid moiety in the Q_nNS surface ligands was introduced to achieve the redox-switchable fluorescence properties for signal multiplexing. The 1,2,3-triazole groups can enhance the compatibility of Q_nNS-QDs in biological system because of the similarity with histidine. Three alkyl spacers (C_2, C_5, C_{10}) confer various electron-transfer abilities either the core or the surface of QDs. As a final point, the disulfide group facilitates modification of Q_nNS ligand to the surface of QDs. Using the Q_nNS-QD bioconjugates, enhancement or quenching of the fluorescence of QD bioconjugates can also be switched by modulating the redox state of surface-capping ubiquinone ligands (**Figure 3b**) [38].

Interestingly, the emission of QD bioconjugates was enhanced when the surface-attached ubiquinone layer was reduced to ubiquinol in the presence of NADH and complex I in an effort to mimic the initial stages of mitochondrial respiration. The fluorescence intensity of ubiquinol-QDs was decreased gradually when the $O_2^{\bullet -}$ was added. As the concentration of $O_2^{\bullet -}$ is higher, the luminescence of the QDs is quenched to a higher extent, consistent with the formation of a higher coverage of the oxidized ubiquinone-modified QDs. Moreover, these systems provide the general framework for the creation of probes to monitor the reactive oxygen species in living cells according to their redox state, suggesting that this principle can be generalized to many different biological systems and applications. To demonstrate 1,2,3-triazole groups incorporated into the ubiquinone ligands can enhance the compatibility of QD bioconjugates in biological systems, we investigated a time-dependent fluorescence process using ubiquinone-assembled QDs with or without 1,2,3-triazole groups. A significant increase in the incubation time was observed for the same enhancement of fluorescence compared to the triazole-linked ubiquinone-QDs in the presence of NADH and complex I (**Figure 3**). This is because that the triazole groups behave similarly to histidine ligands and can be used to cap enzymes through proteins- or peptide-affinity coordination of triazole residues, leading to the triazole ubiquinone ligands efficiently improving binding affinity with complex I. The ubiquinone-QD bioconjugate system could be used for monitoring in vitro and intracellular complex I levels by the fluorescence changes of QD. Epidemiological researches show that the activity of complex I of Parkinson patients is impaired. Therefore, this system can be employed as a potent fluorescence probe for early stage Parkinson disease diagnosis and progression monitoring by observing complex I levels in human neuroblastoma SH-SY5Y cells.

Another novel strategy that uses QDs functionalized with quinonyl ligands was developed [39]. A novel biosensor based on "switch-on" photoluminogenic strategy employing of quinonyl glycosides functionalized QDs for the ingenious and biospecific imaging of human hepatoma Hep-G2 cells that express transmembrane glycoprotein receptors (**Figure 4**). The closely coupled quinonyl glycoside ligands are envisioned to have dual functions: the quinone part acts as a quencher of QDs and the glycoside part as a ligand for targeting a specific receptor. Moreover, self-assembly of quinonyl glycosides to QDs through a sulfide bond may produce QD bioconjugates that expose the glycosides in a clustering manner, enhancing their binding avidity with the target receptors. We observed that the quenched fluorescence

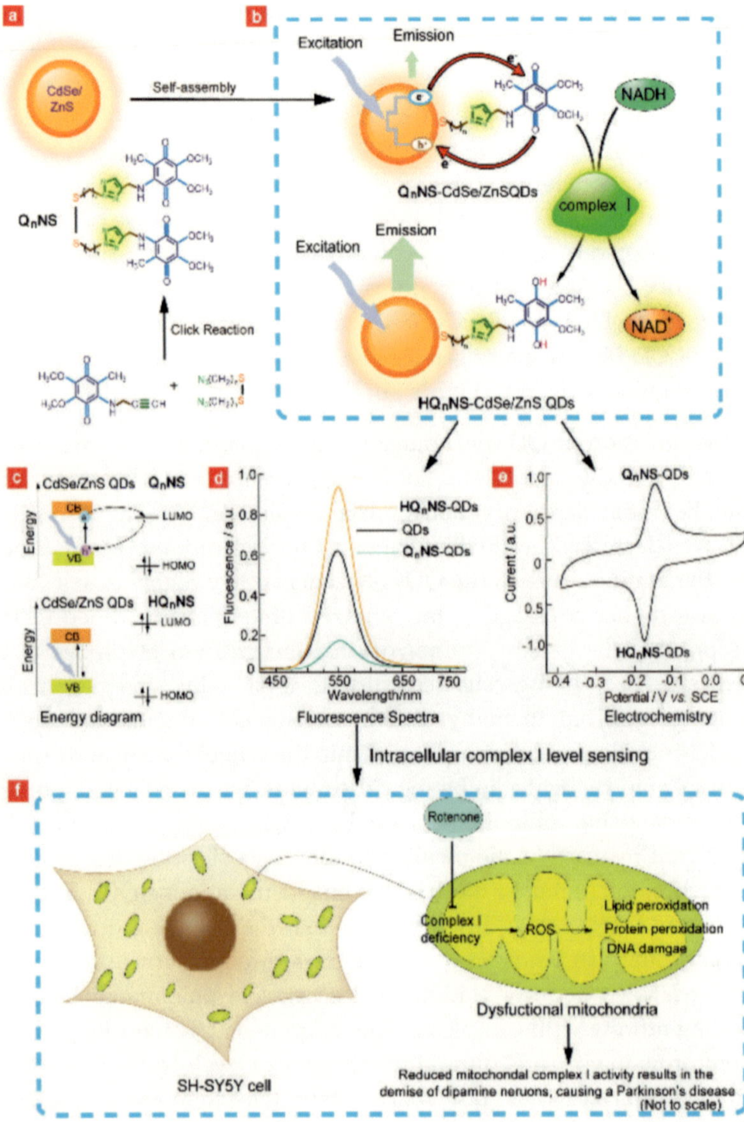

Figure 3. Schematic of ubiquinone-CdSe/ZnS QDs as redox fluorescence biosensor for Parkinson's disease diagnosis. (a) ubiquinone-terminated disulphides (Q_nNS) synthesis and self-assembly of Q_nNS on to CdSe/ZnS QDs. (b) Conceptual visualisation of Q_nNS-QDs as complex I sensor *in vitro*. Under oxidized state (Q_nNS), ubiquinone functions as a favorable electron acceptor, this results in effective QDs' fluorescence quenching. Addition of complex I to Q_nNS-QDs solution in the presence of NADH, ubiquinone coupled electron transfer and proton translocation from NADH, producing reduced ubiquinol (HQnNS) form on the surface of QDs to mimic the initial stages of the respiratory chain. Ubiquinol when in close proximity to the QDs produces fluorescence enhancement. (c) Energetic diagram of the QDs bioconjugates and possible electron transfer processes: electron transfer from the QDs CB to QnNS LUMO, followed by the back QDs VB. HQ_nNS only weakly accepts/donates electrons or energy and the excited QDs can return radiatively to the ground state. Under these conditions, the presence of HQ_nNS results in a significant fluorescence enhancement. (d) Fluorescence spectra of ubiquinone/ubiquinol- functionalised CdSe/ZnS QDs. e, Cyclic voltammetry of Q_nNS-CdSe/ZnS QDs. (f) Visualisation of Q_nNS-CdSe/ZnS QDs as an intracellular complex I sensor. The mitochondrial-specific neurotoxin, rotenone, inhibits complex I and leads to Parkinson's-like pathogenesis. Parkinson's disease is characterized by impaired activity of complex I in the electron-transfer chain of mitochondria. Adapted with permission from [38]. Copyright 2013 Nature Publishing Group.

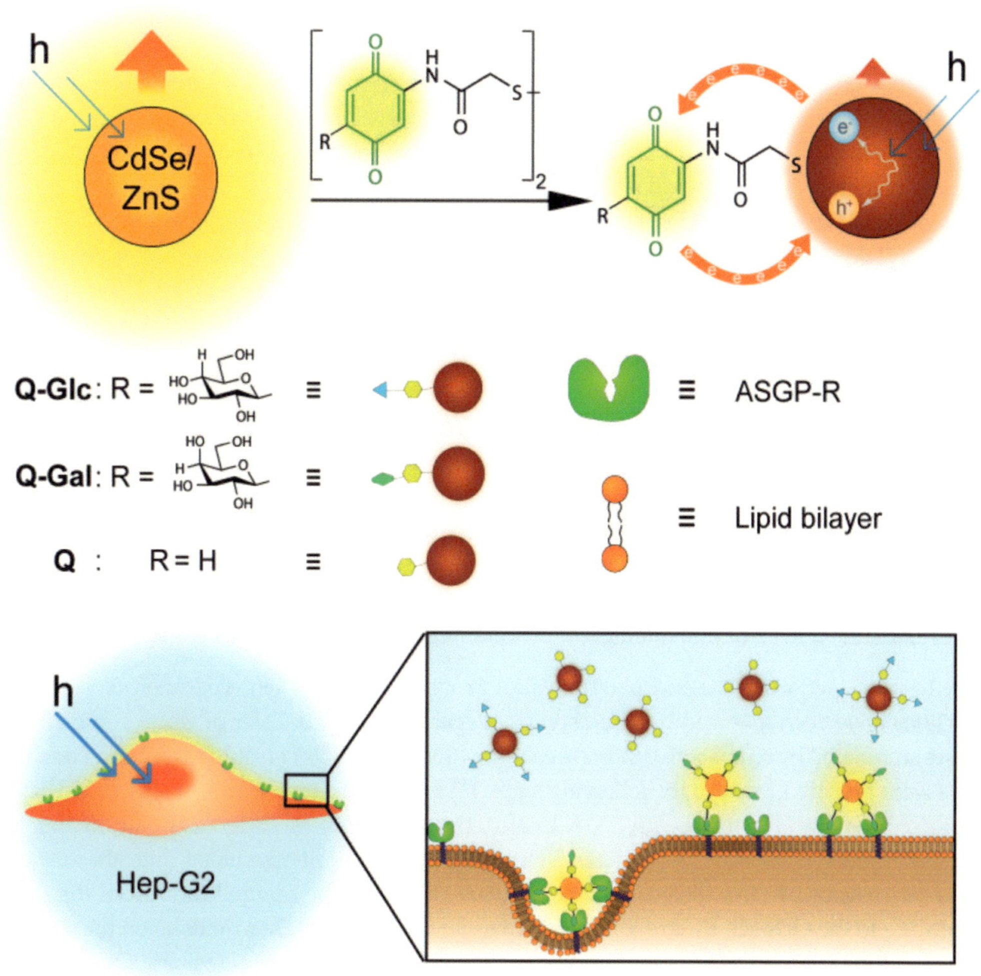

Figure 4. Schematic diagram of quinonyl glycosides functionalized QDs as a novel "switch-on" fluorescence probe for specific targeting and imaging transmembrane glycoprotein receptors of human hepatoma Hep-G2 cancer cells [Q-Glc: quinonyl glucoside disulfide; Q-Gal: quinonyl galactoside disulfide; Q: quinonyl disulfide]. Adapted from [39]. Copyright 2014 American Chemical Society.

of the functionalized QDs (by quinone) could be recovered by a lectin that selectively binds to the quinonyl glycosides clustering the QDs, but showed insignificant fluctuations toward a panel of nonselective lectins. We further determined that QDs coated with quinonyl galactosides could optically image transmembrane glycoprotein receptors of a hepatoma cell line in a target-specific manner (which they showed much weakened imaging ability toward cells with a reduced receptor level). This unique system, by taking advantage of the effective quenching ability of benzoquinone for QDs and natural ligand-receptor pairing on the cell surface (that recovers the signal), paves the way for the development of highly specific and low-background techniques for bioimaging of cancer cells as well as probing of unknown cell-surface receptors.

4. Dopamine-functionalized quantum dots and their application

Dopamine (DA) is an essential neurotransmitter in central nervous system and facilitates various functions in brain. DA-induced neurotoxicity has long been known to be triggered by the oxidation of DA and may play a role in pathological processes associated with neurodegeneration. Under oxidative stress, DA could readily oxidize to produce DA quinone catalyzed by tyrosinase in the presence of O_2 and contribute to the nucleophilic addition with sulfhydryl groups on free cysteine (Cys), glutathione, or Cys residue contained in protein [40, 41]. The interaction between DA quinone and Cys residue yields the formation of 5-Cys-DA in vitro and in vivo. As well-known, Cys residue is particularly critical for maintaining dynamic redox balance of cell and physiological function, which is directly correlated with the level of cellular stress. However, DA inactivation modification between Cys residue and DA quinone may disturb mitochondrial function, scavenge the thiol protein, inhibit protein function, and possibly lead to cell death [42]. Moreover, this modification decreases in the endogenous level of Cys residues, which is often found at the active site of functional proteins. Recent studies suggest that disturbance of Cys residue homeostasis may either lead to or result from oxidative stress in cell, contributing to mitochondrial dysfunction occurs early, and acts causally in neurodegenerative pathogenesis [43]. Therefore, it is of considerable significance to investigate the nature of this interaction process in physiology and pathology.

Due to the superior optical and photophysical properties of QDs, biorecognition or biocatalytic reactions have been followed by fluorescence resonance energy transfer or electron-transfer processes stimulated by redox-active biomolecule-functionalized QDs [44–50]. The DA-functionalized QDs were prepared through the following steps: (1) 596-nm-emitting thiohydracrylic acid capped CdTe/ZnS QDs and a redox-active DA thiol derivative (DAs) as surface-capping ligand were designed and synthesized; (2) the ligand molecule DAs was self-assembled onto the surface of QDs [51]. About 24 DAs molecules per QD were chosen as the optimal ratio from the spectra according to the QDs self-assembled with increasing ratio of DAs. DAs quinone on the surface of QD bioconjugates are generated in the enzymatic oxidation of DAs by tyrosinase/O_2, resulting in the fluorescence quenching (**Figure 5**). With adding a three-fold maximum tyrosinase/O_2, the fluorescence intensity of DAs-functionalized QDs was obviously quenched as expected. However, even much more excess tyrosinase/O_2 did not greatly affect the fluorescence of bare CdTe/ZnS QDs ($\leq$10% quenching). After DAs-QDs catalyzed by tyrosinase/O_2, the resulting product DAs quinone acting as an excellent electron acceptor is efficient for hole trapping of QDs and induces the fluorescence quenching. Fluorescence intensity of DAs quinone-QD bioconjugates recovered gradually upon addition of increasing amounts of Cys. Approximately 96% of the fluorescence was recovered after addition of Cys. It is worth noting that the 5-Cys-DAs containing catechol moiety on the functionalized QDs significantly recovered fluorescence. Here, the photo-excited-functionalized QD bioconjugates decay radiatively to the ground state of QDs because the 5-Cys-DAs ligands could function as poor electron acceptors. This in turn results in a fluorescence recovery due to the transformation from DAs quinone to DAs on the surface of functionalized QDs, blocking the electron transfer from QDs to benzoquinone. Only the presence of Cys residues (Cys or GSH) could induce rapid fluorescence recovery of the DAs quinone-functionalized QDs, confirming the specific coupling of Cys and DAs quinone in this system. In this study, photophysical properties of QDs were used to monitor the redox process of DAs and the formation of 5-Cys-DAs by mimicking the interaction process that DA oxidizes to form DA quinone, which binds covalently to nucleophilic sulfhydryl groups on Cys residues. The enzymatic process catalyzes the transformation of ligand structure between DA quinone and catechol, leading to the fluorescence change of functionalized QD bioconjugates (**Figure 5**). Several lines of evidence suggest that disturbance of Cys residue

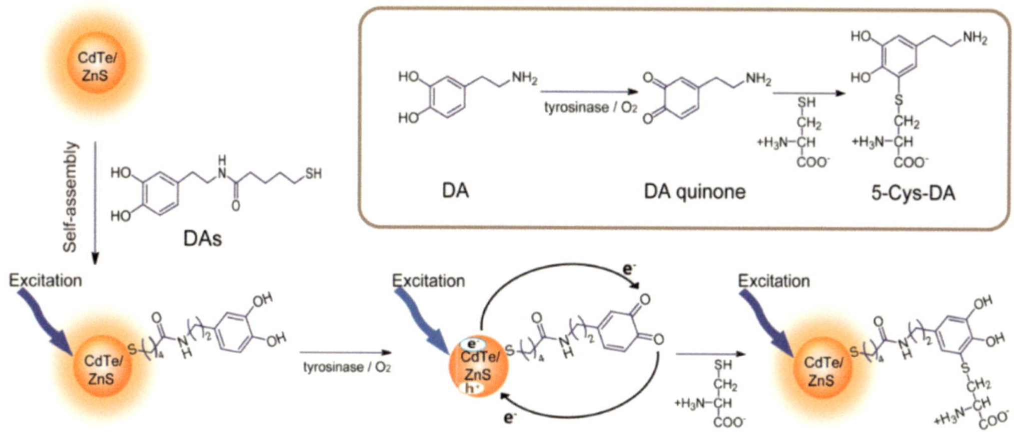

Figure 5. Schematic diagram of self-assembly and FL quenching/recovery characteristics of DAs-functionalized CdTe/ZnS QDs; Inset: schematic of the oxidation of DA and the irreversible interaction between Cys residue and DA quinone. Adapted from [51]. Copyright 2015 American Chemical Society.

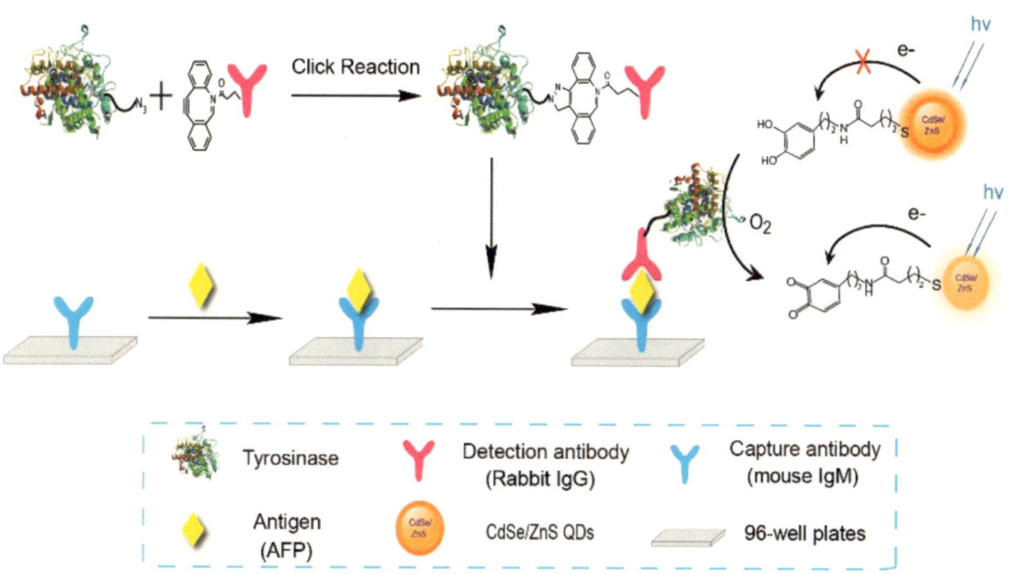

Figure 6. Schematic representation of redox-mediated indirect fluorescence immunoassay for the detection of biomarkers using DAs-functionalized CdSe/ZnS QDs. Adapted from [52]. Copyright 2016 American Chemical Society.

homeostasis may either lead to or result from oxidative stress in cell. In all major examples of neurodegenerative diseases, there is strong evidence that oxidative stress contribute to mitochondrial dysfunction occurs early and acts causally in disease pathogenesis. Thus, this specific fluorescence changes in our proposed system develop a powerful fluorescence sensor to follow the tracks of the neurotransmitter modification.

Inspired by redox-mediated fluorescence strategy, a redox-mediated indirect fluorescence immunoassay was developed for detecting the disease biomarker α-fetoprotein in a model based on DAs-immobilized CdSe/ZnS QDs (**Figure 6**) [52]. In this system, tyrosinase conjugated with the detection antibody was used as a bridge linking the QD fluorescence signals with the concentration of target disease biomarkers; the tyrosinase could catalyze enzymatic oxidation of DA to DA-quinone, resulting in fluorescence quenching in the presence of the analyte. Using this method, the detection limit for AFP was as low as 10 pM. This work provides a new pathway for the detection of disease biomarkers by RMFIA and has good potential for other applications.

5. Conclusion

By using redox-mediated fluorescence strategy, we demonstrated that coupling QDs with redox-active surface ligand is capable of fluorescence detecting of target analytes with high specificity. Ubiquinone-coupled QDs could be used for quantitative detection of ROS and target-specific imaging of transmembrane receptors in living cells. Dopamine as an electron donor could sensitize QDs through different mechanisms for monitoring dopaminergic neurotoxicity. Moreover, the improvement of QD-dopamine bioconjugates as biosensors was used for clinical diagnostic applications. Cumulatively, these results confirm a critical role for redox molecules, and especially quinone, in charge-transfer interactions with QDs for biological application.

Author details

Wei Ma

Address all correspondence to: weima@ecust.edu.cn

Key Laboratory for Advanced Materials, School of Chemistry and Molecular Engineering, East China University of Science and Technology, Shanghai, PR China

References

[1] Moras JD, Strandberg B, Suc D, et al. Semiconductor clusters, nanocrystals, and quantum dots. Science. 1996;**271**:933

[2] Michalet X, Pinaud FF, Bentolila LA, et al. Quantum dots for live cells, in vivo imaging, and diagnostics. Science. 2005;**307**(5709):538-544

[3] Henry CH. Limiting efficiencies of ideal single and multiple energy-gap terrestrialsolar-cells. Journal of Applied Physics. 1980;**51**:4494-4500

[4] King RR. Multijunction cells-record breakers. Nature Photonics. 2008;**2**:284-286

[5] Efros AL, Rosen M. The electronic structure of semiconductor nanocrystals. Annual Review of Materials Science. 2000;**30**:475-521

[6] Murray CB, Kagan CR, Bawendi MG. Synthesis and characterization of monodisperse nanocrystals and close-packed nanocrystal assemblies. Annual Review of Materials Science. 2000;**30**:545-610

[7] Mattoussi H, Cumming AW, Murray CB, Bawendi MG, Ober R. Properties of CdSe nanocrystal dispersions in the dilute regime: Structure and interparticle interactions. Physical Review B. 1998;**58**:7850-7863

[8] Pons T, Uyeda HT, Medintz IL, Mattoussi H. Hydrodynamic dimensions, electrophoretic mobility, and stability of hydrophilic quantum dots. The Journal of Physical Chemistry B. 2006;**110**:20308-20316

[9] Klimov VI, McBranch DW, Leatherdale CA, Bawendi MG. Electron and hole relaxation pathways in semiconductor quantum dots. Physical Review B. 1999;**60**:13740-13749

[10] Klimov VI, Mikhailovsky AA, McBranch DW, Leatherdale CA, Bawendi MG. Quantization of multiparticle auger rates in semiconductor quantum dots. Science. 2000;**287**:1011-1013

[11] Klem EJD, Shukla H, Hinds S, et al. Impact of dithiol treatment and air annealing on the conductivity, mobility, and hole density in PbS colloidal quantum dot solids. Applied Physics Letters. 2008;**92**:212105

[12] Luther JM, Law M, Beard MC, et al. Schottky solar cells based on colloidal nanocrystal films. Nano Letters. 2008;**8**:3488-3492

[13] Koleilat GI, Levina L, Shukla H, et al. Efficient, stable infrared photovoltaics based on solution-cast colloidal quantum dots. ACS Nano. 2008;**2**(5):833-840

[14] Talapin DV, Murray CB. PbSe nanocrystal solids for n- and p-channel thin film field-effect transistors. Science. 2005;**310**(5745):86-89

[15] Larson DR, Zipfel WR, Williams RM, et al. Water-soluble quantum dots for multiphoton fluorescence imaging in vivo. Science. 2003;**300**(5624):1434-1436

[16] Huynh WU, Dittmer JJ, Alivisatos AP. Hybrid nanorod-polymer solar cells. Science. 2002; **295**(5564):2425-2427

[17] Zhang H, Cui Z, Wang Y, et al. From water-soluble CdTe nanocrystals to fluorescent nanocrystal–polymer transparent composites using polymerizable surfactants. Advanced Materials. 2003;**15**(10):777-780

[18] Shim M, Wang CJ, Guyot-Sionnest P. Charge-tunable optical properties in colloidal semiconductor nanocrystals. The Journal of Physical Chemistry B. 2001;**105**:2369-2373

[19] Anderson NA, Lian TQ. Ultrafast electron transfer at the molecule-semiconductor nanoparticle interface. Annual Review of Physical Chemistry. 2005;**56**:491-519

[20] Klimov VI, Mikhailovsky AA, McBranch DW, et al. Mechanisms for intraband energy relaxation in semiconductor quantum dots: The role of electron-hole interactions. Physical Review B. 2000;**61**(20):R13349

[21] Mattoussi H, Radzilowski LH, Dabbousi BO, et al. Electroluminescence from heterostructures of poly (phenylene vinylene) and inorganic CdSe nanocrystals. Journal of Applied Physics. 1998;**83**(12):7965-7974

[22] Schlamp MC, Peng X, Alivisatos AP. Improved efficiencies in light emitting diodes made with CdSe (CdS) core/shell type nanocrystals and a semiconducting polymer. Journal of Applied Physics. 1997;**82**(11):5837-5842

[23] Bruchez M, Moronne M, Gin P, Weiss S, Alivisatos AP. Semiconductor nanocrystals as fluorescent biological labels. Science. 1998;**281**:2013-2016

[24] Chan WCW, Nie S. Quantum dot bioconjugates for ultrasensitive nonisotopic detection. Science. 1998;**281**(5385):2016-2018

[25] Mattoussi H, Mauro JM, Goldman ER, Anderson GP, Sundar VC, Mikulec FV, Bawendi MG. Self-assembly of CdSe–ZnS quantum dot bioconjugates using an engineered recombinant protein. Journal of the American Chemical Society. 2000;**122**:12142-12150

[26] Katz E, Willner I. Integrated nanoparticle–biomolecule hybrid systems: Synthesis, properties, and applications. Angewandte Chemie, International Edition. 2004;**43**:6042-6108

[27] Medintz I, Uyeda H, Goldman E, Mattoussi H. Quantum dot bioconjugates for imaging, labelling and sensing. Nature Materials. 2005;**4**:435-446

[28] Nohl H, Jordan W, Youngman RJ. Quinones in biology: Functions in electron transfer and oxygen activation. Advances in Free Radical Biology & Medicine. 1986;**2**(1):211-279

[29] Sazanov LA, Hinchliffe P. Structure of the hydrophilic domain of respiratory complex I from *Thermus thermophilus*. Science. 2006;**311**(5766):1430-1436

[30] Schultz W. Behavioral dopamine signals. Trends in Neurosciences. 2007;**30**(5):203-210

[31] Roffe L, Schmidt K, Ernst E. Efficacy of coenzyme Q10 for improved tolerability of cancer treatments: A systematic review. Journal of Clinical Oncology. 2004;**22**(21):4418-4424

[32] Li T, Chang CY, Jin DY, et al. Identification of the gene for vitamin K epoxide reductase. Nature. 2004;**427**(6974):541-544

[33] Ernster L, Dallner G. Biochemical, physiological and medical aspects of ubiquinone function. Biochimica et Biophysica Acta—Molecular Basis of Disease. 1995;**1271**(1):195-204

[34] Frei B, Kim MC, Ames BN. Ubiquinol-10 is an effective lipid-soluble antioxidant at physiological concentrations. Proceedings of the National Academy of Sciences of the United States of America. 1990;**87**(12):4879-4883

[35] Ernster L, Beyer RE. Antioxidant functions of coenzyme Q In: Folkers K, Littarru GP, Yamagami T. Biomedical and Clinical Aspects of Coenzyme Q, Vol. 6. Amsterdam: Elsevier; 1991. p. 213

[36] Qin LX, Ma W, Li DW, et al. Coenzyme Q functionalized CdTe/ZnS quantum dots for reactive oxygen species (ROS) imaging. Chemistry-A European Journal. 2011;**17**(19): 5262-5271

[37] Shen R, Shen X, Zhang Z, et al. Multifunctional conjugates to prepare nucleolar-targeting CdS quantum dots. Journal of the American Chemical Society. 2010;**132**(25):8627-8634

[38] Ma W, Qin LX, Liu FT, et al. Ubiquinone-quantum dot bioconjugates for in vitro and intracellular complex I sensing. Scientific Reports. 2013;**3**:1537

[39] Ma W, Liu HT, He XP, et al. Target-specific imaging of transmembrane receptors using quinonyl glycosides functionalized quantum dots. Analytical Chemistry. 2014;**86**(11):5502-5507

[40] Hastings TG, Lewis DA, Zigmond MJ. Role of oxidation in the neurotoxic effects of Intrastriatal dopamine injections. Proceedings of the National Academy of Sciences of the United States of America. 1996;**93**:1956-1961

[41] Van Laar VV, Dukes AA, Cascio M, Hastings TG. Proteomic analysis of rat brain mitochondria following exposure to dopamine Quinone: Implications for Parkinson disease. Neurobiology of Disease. 2008;**29**:477-489

[42] Miyazaki I, Asanuma M. Approaches to prevent dopamine quinone-induced neurotoxicity. Neurochemical Research. 2009;**34**:698-706

[43] Finkel T. Oxidant signals and oxidative stress. Current Opinion in Cell Biology. 2003;**5**:247-254

[44] Liu S, Zhang X, Yu Y, Zou G. A monochromatic electrochemiluminescence sensing strategy for dopamine with dual-stabilizers-capped CdSe quantum dots as emitters. Analytical Chemistry. 2014;**86**:2784-2788

[45] Chou K, Meng H, Cen Y, Li L, Chen J-Y. Dopamine-quantum dot conjugate: A new kind of photosensitizers for photodynamic therapy of cancers. Journal of Nanoparticle Research. 2013;**15**:1348

[46] Liu J, Bao C, Zhong X, Zhao C, Zhu L. Highly selective detection of glutathione using a quantum-dot-based OFF-ON fluorescent probe. Chemical Communications. 2010;**46**:2971-2973

[47] Yuan J, Guo W, Wang E. Utilizing a CdTe quantum dots-enzyme hybrid system for the determination of both phenolic compounds and hydrogen peroxide. Analytical Chemistry. 2008;**80**:1141-1145

[48] Gill R, Freeman R, Xu J-P, Willner I, Winograd S, Shweky I, Banin U. Probing biocatalytic transformations with CdSe-ZnS QDs. Journal of the American Chemical Society. 2006; **128**:15376-15377

[49] Ji X, Palui G, Avellini T, Na HB, Yi C, Knappenberger K, Mattoussi H. On the pH-dependent quenching of quantum dot photoluminescence by redox active dopamine. Journal of the American Chemical Society. 2012;**134**:6006-6017

[50] Li DW, Qin LX, Li Y, Nia RP, Long Y-T, Chen H-Y. CdSe/ZnS quantum dot-cytochrome c bioconjugates for selective intracellular $O_2^{\bullet -}$ sensing. Chemical Communications. 2011;**47**:8539-8541

[51] Ma W, Liu HT, Long YT. Monitoring dopamine quinone-induced dopaminergic neurotoxicity using dopamine functionalized quantum dots. ACS Applied Materials & Interfaces. 2015;**7**(26):14352-14358

[52] Zhang WH, Ma W, Long YT. Redox-mediated indirect fluorescence immunoassay for the detection of disease biomarkers using dopamine-functionalized quantum dots. Analytical Chemistry. 2016;**88**(10):5131-5136

Chapter 9

Enhancement of Photosynthetic Productivity by Quantum Dots Application

Angela Janet Murray, John Love, Mark D. Redwood, Rafael L. Orozco, Richard K. Tennant, Frankie Woodhall, Alex Goodridge and Lynne Elaine Macaskie

Additional information is available at the end of the chapter

http://dx.doi.org/10.5772/intechopen.74032

Abstract

The challenge of climate change promotes use of carbon neutral fuels. Biofuels are made via fixing carbon dioxide via photosynthesis which is inefficient. Light trapping pigments use restricted light wavelengths. A study using the microalga *Botryococcus braunii* (which produces bio-oil), the bacterium *Rhodobacter sphaeroides* (which produces hydrogen), and the cyanobacterium *Arthrospira platensis* (for bulk biomass) showed that photosynthetic productivity was increased by up to 2.5-fold by upconverting unused wavelengths of sunlight via using quantum dots. For large scale commercial energy processes, a 100-fold cost reduction was calculated as the break-even point for adoption of classical QD technology into large scale photobioreactors (PBRs). As a potential alternative, zinc sulfide nanoparticles (NPs) were made using waste H_2S derived from another process that precipitates metals from mine wastewaters. Biogenic ZnS NPs behaved identically to ZnS quantum dots with absorbance and emission maxima of 290 nm (UVB, which is mostly absorbed by the atmosphere) and 410 nm, respectively; the optimal wavelength for chlorophyll a is 430 nm. By using a low concentration of citrate (10 mM) during ZnS synthesis, the excitation wavelength was redshifted to 315 nm (into the UVA, 85% of which reaches the earth's surface) with an emission peak of 425 nm, i.e., appropriate for photosynthesis. The potential for use in large scale photobioreactors is discussed in the light of current PBR designs, with respect to the need for durable UV-transmitting materials in appropriate QD delivery systems.

Keywords: photosynthetic enhancement, bioenergy, quantum dots, zinc sulfide, *Botryococcus braunii, Arthrospira platensis, Rhodobacter sphaeroides*, bio-oil, bio-hydrogen, biomass

1. Introduction

1.1. Photosynthetic biotechnologies for biomass and fuels

The term "bioenergy" is used to describe the conversion of materials of biological origin into fuels and also includes the use of living organisms to produce a material that is a fuel or fuel precursor. This chapter focuses on the use of photosynthetic microorganisms: bacteria, cyanobacteria, and algae. These all grow at the expense of sunlight, while at the same time fixing carbon dioxide from the atmosphere or dissolved in water (algae) or converting organic waste into biomass material (bacteria). Traditional photobiotechnologies have used algae which range from seaweeds to small unicellular organisms within the "kingdom" of eukaryotes which also includes all higher forms of life. The "kingdom" of prokaryotes represents a far simpler level of cellular organization and includes single-celled photosynthetic bacteria and also filamentous microorganisms called cyanobacteria (or "blue green algae"). This review will illustrate examples of all three types.

Photosynthesis is achieved via the use of specialized pigments called chlorophylls that trap light energy for conversion into chemical energy to drive microbial processes and growth. Algae and cyanobacteria contain "chlorophyll a," while algae, like higher plants, also have a second chlorophyll, "chlorophyll b." Photosynthetic bacteria have functionally equivalent pigments called bacteriochlorophylls, and also ancillary pigments involved in light trapping.

Photosynthetic microorganisms are united by the need to maximize solar irradiation onto their light trapping centers. Natural growth occurs in, for example, ponds but, focusing on maximizing productivity, biotechnology has developed various strategies using photobioreactors (PBRs) for process intensification. Typical strategies include various PBR formats for optimal growth and production at scale, molecular engineering of light trapping centers to improve light conversion and strategies to convert the unused portions of sunlight into additional light which forms the focus of this chapter. Examples will be presented as a proof of concept, highlighting some of the barriers towards implementation.

1.2. Examples of photosynthetic biotechnologies: three examples

By 2030, the global demand for transport fuel is likely to increase significantly, requiring the production of up to approximately 400–500 billion liters of biofuel per year [1, 2]. Biofuel production could rise to 165 billion liters by 2030, if the US, Canada, and Europe adopt a common E15 blending standard [3], but clearly there will be a shortfall. Biofuel production by photosynthetic microbes is an alternative to crop-based biofuels as fertile soil and a hospitable climate are not required, and hence, biofuels could be produced using contaminated land, steeply sloping hillsides, deserts, urban areas, or rooftops. Therefore, unlike crop-based biofuels, microbial biofuels would not necessarily impact upon agricultural food production.

The microscopic alga *Botryococcus braunii* is potentially valuable as it secretes long-chain (C_{20-40}) hydrocarbons which can be processed into "drop-in" liquid fuels [1, 4]. As an alternative approach, algae have been grown as a source of biomass for production of another form of bio-oil. Thermochemical treatment (pyrolysis) produces oil, which is akin to fossil oils when suitably processed via upgrading and refinery processes [5].

Cyanobacteria (historically misnamed "blue-green algae") are functionally similar to algae but distinct in many ways. The filamentous cyanobacterium *Arthrospira* ("spirulina") *platensis* is grown as a high-value food supplement and also (under less stringent production standards) as animal feed due to its high content of protein and other nutrients [6, 7]. Spirulina production is highly practical, as the alkaline medium it prefers suppresses contaminants that can impact on algae production. Notably, too, it can utilize soluble bicarbonate ion which forms in alkaline solution, following the dissolving of gaseous CO_2.

The anoxygenic photosynthetic "purple nonsulfur bacteria" (e.g., *Rhodobacter sphaeroides*) offer potentially both fuels and chemicals, producing C_4-C_5 polyhydroxyalkanoates (bioplastic precursors; 50–80% w/w) [8] and high-purity hydrogen gas (typically 90% v/v) as part of an integrated biohydrogen refinery, which could exceed the delivered energy densities of mainstream renewable energy systems, such as photovoltaic cells and on-shore wind turbines [9, 10]. Unlike cyanobacteria and higher algae, *R sphaeroides* utilizes organic acids which are almost ubiquitously produced as by-products from various fermentations and wastewater treatment processes.

Lacking complex structures, microbes can achieve much higher productivity than crop plants. The efficiencies of light conversion to fuel are 0.4–0.8% for algal oil and ~1–5% for purple bacterial H_2 [11], whereas for higher plants, the value is at most 0.16% and normally much less [12]. Significantly higher photosynthetic productivities are needed to make significant progress toward supplanting fossil fuels. As well as improving the microorganism, the "value" of sunlight and its delivery can also be improved, which forms the focus of this chapter.

2. Overview of photobiotechnologies

Bacterial photobiotechnologies are not yet developed at scale, and in some cases, as for the biohydrogen process noted above, the photobioreactor design can be complex due to the need to exclude air. In contrast, algal biotechnology is relatively well developed [13], even though predictable algal culture at industrial scales (10^5–10^6 l), for extended periods, remains problematic [14].

Although the basic requirements for algal culture are simple—water, dilute inorganic nutrients (nitrate, phosphate and trace elements), CO_2, and light—a number of physical and biological factors limit the basic engineering designs of algal culture platforms which, as a consequence, have changed little in the last 50 years. These limiting factors include mainly light attenuation in water (notably of the photosynthetic, red wavelengths) due to absorbance and scattering, CO_2 dissolution, water temperature, and, often overlooked, the fact that algal cultures typically comprise unicellular organisms that are fundamentally "selfish" and are in a permanent competition with all other individuals in the culture [15]. This latter point means that algae are superbly adapted to acquiring more photons than are actually required for their photosynthetic processes and dissipate the surplus as nonphotosynthetic radiation. In algal cultures, illumination typically follows the Beer-Lambert law, with light intensity decreasing exponentially depending on the biomass concentration [16]. Consequently, in static cultures, cells at the surface of the photic zone experience high intensities of light and temperature, while the majority of the culture is in complete darkness [17–19], the consequence of which is that static cultures rapidly become light limited, and overall growth slows or reaches a plateau.

Engineering solutions to these problems typically include: constructing short light paths within the culture system or using high-intensity illumination; mixing cultures using pumps, impellers, paddle-wheels, or bubbles to maintain an overall average illumination experienced by all cells in the culture; controlling temperature; and increasing the concentration of dissolved CO_2. Fundamentally, engineering algal culture systems is a complex problem [20], involving multiple possibilities, and compromises that must be aligned with the final application, as any solution invariably has a cost that will be reflected in that of the product.

Algal culture platforms are conventionally divided into two categories, open or closed systems, each of which has different advantages, uses, and productivities.

Open algal cultures (**Figure 1a** and **b**) are typically shallow ponds, or "raceways", in which mixing is performed by direct displacement of the liquid using impellers or paddle wheels or by bubbles in airlift systems [14]. Raceways are designed to provide predictable, circulatory patterns (**Figure 1c**), enabling a more or less homogenous distribution of nutrients and access to light for all individual algal cells [22]. Although photosynthetically active radiation may not penetrate dense cultures, a combination of shallow ponds (20–50 cm deep) and mixing allows the algae sufficient time in the photic zone to grow (**Figure 1d**). The larger the installation, the more energy is required for mixing, increasing hydrodynamic shear, and the possibility of localized "dead-zones," where mixing is sub-optimal, and resulting in sub-optimal productivities [23]. Moreover, open ponds require large expanses of flat land which, in certain locations, is sought after for other, more lucrative uses, thereby increasing the capital cost of the installation. The addition of CO_2 to open systems is also problematic. Finally, open ponds carry the possibility of culture contamination by undesirable organisms such as other algal species or algal predators. While some applications, notably bioremediation, might benefit from a diverse population of different algal species with regard to resilience, stability, and performance [24–26], when the culture of a single algal species is preferred in an open setting, a limited number of extremophiles and rapidly growing algal species are used to minimize contamination [27].

Closed systems (also termed "photobioreactors" or PBRs), in which there is no direct exchange of culture media, gases, and potential contaminants with the environment, offer a number of advantages for algal culture, including better control over culture conditions (light intensity, temperature, pH, oxygen concentration, and CO_2), higher levels of reproducibility, higher biomass productivity, a lower risk of contamination, enabling culture of a wider variety of species, and, because they are contained, the use of genetically modified algal strains. Several types of PBRs have been devised [28, 29] (**Figure 2**) that can be located either outdoors or, for more accurate temperature control, in greenhouses or in artificially lit chambers. Apart from shaken flasks in an illuminated incubator, the simplest PBR design is a hanging, translucent, or transparent plastic bag or vertical, transparent tube, in which algal cultures are mixed by gas sparging ("bubble columns" or "airlift columns"). Such PBRs have a high surface area to volume ratio suitable for light transmission and satisfactory heat and mass transfer, providing a homogenous culture environment and efficient release of gases. Other advantages include low shear; the lack of moving parts makes bubble columns relatively inexpensive and easy to maintain. Alternatively, algal and media mixing may be achieved by an impeller (so-called "stir-tank" reactors; conceptually similar to an illuminated bacterial fermenter); here, the effectiveness of mixing depends upon the design of the impeller blades, the speed of rotation,

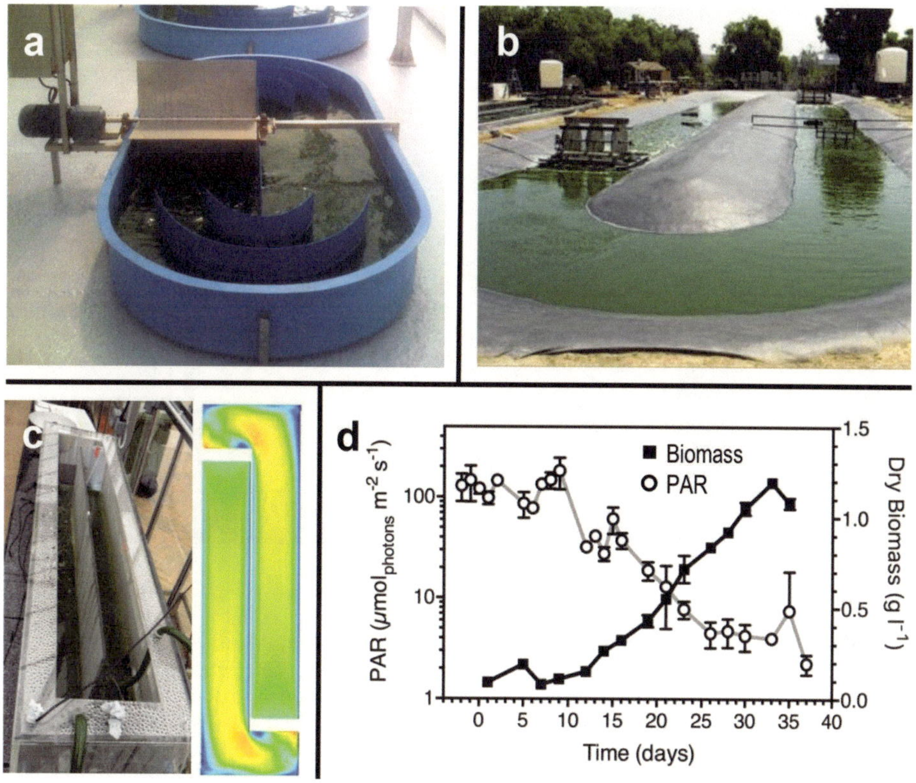

Figure 1. Open algal culture systems. Open-pond systems used for (a) small-scale (≈1000 l) and (b) commercial scale (>100,000 l) algal culture (image courtesy of the South Australian Research Institute (SARDI)), where the medium is displaced by paddle-wheels that are easy to service and cause low hydrodynamic shear. Panel (c) shows a model raceway (200 × 50 and 20 cm water depth) in which currents are driven by impellers. Computational fluid dynamic (CFD) modeling of this system topography (panel c, right) shows the distribution and strength of currents, with regions of low water movement in cold colors and faster water movement represented by warmer colors. The CFD was performed by Robert Rouse and Gavin Tabor (University of Exeter Department of Engineering) using empirical data. The graph in (d) shows the growth of a *Botryococcus braunii* culture (closed squares representing the mean of 3 replicates) in a 20 cm deep raceway and the reduction in photosynthetically active radiation (PAR; open circles representing the mean of readings from 4 sensors placed under the tank and 49 cm intervals) at the bottom of the pond, as the culture grows. Note that after approximately 15 days, PAR is only 10% of the starting level but, due to mixing, the culture continues to grow for a further fortnight. Bars represent the standard error of the mean.

and the depth of liquid. Vertical or horizontal tubular reactors in which media and algae are pumped from a main sump through the structure ("biofence") provide a scale-up capacity to several hundred liters. Shorter light paths are achieved using flat-panel reactor designs.

Illuminating plants with light emitting diodes (LEDs) leads to higher biomass productivity per unit of irradiance [30]. LEDs have several advantages over conventional, incandescent, or fluorescent lights, including small size, durability, long lifetime, cool-emitting temperature, and the option to select specific wavelengths, notably in the photosynthetic red and blue wavelengths [31, 32].

Despite the concomitant reduction in energy use from LEDs compared to other forms of illumination, and the effectiveness of different PBRs at laboratory scale, the mass production of

Figure 2. Examples of closed algal culture systems. (a) Flasks containing 100 ml of algal culture in a shaking, lighted incubator with CO_2-enriched atmosphere. (b) Polyethylene bag containing 100 ml of algal culture, located in a greenhouse with natural and complementary artificial lighting. Mixing is achieved by an air-stone (aquarium) bubbler. (c) Translucent polyethylene, conical bucket for airlift culture, containing 100 l for media. Air is provided by a simple tube at the bottom of the vessel. The conical shape and tap towards the bottom of the container enable simple harvest of the algal suspension. Photograph courtesy of Dr. Mike Allen, Plymouth Marine Laboratory. (d) Airlift column photo-bioreactor (perspex; 2 m in length; 10 cm in diameter), containing 10 l of culture and lit by a combination of white, blue and red LED's optimized for algal growth. An air inlet at the base of the column provides mixing. (e) Stirred photo-bioreactor with white LED light jacket, containing 2 l of algal culture. (f) Horizontal, tubular photo-bioreactor (or "biofence") containing 600 l of *Phaeodactylum* culture, located at Swansea University (Wales, UK). The culture is pumped through the transparent tubes from a sump enabling control of media composition and temperature. (g) Flat-panel photobioreactor containing 100 ml of algal culture. The light path is 5 mm and the algae are pumped through the reactor from a sump.

microalgae in closed systems remains expensive in terms of construction costs, materials, and energy. Moreover, up-scaling is problematic, as most PBR designs suffer from a number of limiting factors, including poor gas exchange, difficulties in nutrient delivery, heat balance, and, in locations where seasons are marked, available light [33, 34].

3. Process intensification: Limitations of light availability

The problems of light delivery to a photobioreactor are 2-fold. In equatorial regions, the photoperiod (day length) and seasonality are reasonably constant. However, at higher and lower latitudes, the day length and incident light are seasonally variable, and hence, for a proportion of the year, a PBR cannot operate during significant periods of darkness or is impaired by low light intensity. A pilot scale tubular photobioreactor using *R. sphaeroides* to produce hydrogen (**Figure 3a**) was programmed to operate at UK latitude (~55°N) at equinox and maintained on that diurnal cycle for 3 months, with the light intensity varied day to day. Saturation occurred at ~400 W/m^2 (**Figure 3c**). **Figure 3d** shows that in spring and autumn, a

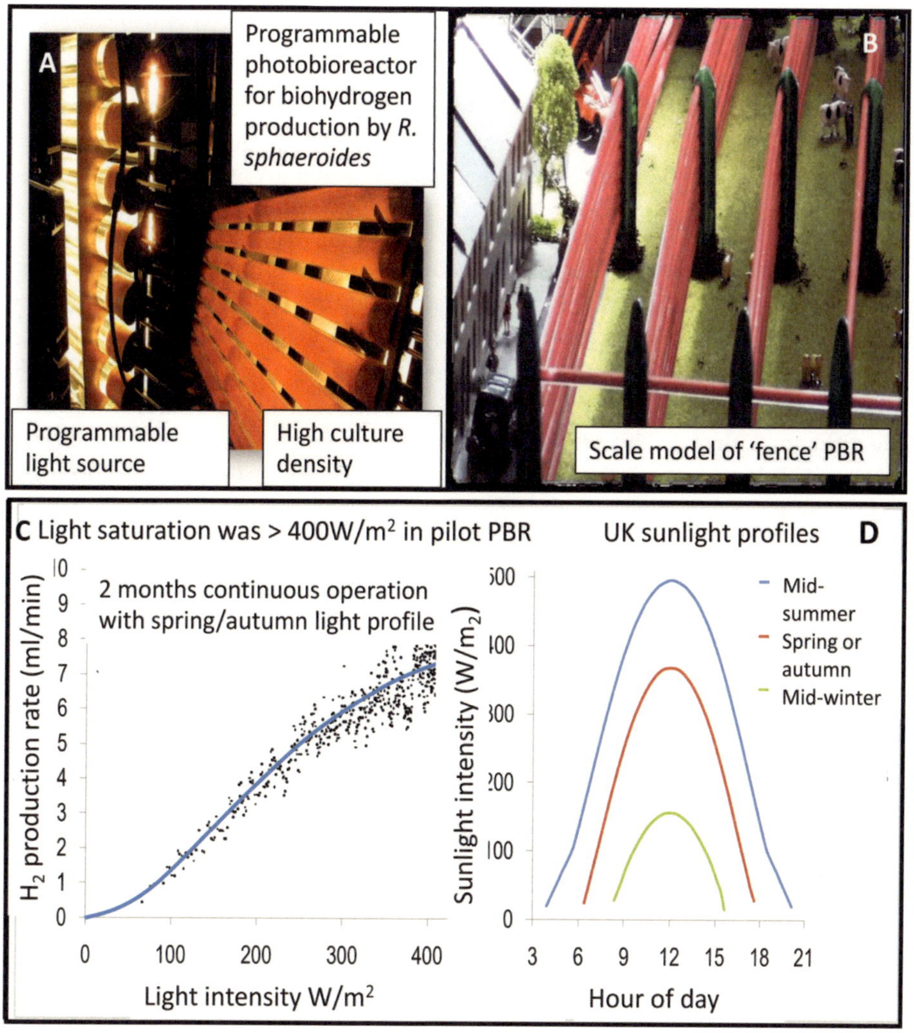

Figure 3. Performance of a programmable photobioreactor. The PBR was operated over 3 months of continuous diurnal operation set to spring and autumn equinox (12 hour days/nights plus dawn/dusk periods). The light intensity was varied from day to day (at random) and parallel rooftop tests confirmed the pilot scale data. A. PBR tubular construction and orange-pigmented *R. sphaeroides*. B. Scale model of full scale PBR constructed on the basis of the pilot data C: Determination of light saturation in terms of biohydrogen productivity. D: Light intensity (W/m^2) as typical UK profiles.

PBR will not reach light saturation, while in winter, the productivity would reach only ~33% of its potential maximum at midday (**Figure 3c**, **d**). Saturation would only be reached in mid-summer (**Figure 3d**); hence, an increase in light intensity of up to 4-fold would be required for maximum productivity.

To attempt to overcome the limitation, the culture biomass intensity/ml can be increased, but this results in significant "self-shading". In illustration, using the spring/autumn illumination profile (**Figure 3c**) of the biohydrogen PBR using low, intermediate and high density cultures gave a hydrogen yield of 11.5, 7.0, and 3.5 ml/min, respectively, i.e., simply introducing more bacteria is counterproductive, and this also increases the running costs with respect to both the make-up feed (trace nutrients) and the final biomass for waste disposal, if the biomass is not used for biofuel.

The second limitation is that the solar spectrum is very wide, yet the wavelengths captured by photosynthetic pigments are quite conservative (**Figure 4**), with algal/cyanobacterial chlorophyll utilizing visible wavelengths, while bacteriochlorophyll utilizes light in the visible/near infrared (NIR) region.

A novel study used a beam splitting approach to supply an algal and a bacterial system (similar to that shown in **Figures 5** and **6**, without quantum dots), taking advantage of their respective preferred wavelengths and giving the potential to operate two parallel PBRs. This enhanced the microbial productivity per incident photon [35], an approach that could be useful in, for example, biohydrogen production, where bacteria and algae both make bio-H_2 but by using different pathways [36]. Hence, it may be possible to produce bio-H_2 by bacterial and algal reactors side by side, with the additional advantage of providing a "sink" for bacterially produced CO_2 into algal biomass.

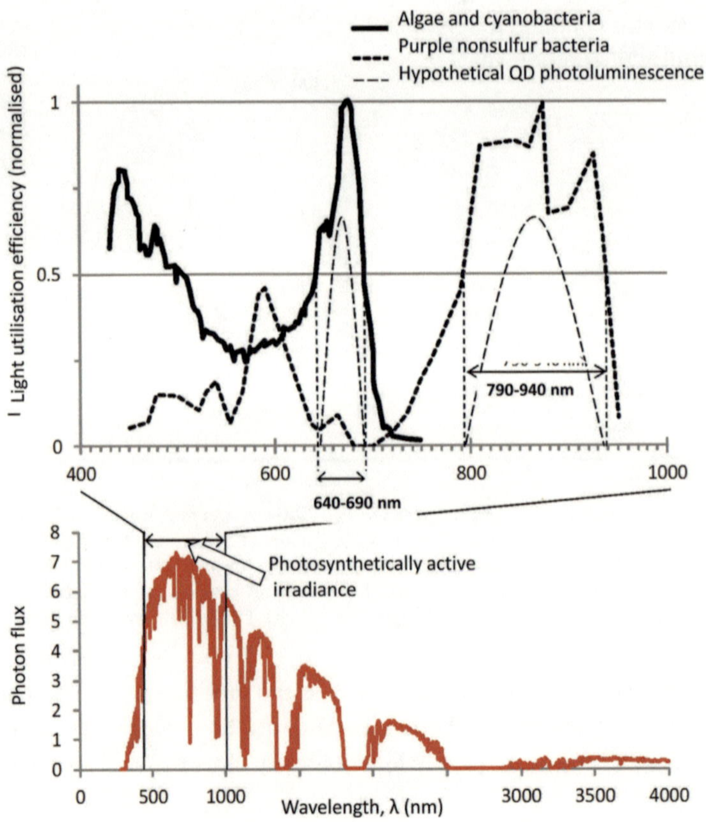

Figure 4. Action spectra and identification of targets for spectral enrichment. Generic action spectra were adapted from [35]. Note that action spectra differ substantially from whole-cell absorption spectra, which show strong wavelength-nonspecific attenuation due to the scattering effect of cells. Above this, small peaks associated with the absorption maxima of chlorophylls can usually be detected. The emission of the desired quantum dots is shown by the dotted line; ideally the emission peak should be narrow and overlap with the absorption maxima of the algal and bacterial chlorophylls.

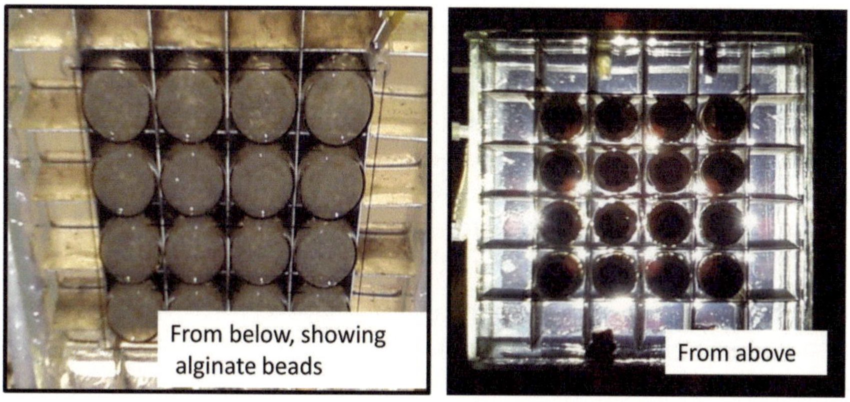

Figure 5. Boosting photosynthetic activity of *Rhodobacter sphaeroides*. Qdot'792 (to 20 nm) was encapsulated in 2% alginate beads (diam. 2.6 ±0.04 mm). Beads were prepared by mixing QD or blank (50 mM sodium borate buffer, pH 9.0) with 2% sodium alginate and dropping into 100 mM $CaCl_2$ through an 18G needle. After curing (60 min), the beads were washed with deionized water and used immediately. Concentrated cell suspension was diluted to 0.547 gdry weight/l with fresh butyrate medium [35]. Vials (4 ml bacterial suspension, 3 ml beads containing Qdot'792 and ~5 ml headspace) were sealed with gastight stoppers and purged with argon (30 min in darkness) and incubated (30°C, 3d, 10.0W/m^2 simulated sunlight). H_2 formation was measured as described previously [35]. Each vial contained 0.056 nmol Qdot'792 distributed over an illuminated surface of 3.14 cm^2. Optical dividers prevented optical interactions between vials and ambient light was excluded by covering the assembly with black cloth.

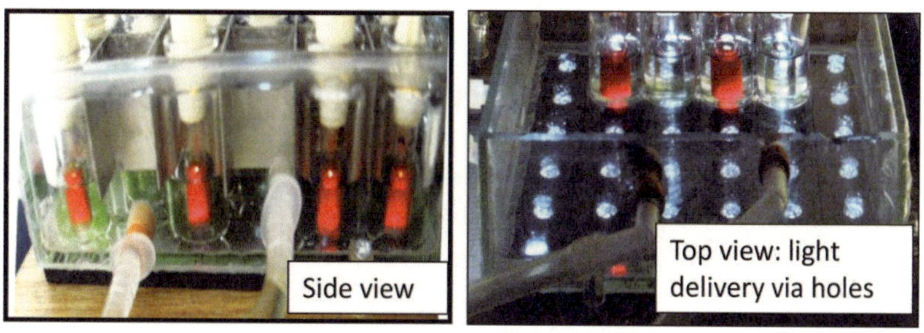

Figure 6. Boosting photosynthetic activity of *Arthrospira* ("spirulina") *platensis*. Growth was as in [35]. Inocula for photonic enhancement experiments were taken 2 days after subculturing (to ensure active growth) and diluted with fresh medium to an OD 660 of 0.364.4 ml was transferred into each vial. Cultures were illuminated by a close-match solar simulator undergrowth-limiting irradiance (10W/m^2) (supplementary material in [35]. Photosynthetic action was inferred from growth measured after 22 h [35]. To isolate optical effects, QDs were isolated in a glass insert, each containing 1 ml of Qdot'652 in 100 mM Na_2SO_3 with controls of QD-free Na_2SO_3. The inserts aligned with 10 mm openings in an opaque sheet to allow illumination from beneath with artificial sunlight. Ambient light was excluded by covering the assembly with black velvet and opaque barriers were placed around each mini-reactor. The temperature was controlled by circulating water at 30°C.

However, in practice, other than to make a common product in an integrated process, it may be impractical to co-locate both types of PBR due to other requirements; for example, organic acid feedstock for bacterial hydrogen production can be supplied by urban wastewater treatment plants [37], whereas algal biotechnologies typically require a large land area which can be waste or nonarable land. Hence, a generic method is required to "upgrade" solar light by converting unused wavelengths into used wavelengths for a particular process, thereby increasing the usable light and productivity without increasing biomass density and "self-shading". This forms a goal for photobiotechnology process intensification.

The wavelength dependence of photosynthesis by purple bacteria and microalgae has been known since the early twentieth century and confirmed many times in different species. As shown in **Figure 4**, green algae/cyanobacteria/higher plants show the greatest activity with red light, whereas purple bacteria are most active under near-infrared (NIR) [38–42]. The effect is so powerful that these organisms have developed an apparent "phototaxis" response, accumulating in the optically optimal part of a natural water column [38]. This ability is very important because red light is absorbed strongly by water, and hence, the availability of useful light drops markedly with depth. Blue light has a far greater penetration (see later).

The action spectra of photosynthetic microorganisms have been extensively surveyed. Green microorganisms (and plant chloroplasts) conform to a generic action spectrum, while purple bacteria conform to a distinctly different generic action spectrum (see [35] and **Figure 4**) attributable to the different chlorophylls evolved in the taxanomic groups.

4. Quantum dots as a potential means of "upgrading" light

One method of "upgrading" "waste" light of a particular wavelength is to use the light-emitting properties of quantum dots (QDs). QDs are single crystals of uniform size and shape of ~2–10 nm diameter and usually comprising pairs of semiconductors (e.g., CdSe, PbSe). QDs are replacing fluor dyes in cell biology due to their high brightness and photostability [43]. The properties and potential applications of QDs are described elsewhere in this volume, and indeed, QDs are commercially available in appropriate delivery systems for boosting horticulture and small-scale crop production [44] but have yet to find application in large-scale photobioreactor systems. However, for bioenergy applications and bulk-scale animal feed production, large scale constructions would be required (e.g., see **Figure 3b** and **Table 1**). Hence, a feasibility study was undertaken using the three microorganisms described above to indicate whether photosynthetic boosting via QDs is feasible for algal and bacterial growth systems. The use of LEDs to supply additional lighting at the optimal wavelengths is well-established technology [44, 45], and it is assumed to be intrinsically scalable, although a full cost-benefit analysis is required for applications in biofuels production.

4.1. Boosting of three photosystems using quantum dots

The concept of photonic enhancement is to increase the proportion of the solar spectrum that corresponds to the major peak(s) of the organismal action spectrum (**Figure 4**), at the expense of other irradiance at less active wavelengths. The part of the spectrum to be intensified is referred to as the target band. **Figure 4** (top panel) shows the boundaries of the target band corresponding to the half maximum of the major peak in the organismal action spectrum. Using generic action spectra derived previously [43], the target bands of 640–690 nm and 790–940 nm were determined for algae/cyanobacteria and purple bacteria, respectively. These bands account for 25 and 67% of the total action, respectively.

The study used test quantum dots purchased from Invitrogen: Qdot'792 (ITK carboxyl, no Q21371, lot 834,674; quantum yield (QY) 72%; full width height maximum (FWHM): 82 nm) and QD'652 (ITK carboxyl, no. Q21321MP, lot 891,174; QY 78%; FWHM 26 nm) for cultures of *R. sphaeroides*

PBR type	Dimensions; vol, m^3	Algae production kg dry solids/yr	Microalgae cost price (€/kg dry solids)	Ratio electricity*/variable cost# (%)
Open pond	1000 m^2; 0.03 m water depth; 300 m^3	1538	36	35
Tubular	1000 m^2; 0.06 m tube diam.; 45 m^3	3076	18	50
Flat panel	1000 m^2; 0.03 m plate spacing; 60 m^3	5127	12.50	68
Solar LEDs/flat panel*	1000 m^2; 0.03 m plate spacing: 60 m^3	12,818*	~ 4.20*	~30*

*Electricity is taken from the grid (@ €0.107/kWh) and includes "parasitic energy" consumption (pumps for culture and water heating/cooling circulation, centrifuge and blower for the flue gas supplying CO_2).

*Projection via use of solar cells, efficient battery technology and LED supplementary illumination. Calculations by R.L. Orozco (unpublished work).

#variable cost includes cost of water use, electricity (parasitic energy), labor, fertilizers (N & P) and waste water. V: culture volume.

Table 1. A comparative study [21] on algal (***Chlorella vulgaris***) cultivation technologies which include open pond and closed photobioreactors (PBRs: Tubular and flat panels) and economics of algal biomass production. The high productivities of flat panels compared to the other systems are reflected in the lower cost price.

and the cyanobacterium/green alga spirulina and *B. braunii*, respectively. The loading densities in the tests were 0.0178 (*R. sphaeroides*), 0.0792 (spirulina), and 0.050 (*B.braunii*) nmol/cm^2.

4.2. Experimental test systems

4.2.1. *Rhodobacter sphaeroides for biohydrogen production*

R. sphaeroides was used in a test system of mounted vials as shown in **Figure 5**, using simulated sunlight.

4.2.2. *Spirulina for biomass production*

Arthrospira platensis (spirulina) was used in a test system (**Figure 6**) using a close-match solar simulator (supplementary material in Redwood et al. [35]).

4.2.3. *B. braunii: A single-celled alga for bio-oil production*

B. braunii was cultured routinely in shake flask cultures. Q dot'652 was added directly into small 25 ml cultures to 10 nm. Cultures were shaken in a temperature-controlled greenhouse (average solar photon flux was 11 μmol/m^2/s). Photosynthetic action was inferred from growth at 21 days as estimated by OD_{600}.

4.3. Photosynthetic enhancement using commercial quantum dots

The first test, using *R. sphaeroides* to produce H_2, showed a photonic enhancement of ~10% (**Table 2**; **Figure 8c**). This was a close fit to the increase predicted by the known QD quantum efficiency and the QD loading/cm^2. Photonic enhancement of the growth of *A. platensis* (spirulina) doubled the biomass yield, which was ~25% higher than a predicted stimulation on the basis of quantum yield and QD loading density (**Figure 8d**). In this example, the QDs were held separate from the culture (**Figure 6**), which rules out stimulation via components leaching from the QD preparation. Finally, using *B. braunii* with QDs added directly into the culture and incubated in sunlight, the biomass yield was increased by 2.4-fold (**Figure 8b**). This photonic enhancement, 2.4-fold with respect to optical density, was >50% higher than that predicted on the basis of quantum yield and loading. A growth stimulatory effect of contaminants was largely ruled out on the basis of the test using spirulina, which was held separate from the cells (above). However, as *B. braunii* becomes heavily loaded with oil globules during growth (**Figure 7**), their contribution to increasing the size (and hence OD_{600}) of the cells cannot be precluded. The effect of photosynthetic enhancement on oil production was not examined in this study. It was concluded that the use of QDs as photonic enhancers has potential, but the light emission from the commercial QDs was not at the ideal wavelength (**Figure 4**), while the high cost of commercial QDs would currently be prohibitive in large scale systems, although the potential cost reduction at bulk scale is not known.

Microbial group	Organism	QD-free controls	Experiment with QDs*	Photonic enhancement*
Purple nonsulfur bacteria	*Rhodobacter sphaeroides*	15.54 ± 0.31 (13)	17.00 ± 0.16 (13)	1.1-fold
Cyanobacteria	*A. platensis*	0.025 ± 0.002 (6)	0.052 ± 0.011 (4)	2.1-fold
True algae#	*B. braunii*	0.106 ± 0.032 (3)	0.251 ± 0.011 (3)	2.4-fold

Data are means ± SEM for the number of experiments shown in parentheses. Photonic enhancements are modest due to the low dose of QDs used but were in accordance with theoretical predictions. The maximum enhancement was not tested. Enhancements are statistically significant at P = 0.95.

*The criterion for the algae was biomass content/ml (OD_{600}) that for *R. sphaeroides* was production of hydrogen.

#Data from shake flask tests.

Table 2. Photosynthetic boosting using quantum dots.

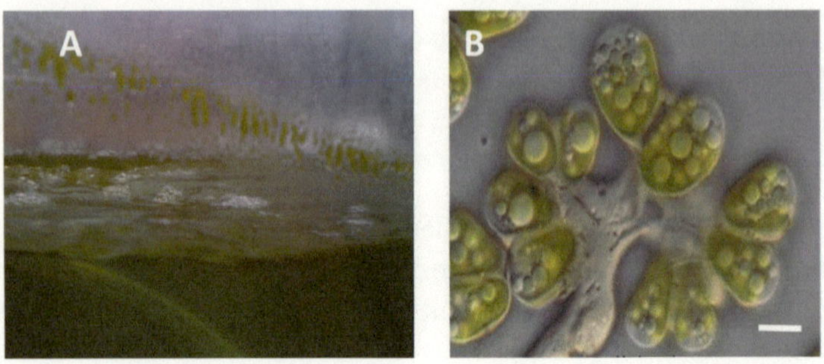

Figure 7. *Botryococcus braunii*. (A) Twenty-day-old *Botryococcus braunii* culture in uplift photobioreactor. (B) Bright field image of *Botryococcus braunii*, Race B. Pyriform *B. braunii* cells held together by a hydrocarbon-polysaccharide matrix. Oil containing vesicles are clearly visible inside the cells, which contain a single chloroplast. Images were acquired using an Olympus BX51 System Microscope with an attached DP71 digital CCD camera. Image processing and analysis software used was Cell F version 2.8 from Olympus).The scalebar represents 5 μm.

http://dx.doi.org/10.5772/intechopen.74032

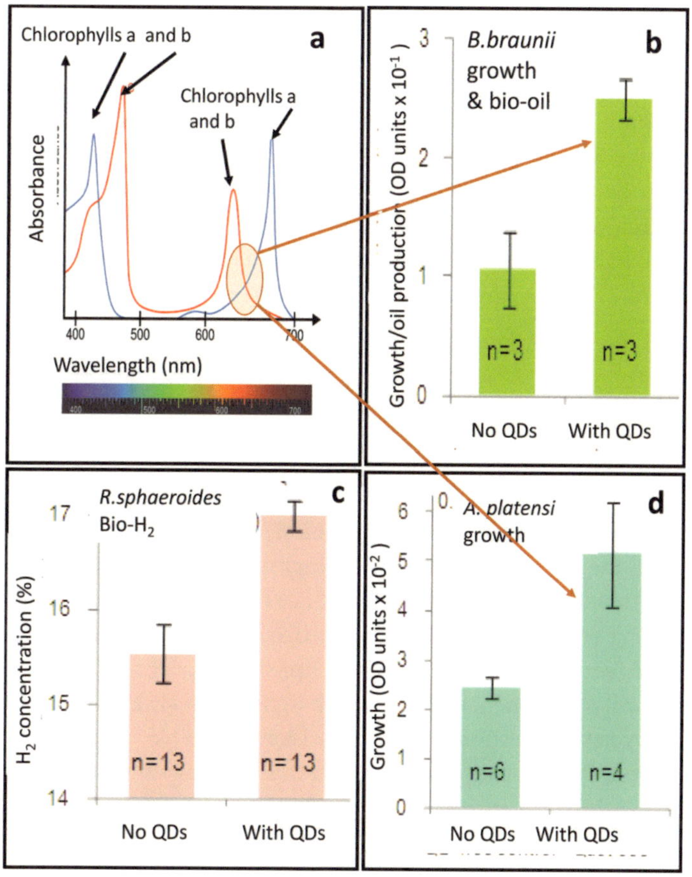

Figure 8. Enhancement of photosynthetic activity using QDs in three test systems. (a) Spectrum showing absorbance of chlorophylls a and b, and QD emission. Red emission was used (circled) for the algae and near infra red emission for *R. sphaeroides* (not shown). (b), (d) Photoproductivity of *B. braunii* (b) and *A. platensis* (d) with QDs (c) Hydrogen production by *R. sphaeroides* with QDs. N is the number of tests in each case.

5. Factors affecting development of quantum dots for enhancement of photosynthetic biomass and biofuel processes

This feasibility study (carried out in 2012) was limited by the suitable but nonideal properties of commercially available QDs. Small shifts in the emission peaks could improve the photonic enhancement, particularly in purple bacteria.

For cyanobacteria and algae, the available Qdot'652 was suitable as the major peak in the action spectra of cyanobacteria and algae occurs inside 640–690 nm (**Figure 4**), while the main part of the emission of Qdot'652 was within 639–665 nm (boundaries placed at half-maximum). Further development would aim to adjust the emission peak to ~655 nm while maintaining high quantum yield (QY) and full width height maximum (FWHM) ≤35 nm. For purple bacteria, the major peak in the action spectrum occurs inside the 790–940 nm region (**Figure 4**), while the main part of the emission of Qdot'792 was within 751–833 nm (boundaries placed at half-maximum). Therefore, a significant fraction of the emission fell outside the target band. Further development using this method would aim to adjust the emission peak to ~855 nm while maintaining high QY and FWHM ≤83 nm, thereby placing almost all QD emissions within the target band. For cyanobacteria and algae, this would require a FWHM for the QD of ≤35 nm, which matches the manufacturer specifications for Qdot'655. However, for purple bacteria, this could be more challenging as the longer wavelength emitting QDs typically produce broader emission peaks. However, the major action peak for purple bacteria is also broad (770–940 nm; half-maximum; **Figure 4**), suggesting an ideal emission peak of ~855 nm with FWHM ≤83 nm, which is similar to published specifications.

The discussion does not consider other potential impacts of QDs on the photobiological apparatus. The absorbance of less useful solar wavelengths by QDs could protect against damage from heat and UV irradiation, a benefit that would not be apparent from the experiments described here, as the temperature was actively controlled and much of the UV element of sunlight was absorbed by several layers of glass before reaching the QDs or the culture.

In this feasibility study, a single type of QD was selected to align as closely as possible with the major action peak of the organism. Further development could combine different QDs to further enrich the solar spectrum, according to the minor action peaks (**Figure 4**). There is also further potential in using combinations to further enrich the spectrum at ~680 or ~850 nm above the model presented here. Nature has evolved complex but optimal systems, for example, purple bacteria have ancillary pigments which absorb light in the visible region (e.g., 400–500 nm) and transfer energy very rapidly onto the bacteriochlorophylls [46, 47]. A biomimicry approach could use alternative QDs to construct a spectrum that precisely mirrors the action spectrum.

One important technical factor affecting practical photonic enhancement would be the stability of QDs. QDs can be affected by photobleaching [48], and the leaching of QD degradation products could have a potential negative environmental impact. Therefore, further investigations should focus on QD immobilization methods, aiming to make QDs a permanently encapsulated part of a photobiological installation and to enable low-risk handling in large quantities and recovery for multiple uses.

Finally, since the photobiotechnologies envisaged here would necessarily be at large scale to supply energy carriers for replacement fuels (as compared to the relatively small PBRs used for high-value products), maximum light transfer from sunlight to the QDs and light upgrading to the cells is essential, while maintaining a minimum QD loading for economy.

Color	Wavelength, nm	Percentage absorbed in 1 m of water (%)
Violet	400	4.2
Blue	475	1.8
Green	525	4.0
Yellow	575	8.7
Orange	600	16.7
Red	725	71.0
Infrared	800	82.0

Adapted from [49]

Table 3. Light attenuation in water.

However, given poor penetration of the red component of sunlight in water (**Table 3**), it is apparent that a deep QD-reactor system with irradiation from above would be unsuitable for purple bacteria as they use red-infrared light. An algal system is less sensitive to culture depth, as it can utilize blue light; the loss of light at 655 nm was calculated to be ~30%, which would still be a factor to consider in photobioreactor design. However in the blue region, corresponding to an absorption maximum of chlorophyll b (575 nm: **Figure 4**), very little light is lost, while at 430 nm (optimum for chlorophyll a), the available light intensity is still acceptable with depth, meaning that "point" sources of QD light could be used (insets or roof panels).

6. Potential alternative strategy for economic production of quantum dots at scale

For incorporating quantum dots into photosynthesizing cultures, some forms of QD encapsulation or barrier method are likely to be required (see above), while the use of toxic materials *per se* is unattractive for manufacturing, even assuming that the QDs are held separate from the cells, are easily recovered and are re-usable. Given the high cost of commercial quantum dots, the possibility to use more traditional metallic-based semiconductors was revisited, since these can be made economically at scale, but the use of highly toxic metals such as Cd should still be avoided. The waste hydrogen sulfide off-gas from an (unrelated) bioremediation process was considered for use to promote the formation of zinc sulfide nanoparticles which are well-known QDs. Using a waste from a remediation process (which is, in itself, used to recover Zn and Cu from acidic mine wastes [50, 51]) is a paradigm example toward realizing a circular economy. The liquid minewater wastes are obtained via the activity of microorganisms that leach the metals out of ore residues and closed mines. They also lower the pH (by formation of sulfuric acid), and hence, they are acid-loving bacteria (acidophiles). The acidophilic bacteria are fed by using additional nutrients derived from an algal source, *Coccomyxa onubensis*, and hence, development of a method for enhancing growth of this alga via a QD-enhancement approach would impact positively on the economics of the primary metal recovery process (combined metal bioleaching and recovery as metal sulfides), which produces excess waste H_2S from the activity of sulfate-reducing bacteria. These convert sulfate (dilute H_2SO_4) to sulfide, which is available to form ZnS quantum dots by combination with Zn^{2+} ions. This strategy was tested in principle.

Zinc sulfide has a bandgap varying from, in bulk material, 3.7 eV to, in nanoparticles, 4.2 eV [52, 53]. It has large exciton energy (~ 40 meV) and has been used in light-emitting diodes and, for example, flat panel displays [54]. The nanoparticles have to be stabilized during synthesis in order to minimize extensive agglomeration. This is important because the quantum yield is lower in larger particles [54].

Methods of QD nanoparticle synthesis commonly use organic solvent [55], capping agent, and/or surfactant in order to control agglomeration. These methods may introduce problems of reproducibility as well as complexity and cost, as well as leaving residual chemicals and hence being nonsustainable (see [56] for overview). Looking toward large scale manufacturing, various "traditional" methods could reduce the high cost of ZnS NP-synthesis. Khani et al. [57] incorporated 2-mercaptoethanol as a capping agent; Na_2S and mercaptopropionic acid have also been used [58]. Here, refluxing with tetrapropyl ammonium hydroxide resulted in QDs of nanoparticle size 4.5 nm, and the respective absorption and emission peaks were 315 and ~415 nm [58]. Other work reported QDs with absorbance and emission peaks at 279 and 435 nm, respectively; this method utilizes thiolactic acid with Zn^{2+} solution and Na_2S [59]. Being very close to the absorption peak of chlorophyll a at 430 nm (**Figure 4**), this raises the possibility to use ZnS NPs as a quantum dot ancillary to Qdot'655 (above) or, indeed as a substitute for the latter, using emitted blue light via the other absorbance region for chlorophyll a (see above).

A first report [60] showed that the characteristics and the light emitting properties of ZnS quantum dots made by use of bacterially made waste H_2S left over from the metal bioremediation process [50, 51] were comparable to those made by "classical" methods, which required more complex procedures. As a potential synthesis method at scale, this shows potential for commercial QD production and introduces the possibility to use these biogenic ZnS QDs to promote algal growth for the applications described above and also to provide algal feedstock as a nutrient source for other processes (e.g. high-value chemicals); algae as biomass feedstock *per se* for pyrolysis oil production has also been reported (e.g., [5]).

The price of commercial QDs discourages development above small-scale and a full-scale energy plant is probably currently unfeasible. However, QDs are rapidly developing from niche markets into consumer electronics [61], which is expected to bring substantial increases in production scale, and hence, reduction in cost may be expected in the future. From the tests and data shown above, the QD cost would have to be reduced by up to 100-fold in order for photonic photobioreactors to achieve parity with standard PBRs in terms of capital cost (M.D. Redwood, unpublished). This estimation was based on early published values for capital costs of different photobiological systems [62–65] and a survey (in 2012) of market prices for commercial QDs. Because open systems or raceways present much lower capital costs (per unit area) than enclosed PBRs, the estimated minimum QED cost reduction would be ~10 fold more attractive for raceways, suggesting that such enhancement would be first tested in PBRs then developed in raceways at scale as costs fall.

However, these estimations do not consider the reduced land requirement and reduced running costs of photonically enhanced photobioreactors, which may lessen the cost impact. On the other hand, end-of-life decommissioning may be more costly if potentially toxic metals have been used. QD retention via immobilization/encapsulation and re-use would be a key strategy. The extent to which biofouling of transparent surfaces in contact with the culture may impact adversely on QD-enhanced PBR useful life has not been taken into account (nor tested). Common methods to remove biofouling deposits (e.g., scraping) may damage surfaces that have been precision-machined or polished for optical transmission. Hence, an air gap between the QD enclosure and the culture liquid may prove beneficial. In practice, as long as there is sufficient stirring, the shear force is sufficient to prevent fouling problems. This means that sufficient shear force being produced by sparging of the PBRs can prevent the algae being able to settle on the (e.g., perspex) surface. However, if the perspex is scratched, then algae will adhere more readily. In some cases, fouling can be a major problem; some algal species are more adherent than others, but if the circulation in the PBR is sufficiently high, the algae will not adhere. Conversely, if the shear forces are too high, this may damage the algae. Most of the species that are grown commercially are fairly robust, but some species are shear sensitive; hence, this would need to be tested on a case by case basis (D. McKenzie, Xanthella Ltd., personal communication).

Based on this discussion, and the ease and potential scalability of bimanufacture of ZnS quantum dots, these were considered as a possible alternative to boost photosynthetic output via irradiation of chlorophyll a in the blue region, as an alternative to visible-red wavelengths, also noting the preferred use of blue light for deep culture (above). The emission of biogenic ZnS QDs prepared in 50 mM citrate buffer, pH 6, was reported at 410 nm [60], whereas the optimal absorbance wavelength of chlorophyll a is ~430 nm (**Figure 4**); respective molar extinction coefficients at 410 and 431.66 nm were calculated as 70,733 and 110,789 cm^{-1}/M, respectively [66, 67]. At ~425 nm, this was given as 93,099 and 98,874 cm^{-1}/M at 424.8 and 426.15 nm, respectively [66]. This illustrates the need to redshift the QD emission of the biogenic ZnS QDs by up to 15–20 nm to realize the full potential.

7. Toward realizing useful quantum dots from biogenic ZnS

The early study [60] used 50 mM citrate buffer (pH 6) to prevent uncontrolled precipitation of ZnS by chelating the Zn^{2+} in solution and acting as a passivant for the ZnS nanoparticles.

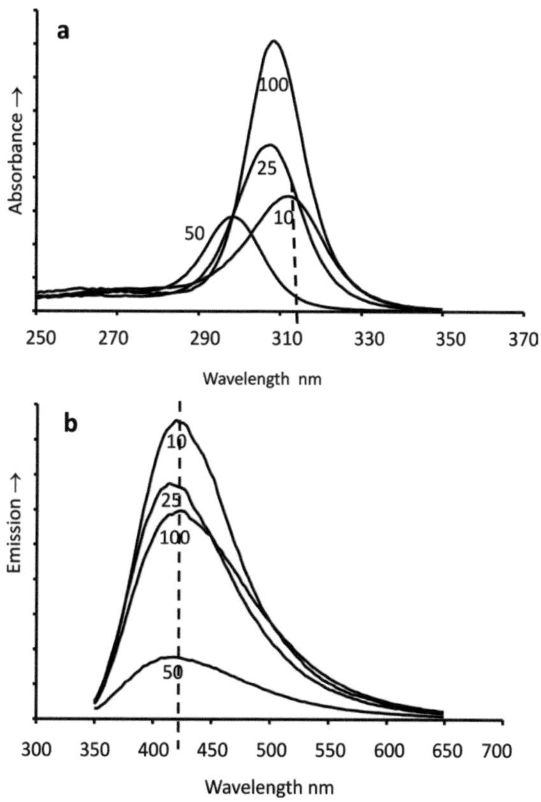

Figure 9. Excitation (a) and Emission (b) Spectra of ZnS Quantum dots. QDs were synthesized using excess waste biogenic H_2S from a metal bioremediation process [60]. Samples were sparged with the culture off-gas for 30 min (flow rate 132 ml/min). Example scans are shown. ZnS QDs were made in the presence of 10 mM, 25 mM, 50 mM and 100 mM citrate buffer (pH 6) as shown. Note excitation redshift to ~ 310 nm and emission from ~ 410 to ~425 nm (dotted lines) with decreased citrate concentration from 50 to 10 mM. Buffering at pH 6 without citrate (MES-NaOH buffer) gave material with negligible emission.

However, the use of citrate should be minimized for process economy. Omission of citrate or its substitution by 50 mM MES-NaOH buffer gave a ZnS nanomaterial with poor light emission at 410 nm. By using lower concentrations of citrate in preparation (10 and 25 mM), the light emission was observed to increase by up to 5-fold, together with a redshift from 410 to

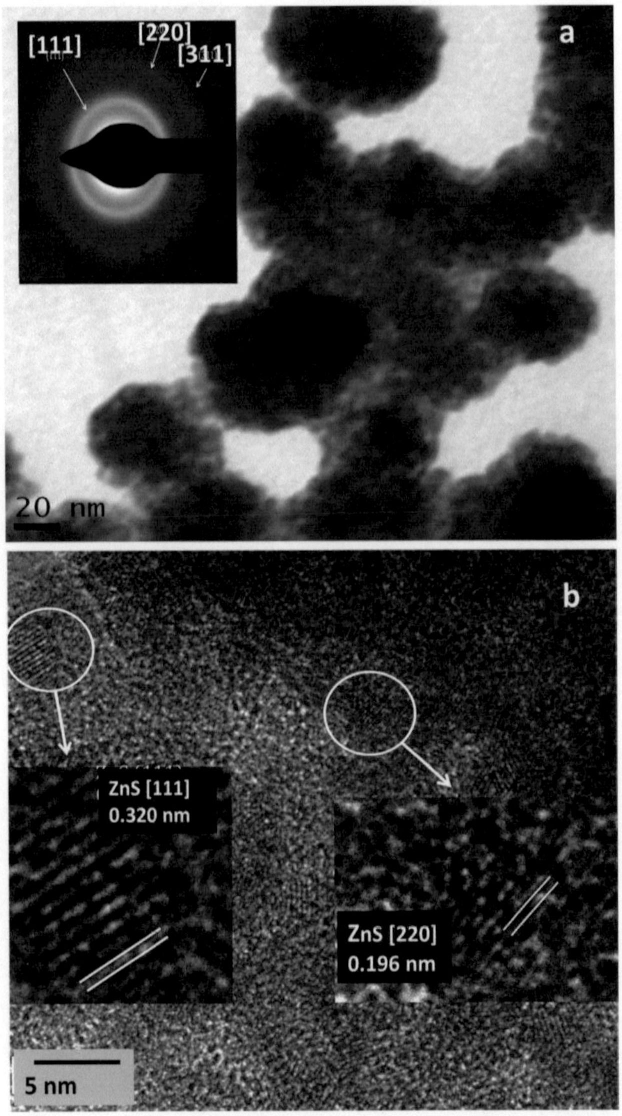

Figure 10. Transmission electron microscopy of ZnS Quantum Dots. QDs were synthesized in 50 mM citrate buffer, pH6 using biogenic H_2S [60]. Accelerating voltage was 80 kV which is optimal for contrast and shows (a) agglomerations of ~60 nm containing discrete small nanoparticles. (b) High-resolution TEM study (300 kV) of a single area of an agglomerations how nina.Lattice details are visible showing facets of ZnS:(111) and (220) corresponding to interplanar spacings of 0.320 nm and 0.196 nm. a (inset): Selected area diffraction of a single nanoparticle at accelerating voltage of 300 kV. The diffraction rings correspond to the (111), (220), and (311) facets of ZnS by reference to the JCPDS database. Calculations from Image J software, in collaboration with J. Gomez-Bolivar.

425 nm, i.e., into the absorption peak for chlorophyll a (**Figure 9**). Increasing the concentration of citrate (to 100 mM) during QD synthesis gave a similar effect; the reason for this was not investigated but future development would require the minimum amount of citrate. Further tests showed that further reduction of the citrate concentration to 7 mM retained the emission peak at 425 nm.

It is well known that increasing the size of quantum dots produces a redshift in the emission spectrum [68]. Hence, the ZnS QD material produced from the biogas from Zn^{2+} solution using high (50 mM) and low (7 mM) concentrations of citrate was examined using two methods: high resolution transmission electron microscopy (HRTEM) and differential centrifugation analysis for determining the size distribution of native nanoparticles.

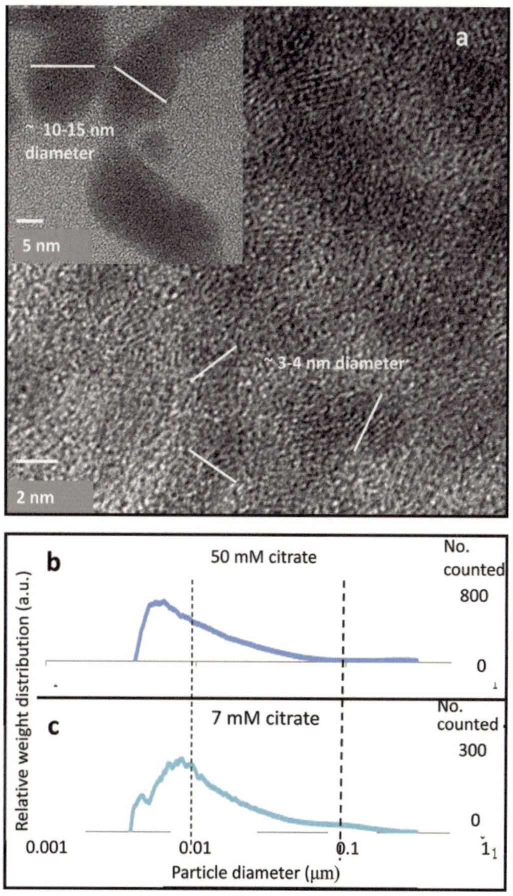

Figure 11. High resolution TEM study and population size analysis of ZnS QDs made under two citrate concentrations. Biogenic H_2S was used to make ZnS in 50 mM citrate buffer (a, main image) and 7 mM citrate buffer (a, inset). Approximate respective nanoparticle sizes are 3–4 and ~ 10 nm as shown. The population from 7 mM citrate solution had additional small nanoparticles of size ~5 nm (a, inset). Bars are 2 and 5 nm as shown. b,c: Estimation of nanoparticle sizes using an analytical disc centrifuge as described in [60] using light scattering. The nanoparticle size in 50 mM citrate buffer was calculated (from log plots) as ~4–5 nm and in 7 mM citrate buffer was ~ 8–10 nm with a smaller population of size <5 nm. The small nanoparticles are too close to the lower size cutoff of the instrument [60] to be an exact measurement, but the two independent methods report the same result. **Figure 11a** was in collaboration with J. Gomez-Bolivar.

Examination of the ZnS material (from 50 mM citrate) revealed agglomerations, within which small NPs were visible [60] (**Figure 10a**). Examination of these confirmed their identity as ZnS (**Figure 10a, b**). Use of 7 mM citrate produced larger nanoparticles (**Figure 11a**) consistent with the redshift observation in **Figure 9** (see above). Electron microscopy can produce artifacts due to drying [60]. Hence, independent confirmation was provided via analytical centrifugation of the liquid suspension in conjunction with light scattering (**Figure 11b, c**).

Examination of nanoparticle sizes by the two methods gave similar results (**Figure 11**). The nanoparticles made with 7-mM citrate were of size in the region of ~10-12 nm with a subpopulation of small NPs of 5 nm or less. In contrast, when made in the presence of 50 mM citrate, the population comprised small NPs of size 3–5 nm. Accurate sizing of the latter was precluded by the limitations of the analytical centrifugation method (see [60] for discussion), but it is clear that this simple method gives the potential to "steer" the ZnS QDs for size optimization.

8. Considerations for large scale process using ZnS quantum dots

A comparison was made of the cost/benefit analysis of electricity production based on the microalga *C. vulgaris* (**Table 1**). Open ponds are feasible and are used routinely, with process intensification achieved using raceways (see earlier), but the best photobioreactor format was concluded to be a flat panel arrangement which, although of volume of 20% of that of an open pond, gave >3-fold more biomass production for ~ twice the cost. (**Table 1**), also outperforming a tubular reactor arrangement. Consideration of algal growth for biofuels production has been reviewed elsewhere [69]. Calculations were made here based on using LEDs to boost light delivery, projecting a system of daytime solar irradiation via solar panels (parallel to the PBR) along with modern battery technology (energy storage) to permit LED-illumination (for culture 'tickover') at night (**Table 1**). A full cost-benefit and life cycle analysis is in progress, but here, it should be noted that ZnS quantum dots have also found application in solar cells [70, 71], giving further scope for cost reduction of solar panels which is not factored into **Table 1**. Similarly, **Table 1** does not take into account any increase in photoproductivity via use of QDs, and the values shown (made on the basis of published values) may reveal further benefits (e.g., 2-fold) via addition of the QD technology described here. Hence, the calculations shown in **Table 1** are taken to form a conservative "baseline."

Although immobilized QDs in suspension in a circulating reactor (raceways) might be appropriate for large scale, well-mixed growth, this may give challenges with respect to stability of the encapsulation material under shear, recovery of the QDs, and importantly, ensuring transparency to UV irradiation in the region of interest (absorbance maximum was ~ 310–315 nm with a peak emission at 425 nm; **Figure 4**). This redshift with respect to early work [60] is very relevant: UV light is divisible into UVA (315–400 nm), UVB (280–315 nm), and UVC (100–289 nm). The latter (the most damaging to living cells due to absorbance by DNA and proteins) is almost all absorbed by the atmosphere. Of the total UV radiance reaching the earth's surface, 85% comprises UVA, while only 5% comprises UVB; At 310 nm, the portion of available sunlight at the earth's surface is ~ half of that obtained at the maximum transmittance of light of ~625 nm [72]. Inspection of **Figure 9** shows that the larger QDs made in 7–10 mM citrate (**Figure 11**) absorb at the interface between UVA and UVB, and further work using lower concentrations of citrate may repay study to redshift both the absorbance and emission peaks by, ideally, a further 5–10 nm. The "preferred" UV light for irradiation also drives the choice of reactor materials (**Table 4**).

From **Table 4**, it is clear that several types of bulk materials would have potential application in large scale photobioreactor technology. LLDPE materials, transparent to both UVA and UVB, are used extensively in bottles and liquid sachets, and a simple approach might involve flotation or suspension of suitable sachet bags into various types of culture as shown in **Figure 2**. However, LLDPE polymers degrade in UV light with a useful life of only about 3 years [73] and would not provide a durable solution, although they would provide a route to easy separation of QDs for re-use. It is also routine to use Perspex™ for photobioreactor materials, e.g., for enclosed inserts. Indeed, Perspex™ is routinely used in numerous applications such as glazing, and its properties are well described, including hardness and scratch resistance; indeed, it is recommended for use as a flooring material [74]. Should scratches occur they can be easily polished out using a proprietary polishing material [74], although the degree of polishing required to achieve a near-perfect optical transmission for optimal use of quantum dots would need to be determined experimentally.

Material	Type	Examples	Transparency UVB	Transparency UVA
Building window materials	Glasses	Clear glass	Opaque	Transparent
		Reflective glass		
		Tinted/wire		
		Tinted glasses		
Nonwindow materials		Quartz glass	Transparent	Transparent
		Perspex	Perspex is the best.	Perspex is the best.
		Furniture glass[1]		
Transparent linear low density polyethylene (LLDPE)	Liquid	Blue crystal	Transparent	Transparent
	Storage bags	"Acqua fil" "Cool Pak"		
	(Sachet bags)	"Ahenpon", "Kenro"		
Polyethylene terephthalate	Plastic bottle	"Standard water"	Opaque	Transparent
	Containers for liquid storage	"Voltic", "Ice Pak"		
		"BelAquah"		
		"Coca Cola"		

Table 4. UV transparency of some common materials used in bulk applications.

9. Conclusions

Photobiotechnologies are maturing rapidly from small-scale high-value applications to large scale operations for biofuels. The major challenge remains optimal use of solar light, since photosynthesis is intrinsically inefficient and effective solar-biotechnologies are currently limited geographically to areas of high and constant solar irradiance. LEDs are already used to supply light into photobioreactors, but their use at large scale requires a careful cost-benefit analysis, especially with regard to the overall energy balance and especially if the biomass is used to make biofuels. Quantum dot technologies, until now used at small scale for niche applications such as imaging, are entering the global commodity market, but traditional QDs are costly. We have shown that commercial QDs can be used to double the photoproductivity, and we also show an economic route to QD manufacture via harnessing a waste from another biotechnology process into the QD manufacturing without compromising quality or performance.

Acknowledgements

We acknowledge the support of NERC (Grant No NE/L014076/1) in the research presented here (AJM and RLO) to develop ZnS-based quantum dots technology via resource recovery from waste. The underpinning evaluation of commercial quantum dots in the three test photobiological systems was supported by the Discipline Hopping Award scheme co-funded by EPSRC, BBSRC, and MRC. The support of BBSRC is acknowledged for MRes studentships (AG and FW). We also thank Drs D.J. Binks and M. Dickinson of the Photonics Institute, University of Manchester, for collaborations, photonics expertise and hosting the secondment of Dr. M.D. Redwood. We acknowledge with thanks the discussions and collaboration with Mr. J. Gomez-Bolivar and Dr. M. Merroun of University of Granada, Spain.

Author details

Angela Janet Murray[1], John Love[2], Mark D. Redwood[1], Rafael L. Orozco[1], Richard K. Tennant[2], Frankie Woodhall[1], Alex Goodridge[1] and Lynne Elaine Macaskie[1*]

*Address all correspondence to: l.e.macaskie@bham.ac.uk

1 School of Biosciences, University of Birmingham, Edgbaston, Birmingham, UK

2 Biocatalysis Centre, University of Exeter, Exeter, UK

References

[1] Hillen LW. Hydrocracking of the oils of *Botryococcus braunii* to transport fuels. Biotechnology and Bioengineering. 1982;**24**:193-205. DOI: 10.1002/bit.260240116

[2] Alfstad T. World Biofuels Study Scenario Analysis of Global Biofuel Markets. BNL-80238-2000. New York NY, USA: Brookhaven National Laboratory; 2008, 2008

[3] Anon. Ethanol Producer Magazine. Global Renewable Fuels Alliance; 2017. Available from: http://www.ethanolproducer.com/articles/14714/grfa-iea-report-highlights-continued-demand-growth-for-biofuels. www.ethanolproducer.com/articles/14714/grfa-iea-report-highlights-continued-demand-growth-for-biofuels.2017. [Accessed December 19, 2017]

[4] Metzger P, Largeau C. *Botryococcus braunii*: A rich source for hydrocarbons and related ether lipids. Applied Microbiology and Biotechnology. 2005;**66**:486-496. DOI: 10.1007/s00253-004-1779-z

[5] Kunwar B, Deilami SD, Macaskie LE, Wood J, Biller P, Sharma BK. Nanoparticles of Pd supported on bacterial biomass for hydroprocessing crude bio-oil. Fuel. 2017;**209**:449-456. DOI: 10.1016/j.fuel.2017.08.007

[6] Qishen P, Baojiang G, Kolman A. Radioprotective effect of extract from *Spirulina platensis* in mouse bone marrow cells studied by using the micronucleus test. Toxicology Letters. 1989;**48**:165-169. DOI: 10.1016/0378-4274(89)90171-9

[7] Samarth RM, Panwar M, Kumar M, Soni A, Kumar M, Kumar A. Evaluation of anti-oxidant and radical-scavenging activities of certain radioprotective plant extracts. Food Chemistry. 2008;**106**:868-873. DOI: 10.1016/j.foodchem.2007.05.005

[8] Mumtaz T, Yahaya NA, Abd-Aziz S, Abdul Rahman NA, Yee PL, Shirai Y, Hassan MA. Turning waste to wealth-biodegradable plastics polyhydroxyalkanoates from palm oil mill effluent – A Malaysian perspective. Journal of Cleaner Production. 2010;**18**:1393-1402. DOI: 10.1016/j.jclepro.2010.05.016

[9] Redwood MD, Orozco R, Majewski AJ, Macaskie LE. Electro-extractive fermentation for efficient biohydrogen production. Environmental Science and Technology. 2012;**107**:166-174. DOI: 10.1016/j.biortech.2011.11.026

[10] Redwood MD, Orozco R, Majewski AJ, Macaskie LE. An integrated biohydrogen refinery: Synergy of photofermentation, extractive fermentation and hydrothermal hydrolysis of food wastes. Hydrolysis of Food Wastes. 2012;**119**:384-392. DOI: 10.1016/j.biortech.2012.05.040

[11] Akkerman I, Janssen M, Rocha J, Wijffels RH. Photobiological hydrogen production: Photochemical efficiency and bioreactor design. International Journal of Hydrogen Energy. 2002;**27**:1195-1208. DOI: 10.1016/S0360-3199(02)00071-X

[12] Anon. Sustainable Biofuels: Prospects and Challenges. London UK: Typesetting Clyvedon Press, Cardiff UK: The Royal Society; 2008. ISBN: 9780854036622

[13] van Beilen JB. Why microalgal biofuels won't save the internal combustion machine. Biofuels Bioproducts and Biorefining. 2010;**4**:41-52. DOI: 10.1002/bbb.193

[14] Lizzul AM, Allen MJ. In: Love J, Bryant JA, editors. Biofuels and Bioenergy. pp. 191-211. Wiley on Line Library ISBN: 9781118350560. 2017. pp. 191-211. Ch12. DOI: 10.1002/9781118350553

[15] Lea-Smith DJ, Bombelli P, Dennis JS, Scott SA, Smith AG, Howe CJ. Phycobilisome-deficient strains of *Synechocystis* sp. PCC 6803 have reduced size and require carbon-limiting conditions to exhibit enhanced productivity. Plant Physiology. 2014;**165**:705-714. DOI: 10.1104/pp.114.237206

[16] Suh IS, Lee SB. A light distribution model for an internally radiating photobioreactor. Biotechnology and Bioengineering. 2003;**82**:180-189. DOI: 10.1002/bit.10558

[17] Lee CG. Calculation of light penetration depth in photobioreactors. Biotechnology and Bioprocess Engineering. 1999;**4**:78-81

[18] Richmond A. Biotechnology at the turn of the millennium: A personal view. Journal of Applied Phycology. 2000;**12**:441-451. DOI: 10.1023/A:1008123131307

[19] Carvalho AP, Silva SO, Baptista JM, Malcata FX. Light requirements in microalgal photobioreactors: An overview of biophotonic aspects. Applied Microbiology and Biotechnology. 2011;**89**:1275-1288. DOI: 10.1007/s00253-010-3047-8

[20] Lee YK. Microalgal mass culture systems and methods: Their limitations and potential. Journal of Applied Phycology. 2001;**13**:307-315. DOI: 10.1023/A:1017560006941

[21] Voort MPJ, van der, Vulsteke E, Visser CLM de. Macro-economics of algae products, public output report WP2A7.02 of the EnAlgae project, Swansea. 2015. DOI: http://edepot.wur.nl/347712

[22] Rosenberg JN, Guzman BJ, Oh VH, Mimbela LE, Ghassemi A, Betenbaugh MJ, Oyler GA, Donohue MD. A critical analysis of paddlewheel-driven raceway ponds for biofuel production at commercial scales. Algal Research. 2014;**4**:76-88. DOI: 10.1016/j.algal.2013.11.007

[23] Ugwu CU, Aoyagi H. Microalgal culture systems: An insight into their designs, operation and applications. Biotechnology. 2012;**11**:127-132. DOI: 10.3923/biotech.2012.127.132

[24] Subashchandrabose SR, Ramakrishnan B, Megharaj M, Venkateswarlu K, Naidu R. Consortia of cyanobacteria/microalgae and bacteria: Biotechnological potential. Biotechnology Advances. 2009;**29**:896-907. DOI: 10.1016/j.biotechadv.2011.07.009

[25] de Godos I, Blanco S, García-Encina PA, Becares E, Muñoz R. Long-term operation of high rate algal ponds for the bioremediation of piggery wastewaters at high loading rates. Bioresource Technology. 2009;**100**:4332-4339. DOI: 10.1016/j.biortech.2009.04.016

[26] Bwapwa JK, Jaiyeola AT, Chetty R. Bioremediation of acid mine drainage using algae strains: A review. South African Journal of Chemical Engineering. 2017;**24**:62-70. DOI: 10.1016/j.sajce.2017.06.005

[27] Borowitzka MA. Culturing microalgae in outdoor ponds. In: Anderson RA, editor. Algal Culturing Techniques. Amsterdam: Elsevier Academic Press. 2005. pp 205-218. DOI: 10.1016/B978-012088426-1/50015-9

[28] Chisti Y. Biodiesel from microalgae. Biotechnology Advances. 2007;**25**:294-306. DOI: 10.1016/j.biotechadv.2007.02.001

[29] Eriksen NT. The technology of microalgal culturing. Biotechnology Letters. 2008;**30**: 1525-1536. DOI: 10.1007/s10529-008-9740-3

[30] Glemser M, Heining M, Schmidt J, Becker A, Garbe D, Buchholz R, Brück T. Application of light-emitting diodes (LEDs) in cultivation of phototrophic microalgae: Current state and perspectives. Applied Microbiology and Biotechnology. 2016;**100**(3):1077. DOI: 10.1007/s00253-015-7144-6

[31] Lee CG, Palsson B. High density algal photobioreactors using light emitting diodes. Biotechnology and Bioengineering. 1994;**44**:1161-1167. DOI: 10.1002/bit.260441002

[32] Matthijs HC, Balke H, van Hes UM, Kroon BM, Mur LR, Binot RA. Application of light-emitting diodes in bioreactors: Flashing light effects and energy economy in algal culture (*Chlorella pyrenoidosa*). Biotechnology and Bioengineering 1996;**50**:98-107. DOI: 10.1002/(SICI)1097-0290(19960405)50:1<98::AID-BIT11>3.0.CO;2-3

[33] Sorokin C, Krauss RW. The effects of light intensity on the growth rates of green algae. Plant Physiology. 1958;**33**:109-113. DOI: 10.1104/pp.33.2.109

[34] Vonshak A, Richmond A. Problems in developing the biotechnology of algal biomass production. Plant and Soil. 1985;**89**:123-135. DOI: 10.1007/BF02182239

[35] Redwood MD, Dhillon R, Orozco R, Zhang X, Binks DJ, Dickinson M, Macaskie LE. Enhanced photosynthetic output via dichroic beam-sharing. Biotechnology Letters. 2012; **34**:2229-2234. DOI: 10.1007/s10529-012-1021-5

[36] Redwood MD, Paterson-Beedle M, Macaskie LE. Integrating dark and light bio-hydrogen production strategies: Towards the hydrogen economy. Reviews in Environmental Science and Bio/Technology. 2009;**8**:149-185. DOI: 10.1007/s11157-008-9144-9

[37] Stephen AJ, Archer SA, Orozco RL, Macaskie LE. Advances and bottlenecks in microbial hydrogen production. Microbial Biotechnology. 2017;**10**:1120-1127. DOI: 10.1111/1751-7915.12790

[38] Chen H-B, Wu J-Y, Wang C-F, Fu C-C, Shieh C-J, Chen C-I, Wang C-Y, Liu Y-C. Modeling on chlorophyll a and phycocyanin production by *Spirulina platensis* under various light-emitting diodes. Biochemical Engineering Journal. 2010;**53**:52-56. DOI: 10.1016/j.bej.2010.09.004

[39] Chen SL. The action spectrum for the photochemical evolution of oxygen by isolated chloroplasts. Plant Physiology. 1952;**27**:35-48. DOI: 10.1104/pp.27.1.35

[40] French CS. The rate of CO_2 assimulation by purple bacteria at various wave lengths of light. Journal of General Physiology. 1937;**21**:71-87. DOI: 10.1085/jgp.21.1.71

[41] Haxo FT, Blinks LR. Photosynthetic action spectra of marine algae. Journal of General Physiology. 1950;**33**:389-422. DOI: 10.1085/jgp.33.4.389

[42] Nogi Y, Akiba T, Horikosji K. Wavelength dependence of photoproduction of hydrogen by *Rhodopseudomonas rutila*. Agricultural and Biological Chemistry. 1985;**49**:35-38. DOI: 10.1080/00021369.1985.10866684

[43] Michalet X, Pinaud FF, Bentolila LA, Tsay JM, Doose S, Li JJ, Sundaresan G, Wu AM, Gambhir SS, Weiss S. Quantum dots for live cells, in vivo imaging, and diagnostics. Science. 2005;**307**:538-544. DOI: 10.1126/science.1104274

[44] Anon. Availlable from: https://www.led-professional.com/resources-1/articles/more-efficient-plant-growth-with-quantum-dots-by-nanoco-lighting. [Accessed 22 December 2017]

[45] Coppack DL. The effect of changes in irradiance on the growth, biomass, lipid accumulation and pigment composition of *Botryococcus braunii*. MRes Thesis, University of Exeter, UK. 2013

[46] Frank HA, Polivka T. In: Hynter CN, Daldal F, Thurnauer MC, Beatty JT, editors. Energy Transfer from Carotenoids to Bacteriochlorophylls in the Purple Photosynthetic Bacteria. Dordrecht, The Netherlands: Springer. 2009. pp. 213-230. DOI: 10.1016/0005-2728(90)90194-9

[47] Saer RG, Blakenship RE. Light harvesting in phototrophic bacteria structure and function. Biochemical Journal. 2017;**474**:2107-2131. DOI: 10.1042/BCJ20160753

[48] Qin H, Meng R, Wang N, Peng X. Photoluminescence intermittency and photo-bleaching of single colloidal quantum dot Advanced Materials. 2017;**29**:1606923 DOI: 10.1002/adma.201606923

[49] Garrison TS, Ellis R. Oceanography: An invitation to Marine Science Cengage Learning 20 Channel Center St. Boston MA 02210, USA, 2016

[50] Ňancucheo I, Johnson DB. Selective removal of transition metals from acidic minewaters by novel consortia of acidophilic sulfidogenic bacteria. Microbial Biotechnology. 2012;**5**:34-44. DOI: 10.1111/j.1751-7915.2011.00285.x

[51] Ana Laura Santos AL, Johnson DB. The effects of temperature and pH on the kinetics of an acidophilic sulfidogenic bioreactor and indigenous microbial communities. Hydrometallurgy. 2017;**168**:116-120. DOI: 10.1016/j.hydromet.2016.07.018

[52] Huang J, Yang Y, Xue S, Yang B, Liu S, Photoluminescence SJ. Electroluminescence of ZnS:Cu nanocrystals in polymeric networks. Applied Physics Letters. 1997:2335-2337. DOI: 10.1063/1.118866

[53] Khosravi AA, Kundu M, Jatwa L, Deshpande K, Bhagwart UA, Sastry M, Kulkarni SK. Green luminescence from copper doped zinc sulphide quantum particles. Applied Physics Letters. 1995;**67**:2702-2704. DOI: 10.1063/1.114298

[54] Lai CH, Lu MY, Chen LJ. Metal sulfide nanostructures: Synthesis, properties and applications in energy conversion and storage. Journal of Materials Chemistry. 2012;**22**:19-30. DOI: 10.1039/C1JM13879K

[55] Johnston RJ, Wilcoxon JP. Metal Nanoparticles and Nanoalloys, Vol 3. London: Elvesier; 2012. DOI: 10.1016/B978-0-08-096357-0.00006-6

[56] Dixit N, Soni HP. Tuning optical properties of ZnS nanoparticles in micellar medium at different pH. Superlattices and Microstructures. 2014;**65**:344-352. DOI: 10.1016/j.spmi.2013.11.018

[57] Khani O, Rajabi HR, Yousefi MH, Khosravi AA, Jannesari M, Shamsipur M. Spectrochimica Part A. 2011;**79**:361-369. DOI: 10.1016/j.saa.2011.03.025

[58] Senthilkumar K, Ramamurthi K, Kalaivani T, Balasubramanian V. Indian Journal of Advances in Chemical Science. 2013;**2**:1-5

[59] Mandal A, Dandapat A, De G. Magic sized ZnS quantum dots as a highly sensitive and selective fluorescence sensor probe for Ag^+ ions. The Analyst. 2012;**137**:765-772. DOI: 10.1039/C1AN15653E

[60] Murray AJ, Roussel J, Rolley J, Woodhall F, Mikheenko IP, Johnson DB, Bolivar JG, Merroun ML, Macaskie LE. Biosynthesis of zinc sulfide quantum dots using waste off gas from metal bioremediation process. RSC Advances. 2017;**7**:21484-21481. DOI: 10.1039/C6RA17236A

[61] Anon 2016 Quantum dots are impacting the Emarket. Available from: https://www.printedelectronicsnow.com/issues/2016-0301/view_features/quantum -dots-are-impacting-the-market/46392. [Accessed 26 December 2017]

[62] Benemann JR. Hydrogen production by microalgae. Journal of Applied Phycology. 2000;**2**:291-300. DOI: 10.1023/A:100817511

[63] Melis A. Green alga hydrogen production: Progress, challenges and prospects. International Journal of Hydrogen Energy. 2202;**27**:1217-1228. DOI: 10.1016/S0360-3199(02)00110-6

[64] Modigell M, Holle N. Reactor development for a biosolar hydrogen production process. Renewable Energy. 1998;**14**:421-426. DOI: 10.1016/S0960-1481(98)00098-6

[65] Tredici MR, Zittelli GC, Benemann JR. A tubular integral gas exchange photobioreactor for biological hydrogen production: Preliminary cost analysis. In: Zaborsky OR, Benemann JR, Matsunaga T, Miyake J, San Pietro, editors. Biohydrogen. Amsterdam: Pergamon press, Elsevier; 1999. DOI: 10.1007/978-0-585-35132-2_48

[66] Prahl 2017. Optical Absorption and Emission Data of Chlorophyll a. Available from: http://omlc.org/spectra/PhotochemCAD/html/123.html

[67] Dixon JM, Taniguchi M, Lindsey JS. PhotochemCAD 2. A refined program with accompanying spectral databases for photochemical calculations. Photochemistry and Photobiology. 2005;**81**:212-213. DOI: 10.1111/j.1751-1097.2005.tb01544.x

[68] Prasad PN. Nanophotonics. Hoboken, NJ, USA: John Wiley & Sons; 2004. DOI: 10.1002/047 1670251.ch1

[69] Zittelli GC, Rodolfi L, Bassi N, Tredici M. In: Borowitzka MA, Moheimani N, editors. Photobioreactors for Microalgal Biofuel Production. In: Algae for Biofuels and Energy. New York, USA: Springer; 2013. pp.115-131 Ch7. DOI: 10.1007/978-94-007-5479-9_7

[70] Kim JY, Yang J, Yu JH, Baek W, Lee C, Son HJ, Hyeon T, Ko MJ. Highly efficient copper-indium-selenide quantum dot solar cells: Suppression of carrier recombination by controlled ZnS overlayers. ACS Nano 2015;**9**:11286-11295. DOI: 10.1021/acsnano.5b04917

[71] Labiadh H, Slah Hidouri S. ZnS quantum dots and their derivatives: Overview on identity, synthesis and challenge into surface modifications for restricted applications. Journal of King Saud University – Science. 2017;**29**:444-450. DOI: 10.1016/j.jksus.2016.12.001

[72] Sackey SS, Vowotor MK, Owusu A, Mensah-Amoah P, Tatchie ET, Sefa-Ntiri B, Hood CO, Atiemo SM. Spectroscopic study of UV transparency of some materials. Environment and Pollution. 2015;**4**:1-17. DOI: 10.5539/ep.v4n4p1

[73] Burgess G, Fernández-Velasco JG, Lovegrove K. Materials, geometry, and net energy ratio of tubular photobioreactors for microalgal hydrogen production WHEC 16 / 13-16 June 2006 – Lyon France

[74] Anon. Pespex™ for glazing. PXTD 236 13th Edition. Darwen, Lancs UK: Lucite Internaitonal Ltd.; Available from: http://www.qdplastics.co.uk/assets/docs/Acrylic/Perspex_Glazing_236.pdf

Section 2

Magnetic Quantum Dots

Chapter 10

Magnetization Dynamics in Arrays of Quantum Dots

Pablo F. Zubieta Rico, Daniel Olguín and
Yuri V. Vorobiev

Additional information is available at the end of the chapter

http://dx.doi.org/10.5772/intechopen.73008

Abstract

The possibility of preparing materials based on quantum dots with fine-tuned magnetic properties has opened up the door for designing new and more efficient devices where the interplay of different microscopic phenomena balances out in useful ways. Nevertheless, our knowledge of the precise interaction of complex objects built from a great number of such nanometric magnetic components is still limited. The investigation of the spin or magnetization dynamics in such materials represents an important opportunity to better comprehend and predict some missing pieces for the advancement of a great deal of promising technologies.

Keywords: magnetism, quantum dots, magnetization dynamics, arrays, macrospin

1. Introduction

Since quantum dots were first discovered [1] and later fabricated, they have attracted a great deal of attention given how, just as single atoms or simple molecules, they depict quantum behavior at the level of their electronic and optical properties, but at the same time allow their tuning as a function of their shape, size, and composition—reason that has led some to refer to them as *artificial atoms* [2].

Depending on whether quantum dots are made of a semiconductor, a metal, or another material, different intrinsic properties of a system based on them can be tailored to exhibit specific values or signatures. By themselves, single quantum dots are of prominent relevance and have found applications in diverse scientific and technological fields [3]. Nonetheless, it is really the properties of arrays of these particles which outline a unique landscape in the design space of new materials. Arrays of regularly ordered *magnetic quantum dots* (MQDs) provide an opportunity to develop materials with characteristics different from those exhibited by traditional solid-state systems.

Arrays—and more generally assemblies—of MQDs comprise all of those systems in which the magnetic nanoparticles are embedded in, dispersed into, or arranged over a different nonmagnetic medium—either a liquid or a solid, e.g., some polymeric material. More concisely, all of them could be simply classified as magnetic nanocomposite or hybrid materials, which depending on the spatial relative placement of the MQDs can form one-, two-, or three-dimensional architectures [4, 5].

Assemblies of MQDs are attractive for their rich spectra of potential technological applications that range from biomedical uses such as magnetic resolution imaging, magnetic hyperthermia, and drug delivery [3–7] to magnetic refrigeration and energy-harvesting devices [8]. Notwithstanding, the major driving forces behind the study of arrays of MQDs are related to the area of information storage technologies [9, 10]. In particular, there is a great hope of achieving through these magnetic storage media of ultrahigh densities [11–13], nonvolatile magnetic random-access memories (MRAM) [14, 15], and logic devices [16–18]. Moreover, the importance of these systems does not stop here; arrays of MQDs are also of special interest as model systems for better understanding interactions and transport processes of magnetic materials in general [19–22].

In this chapter, we present the results from several—mainly theoretical and numerical—works focused on the analysis of magnetization dynamics and interactions of a representative variety of two-dimensional systems of MQDs arrays. Throughout each of the sections, we will review some of the physical models that are used to study, predict, and understand the dynamics and how interparticle interactions affect the overall behavior of these systems. We will also take the opportunity for introducing some of the limitations that have hindered progress on the realization of some of the applications mentioned above and will discuss based on the aforementioned works what possible strategies could be followed in order to steer away the status quo in the field.

While it is not the goal of the current exposition to dive into the experimental aspects that entail the fabrication of arrays of MQDs, it is worth mentioning that numerous efforts have been devoted to the development of different synthetic pathways for MQD arrays, as it can represent a highly challenging task to prepare some of these systems in such a way that the targeted properties of the system are obtained within a desired precision in a predictable fashion. Indeed, at the characteristic length scale of MQDs, i.e., down to less than ten to a few hundreds of nanometers, variations on the shape, size, and distribution of MQDs can significantly impact the magnetic behavior of the whole array.

2. Theoretical modeling of the magnetization dynamics

In general, the starting point for the description of the magnetization dynamics of arrays of MQDs is to consider the relevant fields that modify the magnetization of each individual QD. If we denote the total effective field "felt" at every point in space as $\vec{H}_{eff}(\vec{r})$, then the equation of motion of the magnetization $\vec{m}(\vec{r})$ is the Landau-Lifshitz-Gilbert (LLG) equation [23, 24]:

$$\frac{d\vec{m}(\vec{r})}{dt} = -\gamma\vec{m}(\vec{r}) \times \vec{H}_{eff}(\vec{r}) + \frac{\alpha}{M_S}\left(\vec{m}(\vec{r}) \times \frac{d\vec{m}(\vec{r})}{dt}\right), \tag{1}$$

http://dx.doi.org/10.5772/intechopen.73008

where γ is the gyromagnetic ratio, M_S is the saturation magnetization of the material, and α is the viscous damping coefficient [25]. Here, the magnetization $\vec{m}(\vec{r})$ is normalized over M_S.

Equation (1) can be solved analytically in very few cases and needs, in general, to be integrated numerically, which is usually done by recasting it into the following form:

$$\frac{d\vec{m}(\vec{r})}{dt} = -\frac{\gamma}{1+\alpha^2}\left\{\vec{m}(\vec{r}) \times \vec{H}_{eff}(\vec{r}) + \frac{\alpha}{M_S}\left[\left(\vec{m}(\vec{r}) \cdot \vec{H}_{eff}(\vec{r})\right)\vec{m}(\vec{r}) - \vec{H}_{eff}(\vec{r})\right]\right\}, \quad (2)$$

also referred to as the Landau-Lifshitz equation or explicit form of LLG equation.

The total effective field $\vec{H}_{eff}(\vec{r})$ should be specified for each system of interest, but for most of the systems, it includes magnetic anisotropy contributions, interparticle or substrate-particle interactions, and external field terms.

Depending on the nature, shape, and size of the MQDs, there are two general approaches to solve Eq. (2). The first one, *micromagnetics*, [24] describes the spatial magnetization distribution in the regime where the MQDs are large enough for domain walls and vertices to appear. In this scenario, the interactions between pairs of intra-dot domains render quite involved and time-consuming the computational task of solving the LLG equation, requiring highly technical numerical expertise or specialized software to handle the computational complexity. The second approach, called *macrospin model* [24, 26], is appropriate to describe the magnetization time evolution of assemblies of small nanoparticles, such that each dot magnetization can be described by a single magnetic moment within the QD volume. What we mean by small is unfortunately system dependent, but some known data for spherical MQDs can be found in **Figure 1** as reference.

In the macrospin limit, the magnetization distribution of an array of MQDs becomes $\vec{m}(\vec{r}) = \sum_i \vec{m}_i \delta(\vec{r} - \vec{r}_i)$, where $\vec{m}_i$ denotes the monodomain magnetization of the *i*th particle in the array and $\delta(\vec{r} - \vec{r}_i)$ is the Dirac delta function. Plugging the above expression for $\vec{m}(\vec{r})$ into Eq. (2), we obtain

$$\frac{d\vec{m}_i}{dt} = -\frac{\gamma}{1+\alpha^2}\left\{\vec{m}_i \times \vec{H}_{eff}(\vec{r}_i) + \frac{\alpha}{M_S}\left[\left(\vec{m}_i \cdot \vec{H}_{eff}(\vec{r}_i)\right)\vec{m}_i - \vec{H}_{eff}(\vec{r}_i)\right]\right\}, \quad (3)$$

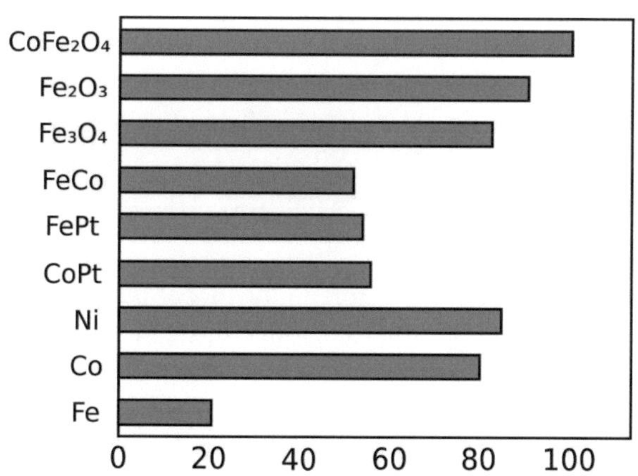

Figure 1. Threshold diameters for maximum monodomain size for spherical nanoparticles. Adapted from Majetich et al. [27].

which is an i-dimensional nonlinear system of ODEs whose solutions can be approximated by traditional numerical integration schemes, e.g., Heun's or Runge-Kutta methods.

Within the same limit, the total effective field $\vec{H}_{eff}$ can be usually written as

$$\vec{H}_{eff} = \vec{H}_{ani} + \vec{H}_{dem} + \vec{H}_{int} + \vec{H}_{ext}, \tag{4}$$

that is, the total field is the sum of the contributions from the magnetocrystalline anisotropy $\vec{H}_{ani}$, the demagnetizing field $\vec{H}_{dem}$, the particle interaction term $\vec{H}_{int}$, and the applied external magnetic field $\vec{H}_{ext}$.

Depending on the crystal structure of a ferromagnetic material, one or more privileged axes—known as *easy axes*—that energetically favor the alignment of the magnetization along their direction may exist. This energy contribution results from spin-orbit interactions and is referred to as magnetocrystalline anisotropy energy.

The demagnetization energy is the energy of the magnetization in the magnetic field created by the magnetization itself. This means that this energy contribution accounts for the dipole-dipole interaction of the elementary magnets.

For MQDs modeled as macrospins and in the absence of other interactions, the field $\vec{H}_{int}$ is simply the dipolar field induced by all of the particles, that is

$$\vec{H}_{int}(\vec{r}_i) = -\frac{VM_S}{4\pi}\sum_{j\neq i}\left(\frac{\vec{m}_j}{r_{ij}^3} - 3\frac{(\vec{m}_j \cdot \vec{r}_{ji})\,\vec{r}_{ji}}{r_{ji}^5}\right). \tag{5}$$

For the micromagnetic description of a system, Eq. (4) should also include the quantum-mechanical exchange interaction.

3. The effect of the geometry of MQD arrays

At this point, it should be clear that when analyzing the magnetic behavior of MQD arrays, one must take into account a lot of different factors. Just to mention a few, let us consider both the chemical composition and the crystalline structure of the nanoparticles themselves, as these factors determine—among other things—the magnetic anisotropy of each MQD; the shape of a single QD gives rise to a demagnetizing field; the spatial distribution among particles in an array establishes how particles will interact via pairs of dipole-dipole potentials; and, equally, any external factor that modifies the array, e.g., some applied external field, will impact the time response of the system.

Given the wide variety of contributing factors to the global magnetic properties of an array of MQDs, let's restrict our attention first to just one of them: the geometric distribution of particles in a two-dimensional assembly.

3.1. Single-domain limit

The possibility to synchronously manipulate the motion of all the magnetic moments in a cluster of magnetic nanoparticles in order to attain the fastest dynamic response in the presence of an external field while achieving some stability for the magnetic moment at the same time has motivated the study of the influence of the geometrical parameters—such as the inter-dot separation and crystal structure of arrays of MQDs—on the magnetic collective behavior of such systems.

In 2014, Meza et al. [28] reported an analysis of the coercivity fields for two-dimensional clusters of ellipsoidal cobalt nanoparticles in two different crystalline configurations (square and hexagonal) as a function of the inter-dot spacing and the frequency of a switching continuous external applied field. For this work oblate nanoparticles of semiaxis lengths of 3 nm × 3 nm × 1.5 nm were chosen. The easy axis of the particles, oriented in plane, was chosen parallel to the boundary of the cluster, and the external field was applied along the same direction. Given that these MQDs lie in the monodomain regime for cobalt, it is safe to assume that they can be properly described by the macrospin model.

By simulating hysteresis cycles for small clusters of cobalt nanoparticles (3 × 3, 5 × 5, and 10 × 10) at different frequencies for the applied oscillating external field (**Figure 2**), it was found that the hexagonal configuration stabilizes the magnetization reversal and narrows the coercivity with respect to the square crystal structure; this is clearly a consequence of the interaction strength promoted by the greater number of the nearest neighbors.

For both crystal systems, it was also observed that as the cluster size gets bigger the coercivity narrows at all frequencies and it is much lower than the coercivity for a single MQD. This is

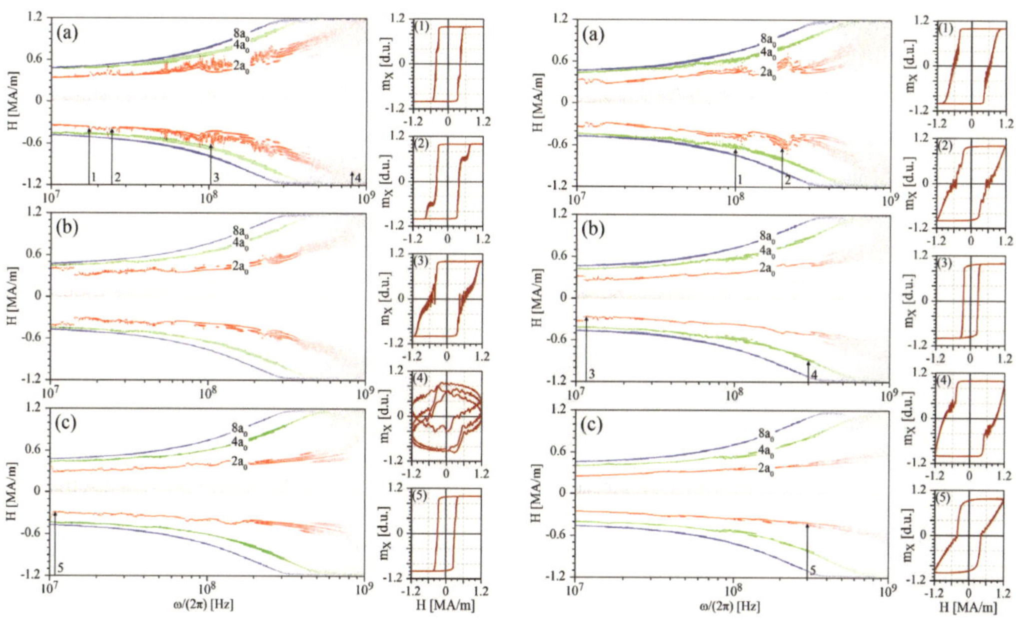

Figure 2. Coercivity histograms for macrospins arranged in a square (left) and hexagonal (right) lattice with different numbers of particles: (a) 3 × 3, (b) 5 × 5, and (c) 10 × 10. The distance between the particles is marked as $d = 2a0, 4a0,$ and $8a0$ (see [28]).

an interesting finding that could be exploited by trying to switch the magnetization only for of small subgroups of MQDs from a bigger array without affecting the rest of the magnetic system.

In all cases, by increasing the proximity of the nanoparticles, i.e., with a tighter packing of the assembly, and the intensity of the dipole interactions becomes more relevant, this in turn stabilizes the reversal at higher frequencies; so, in a macrospin system, the minimum of the switching time for a given frequency of an oscillating external field is limited by the packing of the nanoparticles.

As an additional note, it is noteworthy that experimentally, it is certainly possible to synthesize magnetic nanoparticles of these sizes by chemical methods. Nevertheless, the fabrication of colloidal nanoparticles and its self-assembly via solvent evaporation cannot yet attain the level of control that would be required to fabricate small uniform clusters of the sizes studied. On the other hand, lithographic techniques that allow a more refined control over the position, sizes, and number of particles in an array are still far from reaching the length scales of the system proposed.

3.2. Micromagnetics

To contrast the time evolution of a single-domain small-sized array, let us now consider the work done by Semenova et al. in 2013. In [29], the quasistatic hysteresis of close-packed arrays of NiFe nanodisks is studied. The motivation of this work is quite different from the one presented above and relates to the desire of understanding the phenomena in the field of *magnonics*, where there is a great interest in better grasping the propagation and confinement of spin waves in (among other materials) ordered assemblies of MQDs, which would have potential applications in the fabrication of reprogrammable crystals, magnetoelectronic devices, and metamaterials to name a few.

Although part of the work is focused on obtaining the spin-wave excitation spectra on the arrays of MQDs, we will only discuss the results on the role of the array geometry and packing over the hysteresis loops measured.

The specific set of MQD assemblies studied consisted in two two-dimensional hexagonal arrays of $Ni_{80}Fe_{20}$ nanodisks prepared by etched nanosphere lithography. The disks were of near 350 nm in diameter with an edge-to-edge separation of 65 and 12 nm in average for the two arrays, respectively. Accordingly, there is one tightly packed array, and another one is almost closed-packed array.

The system analyzed requires resolving over much larger distances than it is customary, making the problem of numerically predicting the magnetization dynamics a challenging exercise—mainly because of the complexity required to adequately consider periodic boundary conditions for the long-range interactions of the QDs.

The experimental results and magnetic simulations in [29] indicated that the overall hysteretic behavior in the system is dictated by the nucleation and escape of vortices within each nanodisk. Indeed, the observed sudden drop (see **Figure 3**) in the magnetization near an applied null external field is attributed to the formation of vortices near the center of the QDs; eventually, as the applied field pushes the system toward saturation, it manages to move around the center of every vertex to the edge of its disk. Similar phenomenology has been observed in square arrays of MQDs with some degree of anisotropy on the vortex nucleation [30, 31].

http://dx.doi.org/10.5772/intechopen.73008

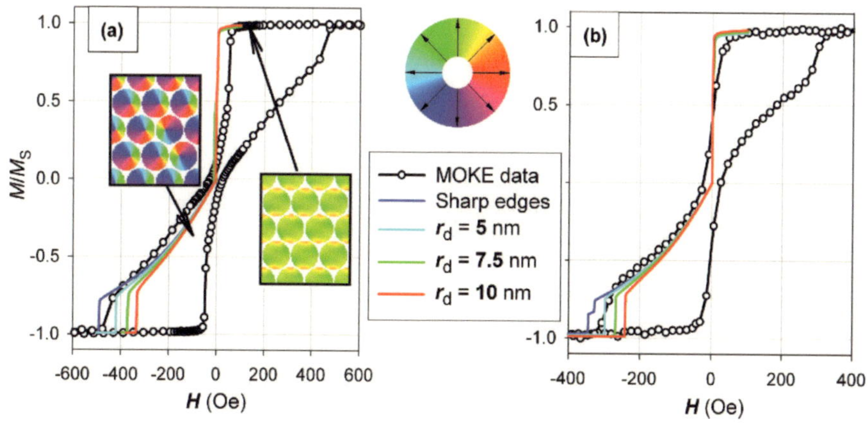

Figure 3. Simulated and measured hysteresis loops for the (a) the tightly packed array and (b) the almost closed-packed array from [29].

In the assemblies described above, the dipolar interactions play a less important role than the internal micromagnetics of each particle near almost null applied fields, and to some extent, the inter-dot dipolar interaction can be thought in an effective way contributing only to the displacement of the vertices out of the disks.

The lesson learned so far is that the net effect of the geometry of arrays of MQDs is of greater importance for monodomain systems where one of the dominant interactions is the dipolar one—for sufficiently packed systems—and that it enters in a much less important fashion for large systems where intra-dot micromagnetics dominate the magnetization reversal properties.

3.3. In between domains

To conclude the section, let us look one final assembly of MQD where the particle's size is of the order of other multi-domain nanostructures but for which the array geometry and the macrospin description might play an important part.

In 2010, Redondo et al. [32] reported experimental observations of the magnetization reversal mechanism for arrays of equally irregularly shaped MQDs aligned and distributed over an ordered square grid. The research team found that for different directions of an applied external field, the magnetization reversal would behave in some cases as if the switching mechanisms were dominated by nucleation and displacement of vortices, but in other directions, for the same system, the behavior would be that of a macrospin array.

For such systems one of the demagnetizing fields or the exchange interactions will benefit from the presence of the external field to balance out the contention.

Conflict of interest

The authors declare no conflicts of interest.

Author details

Pablo F. Zubieta Rico[1], Daniel Olguín[2] and Yuri V. Vorobiev[1*]

*Address all correspondence to: vorobiev@cinvestav.mx

1 Materials Department, Center for Research and Advanced Studies of the National Polytechnic Institute, Querétaro, Mexico

2 Physics Department, Center for Research and Advanced Studies of the National Polytechnic Institute, Mexico City, Mexico

References

[1] Reed M. Spatial quantization in GaAs–AlGaAs multiple quantum dots. Journal of Vacuum Science & Technology B: Microelectronics and Nanometer Structures. 1986;**4**(1):358

[2] Kastner M. Artificial atoms. Physics Today. 1993;**46**(1):24-31

[3] Walling M, Novak J, Shepard J. Quantum dots for live cell and in vivo imaging. International Journal of Molecular Sciences. 2009;**10**(2):441-491

[4] Behrens S. Preparation of functional magnetic nanocomposites and hybrid materials: Recent progress and future directions. Nanoscale. 2011;**3**(3):877-892

[5] Dai Q, Nelson A. Magnetically-responsive self assembled composites. Chemical Society Reviews. 2010;**39**(11):4057

[6] Hergt R, Dutz S, Müller R, Zeisberger M. Magnetic particle hyperthermia: Nanoparticle magnetism and materials development for cancer therapy. Journal of Physics: Condensed Matter. Sep 8, 2006;**18**(38):S2919

[7] Weinstein JS, Varallyay CG, Dosa E, Gahramanov S, Hamilton B, Rooney WD, Muldoon LL, Neuwelt EA. Superparamagnetic iron oxide nanoparticles: Diagnostic magnetic resonance imaging and potential therapeutic applications in neurooncology and central nervous system inflammatory pathologies, a review. Journal of Cerebral Blood Flow & Metabolism. Jan 2010;**30**(1):15-35

[8] Gómez JR, Garcia RF, Catoira AD, Gómez MR. Magnetocaloric effect: A review of the thermodynamic cycles in magnetic refrigeration. Renewable and Sustainable Energy Reviews. Jan 31, 2013;**17**:74-82

[9] Lau JW, Shaw JM. Magnetic nanostructures for advanced technologies: Fabrication, metrology and challenges. Journal of Physics D: Applied Physics. Jul 7, 2011;**44**(30):303001

[10] Weller D, Moser A. Thermal effect limits in ultrahigh-density magnetic recording. IEEE Transactions on magnetics. Nov 1999;**35**(6):4423-4439

[11] Sun S, Murray CB, Weller D, Folks L, Moser A. Monodisperse FePt nanoparticles and ferromagnetic FePt nanocrystal superlattices. Science. Mar 17, 2000;**287**(5460):1989-1992

[12] Ross CA. Patterned magnetic recording media. Annual Review of Materials Research. Aug 2001;**31**(1):203-235

[13] Richter HJ, Dobin AY, Lynch RT, Weller D, Brockie RM, Heinonen O, Gao KZ, Xue J, Veerdonk RV, Asselin P, Erden MF. Recording potential of bit-patterned media. Applied Physics Letters. May 29, 2006;**88**(22):222512

[14] Engel BN, Akerman J, Butcher B, Dave RW, DeHerrera M, Durlam M, Grynkewich G, Janesky J, Pietambaram SV, Rizzo ND, Slaughter JM. A 4-Mb toggle MRAM based on a novel bit and switching method. IEEE Transactions on Magnetics. Jan 2005;**41**(1): 132-136

[15] Gallagher WJ, Parkin SS. Development of the magnetic tunnel junction MRAM at IBM: From first junctions to a 16-Mb MRAM demonstrator chip. IBM Journal of Research and Development. Jan 2006;**50**(1):5-23

[16] Cowburn RP, Welland ME. Room temperature magnetic quantum cellular automata. Science. Feb 25, 2000;**287**(5457):1466-1468

[17] Imre A, Csaba G, Ji L, Orlov A, Bernstein GH, Porod W. Majority logic gate for magnetic quantum-dot cellular automata. Science. Jan 13, 2006;**311**(5758):205-208

[18] Jain S, Adeyeye AO, Singh N. Spin re-orientation in magnetostatically coupled Ni80Fe20 ellipsoidal nanomagnets. Nanotechnology. Jun 18, 2010;**21**(28):285702

[19] Hao R, Xing R, Xu Z, Hou Y, Gao S, Sun S. Synthesis, functionalization, and biomedical applications of multifunctional magnetic nanoparticles. Advanced Materials. Jul 6, 2010;**22**(25):2729-2742

[20] Poudyal N, Liu JP. Advances in nanostructured permanent magnets research. Journal of Physics D: Applied Physics. Dec 14, 2012;**46**(4):043001

[21] Singamaneni S, Bliznyuk VN, Binek C, Tsymbal EY. Magnetic nanoparticles: Recent advances in synthesis, self-assembly and applications. Journal of Materials Chemistry. 2011;**21**(42):16819-16845

[22] Terris BD, Thomson T. Nanofabricated and self-assembled magnetic structures as data storage media. Journal of physics D: Applied physics. Jun 3, 2005;**38**(12):R199

[23] Landau LA, Lifshitz E. On the theory of the dispersion of magnetic permeability in ferromagnetic bodies. Physikalische Zeitschrift der Sowjetunion. Jun, 1935;**8**(153):101-114

[24] Stiles MD, Miltat J. Spin-transfer torque and dynamics. In: Hillebrands B, Thiaville A, editors. Spin Dynamics in Confined Magnetic Structures III. Berlin/Heidelberg: Springer; Jan 1, 2006. pp. 225-308

[25] Gilbert TL. A phenomenological theory of damping in ferromagnetic materials. IEEE Transactions on Magnetics. Nov 2004;**40**(6):3443-3449

[26] Beleggia M, De Graef M. General magnetostatic shape–shape interactions. Journal of Magnetism and Magnetic Materials. Jan 31, 2005;**285**(1):L1-L0

[27] Majetich SA, Wen T, Mefford OT. Magnetic nanoparticles. MRS Bulletin. 2013 Nov;**38**(11): 899-903

[28] Meza MM, Zubieta Rico PF, Horley PP, Sukhov A, Vieira VR. Multi-parameter optimization of a nanomagnetic system for spintronic applications. Physica B: Condensed Matter. Nov 15, 2014;**453**:136-139

[29] Semenova EK, Montoncello F, Tacchi S, Dürr G, Sirotkin E, Ahmad E, Madami M, Gubbiotti G, Neusser S, Grundler D, Ogrin FY. Magnetodynamical response of large-area close-packed arrays of circular dots fabricated by nanosphere lithography. Physical Review B. May 28, 2013;**87**(17):174432

[30] Zhu X, Grütter P, Metlushko V, Ilic B. Magnetization reversal and configurational anisotropy of dense permalloy dot arrays. Applied physics letters. Jun 24, 2002;**80**(25):4789-4791

[31] Natali M, Lebib A, Chen Y, Prejbeanu IL, Ounadjela K. Configurational anisotropy in square lattices of interacting cobalt dots. Journal of Applied Physics. May 15, 2002;**91**(10):7041-7043

[32] Redondo C, Sierra B, Moralejo S, Castano F. Magnetization reversal induced by irregular shape nanodots in square arrays. Journal of Magnetism and Magnetic Materials. Jul 31, 2010;**322**(14):1969-1972

Chapter 11

Dilute Magnetic Semiconducting Quantum Dots: Smart Materials for Spintronics

Jejiron Maheswari Baruah and Jyoti Narayan

Additional information is available at the end of the chapter

http://dx.doi.org/10.5772/intechopen.73286

Abstract

The present day world involved in the fabrication of miniaturized smart devices is in continuous quest of materials with better optoelectronic and magneto-electronic efficiency. Effective incorporation of dopants into semiconductor lattices have been accepted as a primary means of controlling electrical, optical, magnetic and other physico-chemical properties of semiconductors. Manipulations in magnetic spin within a semiconducting material have lead to an effective research for potential ferromagnets with semiconducting properties, leading to an important field, dilute magnetic semiconductors (DMS). On the other hand, quantum dots (QDs) have been registered to be quantum confined nanocrystals with unique optoelectronic properties, having a wide range of potential applications. QDs experienced rapid development leading to the concept of dilute magnetic semiconducting quantum dots (DMSQDs), where transition metals with a few to several atomic percentages, having unpaired d-electrons, are doped in order to manipulate their opto-magnetic properties. These materials are fabricated by alloying transition metals with Group II-VI, III-V and IV-IV elements resulting in multi-component systems. They have tremendous applications in the spintronics industry, where electronic properties are controlled by spin degree of freedom. The present report reveals the significance, electronic origination of the fact, synthesis and their applications toward the fabrication of spintronics devices.

Keywords: dilute magnetic semiconductors, quantum dots, dilute magnetic semiconductors quantum dots, spintronics, opto-magnetic properties

1. Introduction

The dilute magnetic semiconductor quantum dots (DMSQDs) are basically the combination of semiconducting quantum dots, where transition metals are introduced as impurity or dopants. They are symbolically represented by $A_{1-x}M_xB$ or A(M)B or AB(M), where A is often a non-magnetic cation, A and B can be from Group II-VI, Group III-V and Group IV-IV elements. These nanoscale materials play an important role in microelectronics and magnetic storage devices [1–3]. Additionally, these materials have the quality to exist in both Curie temperatures (T_c) as well as in room temperature (RT) with high saturation of magnetization (M_s) [4–6]. These quantum confined materials has unique magneto-optical and optically controlled magnetism properties, which make them essentially important in today's research on materials for spintronics (spin-based electronics) [7, 8] device making. The device making includes miniaturization of electronic devices, magnetic fluids and high density data storage systems [9–11]. The semiconducting quantum dots which are being in research are from the Group II-VI and the dopants are transition metals. The individual potentiality of these materials, generated by coupling of diluted magnetic semiconductor (DMS) and quantum dots (QDs), is expected to be a path breaking one in the future field of optoelectronics and magneto-optoelectronic devices. To understand better, we will first look into the concept of DMS and QDs separately.

2. Dilute magnetic semiconductor (DMS)

Magnetic semiconductors are the semiconducting materials, which can exhibit ferromagnetism. Doping of the transition metals in these materials are said to be dilute magnetic semiconductors (DMS). The DMS are therefore semi-magnetic due to the introduction of magnetic elements in their lattices. Basically, the spintronics property of these materials has attracted the present day research for possible technological applications. By definition, spintronics is a combination of electrons' spin and their associated electronic charge and magnetic moment.

The first generation spintronics devices are derived from passive magnetoresistive sensors [12], but the second generations devices are expectedly achieved with the active spin-based devices, which are manipulated in the host semiconductor with spin-polarized electrons [13, 14]. The thought behind a spintronics device is the presence of spin-polarized electrons which travels through the host. Although the introduction of the ferromagnetic material in the semiconductor material through doping is extensively studied, yet the electronic spin is difficult to preserve throughout the material interface due to the difference in electrical conductivity in both the doped as well as the host material [15]. Hence, to present these materials as expected material, better than both ferromagnetic and semiconductor individually, research is very much crucial and warranted in spin electronic carrier device industries. DMS is next concept to meet the vital applicative operation to establish the spintronics carrier devices, an efficient one. In DMS, the host is non-magnetic semiconductor, whereas, the magnetic material is from transition metal series. These are powerful integrated devices having highly spin-polarized capacity.

3. DMS materials and their applications

3.1. $A^{II}_{1-x}Mn_xB^{VI}$ ternary materials

One of the most extensively studied DMS is alloys of $A^{II}_{1-x}Mn_xB^{VI}$ (*A- Group II element & B- Group VI element*) [16]. This ternary material exhibits both wurtzite and zinc blende crystalline structure as per their compositional range [17]. It is noticed that the change in Manganese (Mn) content can lead the crystal structure to cubic or hexagonal orientation. It has been observed that lower the input of Mn element in the host composition, the resultant structure tends to acquire cubic crystallinity, whereas, with higher the amount of Mn doping wurtzite crystallinity is observed [18]. Crystal structure of the various compositional ranges of the material suggests that, although their symmetry is different, both achieve tetrahedral geometry (s-p^3 bonding) with the involvement of 2*s* valence electrons from Group II element and 6*p* valence electrons from Group VI element. Mn acts as a contributor of its valence $4s^2$ electrons to the s-p^3 bonding arrangement, although Mn differs from Group II material with an exactly half-filled d-orbital. The Hund's rule suggests that introduction of an unpaired electron of opposite spin will require a lot of input of energy (~ 6–7 eV) and hence Mn is acting as complete *3d* shell material [19]. This ability makes Mn eligible for the replacement of Group II elements in the tetrahedral structure. There is another crucial reason behind the establishment of Mn as a replacement of other Group II materials, which is the exact half-filled *3d* orbital configuration of Mn. The fact is as important as, there is a possibility of forming stable phase by other elements too, although, dimer formation is very much a possibility for other than Mn of the Group II element [18].

3.2. TiO_2 base DMS

Among the oxide base DMS materials, Cobalt (Co) intruded Titanium dioxide (TiO_2) system is one of the most consistently researched *n-type* semiconductor, to achieve ferromagnetism far above the RT ($T_c > 650$ K) [20]. The importance of temperature in this kind of system is because the system of magnetic semiconductor is hard to achieve at RT [21, 22]. The reason behind such concept is the faced difficulty in the introduction of both electronic and magnetic dopants in the system and functionalization of the designed material as a good balanced material between dopant spins and free carriers of electrons. Hence, the synthesized material achieves the coupling as a thermally strong dopant spin-carrier coupling [22]. The most recent research suggest that the ternary cited materials of Mn-doped Group II-VI semiconductors are also capable of exhibiting the expected necessary property as Co-doped TiO_2 revealed at RT [23, 24].

TiO_2 is a wide direct forbidden band gap (3.03 eV) material, used for optoelectronic devices and solar cell applications [25–27]. Its crystal symmetry is found to be in tetragonal and rhombohedral orientations [28]. Therefore the thin films of Co-doped TiO_2 can be accommodated in the applicative DMS devices. The first observation of RT ferromagnetism in the Co-doped TiO_2 system was reported at Anatase phase $Ti_{1-x}Co_xO_2$ films ($0 \le x \le 0.08$), on $LaAlO_3$ and $SrTiO_3$ substrates, using laser molecular beam epitaxy, at substrate growth temperatures between 680 and 720°C [29]. Same research group also found the satisfactory results with the thin films of rutile phase TiO_2 with a composition of $Ti_{1-x}Co_xO_2$ ($0 \le x \le 0.05$) onto α-Al_2O_3 substrates, using the same deposition technique [30]. After this achievement, a good number work on this composition was done with various thin film deposition techniques, viz., pulsed laser deposition (PLD) [31–36], laser molecular beam epitaxy (LMBE) [37–39], metal–organic chemical vapor deposition (MOCVD) [40], reactive co-sputtering, oxygen-plasma-assisted molecular beam epitaxy (OPA-MBE) [41] and sol–gel [42] method. Researchers observed that the pressure of Oxygen applied during the thin film deposition is also a very important factor and suggested that at $P_{O2} \ge 1.3 \times 10^{-5}$ mbar, we can have clear streaky RHEED patterns, which suggest two-dimensional smooth surfaces [32].

The ferromagnetism in Co-doped TiO_2 is a topic of interest for the research accompanying spintronics devices. The oxide base DMS materials have extrinsic or intrinsic effect, which is the root of their device driven capability, is still a matter of discussion. The extrinsic effect may be attributed to the interaction of local magnetic moments with magnetic impurities. The intrinsic magnetism may be due to the exchange coupling between the spin of carriers and local magnetic moments. Since, spintronics takes place only in polarized charge carriers, which is possible only when the ferromagnetism is intrinsic. The issue is of great concern because the experimental evidence is not yet available behind the actual reason of magnetism of DMS in TiO_2. Anomalous Hall effect (AHE) and electric field induced modulation by magnetization suggests, for rutile phased Co-doped TiO_2 system, the carrier-mediated ferromagnetism with a value of 13.5% [43, 44].

Recent theoretical studies propose the creation and distribution of oxygen vacancies in Co-doped TiO_2 is responsible for the ferromagnetism in these systems. The ferromagnetism is suppressed when the oxygen content is increased in the unit cell [45]. In a nutshell, for the TiO_2 crystal, in the event of an oxygen vacancy, Ti atoms will give away their electrons to oxygen and hence they will be in the scarcity of electrons to get bind with the oxygen vacancy sites by their own atoms and therefore a situation of hydrogen-like orbital occurs, hence constitutes a Polaron. This phenomenon is supported by a percolation model named bond magnetic polaron (BMP), which was used to study the magnetically doped oxides [46].

In the interaction of the magnetic cations with the hydrogenic electrons in the impurity band, the donors tend to form BMPs, coupling the 3d moments of the ions within their orbits. Depending on whether the cation 3d orbital is less than half filled, or half filled or more, the coupling between the cation and the donor electron is ferromagnetic or anti-ferromagnetic, respectively. Either way, the coupling between two similar impurities within the same donor orbital is ferromagnetic. The polaron radius is a function of the host material's dielectric constant and electron effective mass. If the polaron concentration in the material is large enough to achieve percolation, an entire network of polarons and magnetic cations become interconnected and we observe macroscopic ferromagnetic behavior [47].

Thus, the incorporation of impurities/dopants in the semiconducting lattices have been realized as an important primary means of controlling the magnetic and electrical conductivities, besides having an immense effect on magnetic, magneto-optical and other physical properties of semiconductors.

4. Quantum dots

Quantum dots (QDs) of semiconducting materials have attracted the research community due to their potential application in various fields of humanity, viz., optoelectronics, solar cell, bio-imaging and biosensors, cosmetics, space science, photocatalytic activity, etc. [48–54]. The QDs can be defined with respect to their size, which is supposed to be less than excitons Bohr radius. The material specific Bohr radius also leads to the property of that material. The size factor is supported by differently shaped particles. The size of the QDs leads to the significant change in band gap of the semiconductors than the bulk. The enhanced band energy of the particles is due to the fact of their atom like structures. These particles in this confinement have 10–1000 numbers of atoms within one particle. Therefore, the energy levels of each particle have the merging levels of only some of the atoms in comparison to their bulk entity, where millions of atoms coincide. Because of this fact, very less energy levels can merge with each other in a QD and hence the band gap energy increases drastically. The QDs have another specific property of showing blunt and broad absorption peak. The primary cause behind the phenomenon resides in their size effect. At the atomic level, the slight change in the size of a particle (viz., 0.5 nm) can change their HOMO-LUMO gap drastically. Therefore, whenever there is a solution of QDs, the particle size is never homogeneously uniform in the solution. Hence, for every particle the band gap energy will be different and therefore the absorption maximum shifts accordingly. As a consequence of the presence of differently sized particles in the solution, togetherness of these absorption maxima can be observed and hence a broad peak. Therefore, by tuning the size, we can meet the desired application with these particles. Apart from the size factor, shape phenomenon also plays a significant role in deciding the characteristic features in the field of quantum dot physics. The electrons, which are the driving force behind every electronic transition in a physical matter, have received different orientations in terms of surface of the particle. In the quantum range of physics, the QDs are forced to adapt the required application by modifying their surface. The reason behind such observation is the attachment of the surface electrons for differently shaped particles is different, which is again as an outcome of releasing surface energy of the particle to make it stable. The introduction of capping agents (the ligands) is also having a capability of taming the particle according to their preferred shape. This phenomenon is addressed as surface functionalization. The surface modification can lead us to the fabrication of the particles with better efficacy in different applicative devices.

The property of showing high luminescence by these QD metamaterials is one of the most aspired properties. The generation of double excitons leads the materials toward more promising luminescent material. This extraordinary property blesses these materials to show higher emission range than the traditional dyes and hence they become more appropriate with the fact of getting more emission with the excitation of only one electron. The size and, of course the shape, both have an important role in making them suitable for these applications. Most of the time the tunable size property of these quantum dots is mentioned, due to which one can access the whole light spectrum. The devices such as LEDs and solar cell require these nano dots in such a manner that they have the ability to absorb the whole visible and UV region and emit the same in higher wavelength. Therefore, the luminescence property of these fluorescent dots has to have the tenability to perform in the whole region. Fortunately, researchers found that for different semiconducting quantum dots, we can achieve the luminescence as per our requirement. Another interesting concept of wastage of the solar energy as thermal energy during the absorption of sun light by a photovoltaic cell comes into play, in the present day photo voltaic research. It is observed that a photovoltaic material, such as, QDs (although being the most promising one), cannot absorb the whole sun light as conversion efficiency of the cell becomes less. The reason behind this is the material, that we are using, can absorb the light in the desired range but cannot emit the same in the desired wavelength. To tackle this difficulty, the concept of large stock shift quantum dots has come up. This large stock shift materials can absorb the sun light in short wavelength and emit the same in the long wavelength, which make these functionalized quantum dots more efficient toward these kind of applications [55].

5. Dilute magnetic semiconductor quantum dots (DMSQDs)

Discussions on DMS and QDs have made it easier to understand the concept of DMSQDs. They are quantum dots of semiconducting materials doped with transition metals having magnetic behavior. Due to specific significance of QDs, researchers are tremendously focusing on the ferromagnetic material doped QDs. Since, semiconductors do not possess high magnetism in any level of their atomic growth, it becomes essential to incorporate the magnetic nature of DMS in nanoscale so as to improve its efficiency in the various fields of spintronics applications. It has been observed that the effectiveness of interaction of *sp-d* for the exchange of carrier and magnetic ions in terms of hole energy depends on the high and low magnetic field induces from outside. Hence, it is expected that due to the small size of a quantum dot, the exchange and interaction of *d* electrons with *sp* shelled electrons will be extensive in DMSQDs [56]. Therefore, the spintronics devices developed from DMSQDs are expected to be efficient as well as miniaturized one, probably due to the quantum confinement effect of DMS and therefore better than the single DMS materials.

5.1. Synthesis of DMSQDs

The synthesis procedures are very much similar to those for the synthesis of QDs. The only exception is to incorporate the metallic materials as impurity during the reaction process. Among a vast number of procedures, the chemical route to synthesis DMSQDs is the most commonly use deficient method. The size and shape of such QDs can be easily tailored through this method. Unwanted oxidation can also be prevented during the adopted during the synthesis process. Fe, Co, Ni and Mn are the main doping elements used for the preparation of DMSQDs of semiconductors of Group II-VI [56].

Clustering and surface doping are two main issues that are faced during the synthetic process for obtaining uniformed DMSQDs. To eliminate these key issues, one has to overcome self-purification [57] of the materials and understand the reactivity between the host-guest materials [58, 59]. The self-purification is a process where host molecule expels the guest molecule from the surface to attain a thermodynamically stable state by reducing its defect energy. Self-purification can be resolved by making the magnetic core at first, followed by coating with the semiconducting material and then annealing at higher temperature for a longer time to diffuse the dopant inside the host properly before it get expelled by the host. The reactivity issue can be sorted out by two ways: a) Nucleation doping and b) Growth doping. Successive ionic layer adsorption reaction (SILAR) method was used to dope Fe in CdS in one of the methods of its preparation. This was attained at high temperature. This method showed excellent result with the homogeneous diffusion of Fe in CdS shell. It was also reported that the oxidation state of Fe was reduced to 2 from 2.44 due to the presence of reducing reagent and replaced the Cd site with substitutional doping.

5.2. Properties and applications of DMSQDs

DMSQDs possess unique properties which make them suitable for wide range of applications. Their properties are primarily divided into magnetic and magneto-optical as well as magneto-electrical properties. These properties are attributed to the exchanged interaction of *sp-d* between the dopants magnetic material and the host semiconductor, although proper mechanism of origin of the effect and the governance of ferromagnetism are not yet confirmed. Secondly, it has been observed that due to the presence of quantitatively unknown weights of ligands within the synthesized material makes it difficult to calculate the conversion of magnetic moments from magnetic ions, [60] however there have been improvement toward the production of DMSQDs from time to time. Early reports on quantification showed the presence of a few magnetic moments in emu/gram (memu/g) [61] due to the doping of magnetic ions, instead of much more as expected. The effect of unknown amount of magnetic moment hinders the knowledge of comparison between the absolute magnetism of bulk and nano materials. The plausible reason to this effect may be the clustering of magnetic dopants or/and inherent *sp-d* exchange interactions.

One of the most advantageous finding on DMSQDs shows an exceptionally different nature of magnetism. It is the co-doping of ZnO with Cu and Fe [62]. Interestingly, ZnO individually doped with Fe or Cu showed an anti-ferromagnetic behavior without a trace of ferromagnetism. Whereas, the co-doping of both the transition metals in ZnO showed high quality ferromagnetism with magnetic moment as high as 600 memu/g. This work has proved the anti-ferromagnetism of Cu doped ZnO with the help of *M* versus *H* (Magnetization versus Magnetic field intensity) plot and anti-ferromagnetism of Fe doped ZnO with the inverse susceptibility plot as function of temperature, showing a negative intercept. But, in co-doped ZnO with Cu and Fe, X-ray absorption spectroscopy (XAS) clearly showed the presence of both Fe^{+2} and Fe^{+3} and its relative percentage is dependent on the presence of Cu as a dopant. Another research revealed that the size of Fe doped CdS QD was responsible for the magnetic moment [57]. They have achieved a magnetic moment of 80 memu/g at RT for doped CdS and undoped CdS showed negligible amount of magnetic moment with the same scale reaction. It has been observed that in the presence of an external magnetic field, a non-magnetic substance showed a small internal magnetic field due to Zeeman splitting (having an order of 2), whereas, materials like DMSQDs, the intensity of internal field is very high in the presence of external magnetic field [56]. It is also observed that along with the large internal field, a small external magnetic field also gets generated during this process. This happens due to the presence of the magnetic ions inside the material and the tendency to align themselves in the direction of the applied magnetic field. Theoretical modeling of magneto-optical and electronic property of core-shell nanoparticles of CdS-ZnS, doped with magnetic impurities of Mn showed that, these nanocrystals can give an attuned value of g over a wide range and make them suitable for spintronics devices, if the position of the magnetic impurities can be controlled [63]. Spectral fingerprints of the spin–spin interactions between the host excitons and the dopant is also revealed by single particle spectroscopy with discrete projections of individual Mn^{+2} ions observed from emission peaks. These QDs showed enhancement in exchange splitting at elevated temperatures by an order of magnitude compared to their epitaxial counterparts, which is useful for solotronics applications. The circularly polarized photoluminescence in the presence of magnetic field (MCPL) for bulk DMS is very much different than the QDs. In case of DMS material, the emission band edge of the host material showed a polarization due to the splitting of the band, but doped material (Mn^{+2}) do not show any band polarization due to spin and orbital forbidden emission [64]. But, in DMSQDs, along with the host, the dopant also showed polarized emission band edge in the presence of magnetic field. This surprised effect was although not yet properly understood, but expected to be due to quantum confinement, where wave functions are overlapped extensively [65]. Magneto-optical response in Cu doped chalcogenide QDs is also a tremendous effect observed in DMSQDs. This photo-excitation phenomenon in these DMSQDs has come as a result of strong spin-exchanged interaction between the valence band-conduction band (VB-CB) of the host and the paramagnetic Cu dopant. The magnetic circular dichroism (MCD) studies revealed the enhancement of paramagnetism up to 100% in these Cu doped ZnSe/CdSe QDs under the UV light excitation. Again, in dark, these materials retained a photo-magnetization memory for timescales of hours [66]. Another application is reported for Mn-doped CdSe QDs as light-induced spontaneous magnetization, where spin effect is controlled to generate, manipulate and read out spins [67]. In this case, no external magnetic field was applied but still showed large Zeeman splitting as a result of photo-excitation. The reason behind these giant splitting is the generation of large dopant-carriers exchange fields. These materials are having potential applications in the field of magneto-optical storage and optically controlled magnetism. DMSQDs are also known to respond to charged carriers. The carrier-mediated ferromagnetic interaction in Mn-doped CdSe QDs are also reported, which arise due to photo-excited carriers from surface defect states of smaller QDs (~3 nm) [68]. Mn-doped ZnO QDs also exhibited ferromagnetic exchange interaction due to photo-excitation in the absence of oxygen [69] and as is reported in air-stable Fe–Sn co-doped In_2O_3 [70]and Mn–Sn co-doped In_2O_3 [71]. Research has proved the conduction band electron–dopant ferromagnetic exchange interaction, offers magneto-electric and magneto-plasmonic properties which helps in wide scale spintronics applications.

Author details

Jejiron Maheswari Baruah and Jyoti Narayan*

*Address all correspondence to: jnarayan.nehu@gmail.com

Department of Basic Sciences & Social Sciences (Chemistry division), Synthetic Nanochemistry Laboratory, School of Technology, North Eastern Hill University, Shillong, India

References

[1] Mandal SK, Mandal AR, Banerjee S. High ferromagnetic transition temperature in PbS and PbS:Mn nanowires. ACS Applied Material Interfaces. 2012;**4**:205-209

[2] Zhou Y, Liu K, Xiao H, Xiang X, Nie J, Li S, Huang H, Zu X. Dehydrogenation: A simple route to modulate magnetism and spatial charge distribution of Germanane. Journal of Materials Chemistry C. 2015;**3**:3128-3134

[3] Odio OF, Lartundo-Rojas L, Santiago-Jacinto P, Martínez R, Reguera E. Sorption of gold by naked and Thiol-Capped Magnetite nanoparticles: An XPS approach. Journal of Physical Chemistry C. 2014;**118**:2776-2791

[4] Dietl T. A ten-year perspective on dilute magnetic semiconductors and oxides. Nature Materials. 2010;**9**:965-974

[5] Felser C, Fecher GH, Balke B. Spintronics: A challenge for materials science and solid-state chemistry. Angewandte Chemie International Edition. 2007;**46**:668-699

[6] Gao D, Yang G, Li J, Zhang J, Zhang J, Xue D. Room-temperature ferromagnetism of flowerlike CuO nanostructures. Journal of Physical Chemistry C. 2010;**114**:18347-18351

[7] Sinova J, Zutic I. New moves of the Spintronics tango. Nature Materials. 2012;**11**:368-371

[8] Wong PKJ, Zhang W, Wang K, van der Laan G, Xu Y, van der Wiel WG, de Jong MP. Electronic and magnetic structure of C60/Fe3O4(001): A hybrid Interface for organic Spintronics. Journal of Materials Chemistry C. 2013;**1**:1197-1202

[9] Schwartz DA, Norberg NS, Nguyen QP, Parker JM, Gamelin DR. Magnetic quantum dots: Synthesis, spectroscopy, and magnetism of Co2+– and Ni2+–doped Zno Nanocrystals. Journal of the American Chemical Society. 2003;**125**:13205-13218

[10] Schwartz DA, Kittilstved KR, Gamelin DR. Above-room-temperature ferromagnetic Ni2+–doped ZnO thin films prepared from colloidal diluted magnetic semiconductor quantum dots. Applied Physics Letters. 2004;**85**:1395-1397

[11] Ohno H. Making nonmagnetic semiconductors ferromagnetic. Science. 1998;**281**:951-956

[12] Chappert C, Fert A, van Dau FN. The emergence of spin electronics in data storage. Nature Materials. 2007;**6**:813-823

[13] Bland T, Lee K, Steinmuller S. The spintronics challenge. Physics World. 2008;**21**:24-28

[14] Awschalom DD, Flatté ME. Challenges for semiconductor spintronics. Nature Physics. 2007;**3**:153-159

[15] Wolf SA, Awschalom DD, Buhrman RA, Daughton JM, von Molnár S, Roukes ML, Chtchelkanova AY, Treger DM. Spintronics: A spin-based electronics vision for the future. Science. 2001;**294**:1488-1495

[16] Furdyna JK. Journal of Vacuum Science and Technology A;**4** 2002 (1986)

[17] Pajaczkowska A. Physicochemical properties and crystal growth of $A^{II}B^{VI}$-MnB^{VI} systems. Progress in Crystal Growth Characteristics. 1978;**1**:289

[18] Furdyna JK. Diluted magnetic semiconductors. Journal of Applied Physics. 1988;**64**: R29-R64

[19] Hass KC, Ehrenreich H. Journal of Crystal Growth. 1988;**86**:8

[20] Fukumuraa T, Yamadaa Y, Toyosakia H, Hasegawab T, Koinuma H, Kawasakia M. Exploration of oxide-based diluted magnetic semiconductors toward transparent spintronics. Applied Surface Science. 2004;**223**:62

[21] Seife C. 125 big questions that face scientific inquiry over the next quarter-century. Commemorative Issue Celebrating the 125th Anniversary of the Science Magazine. Science. 2005;**309**:82

[22] Chambers SA, Droubay TC, Wang CM, Rosso KM, Heald SM, Schwartz DA, Kittilstved KR, Gamelin DR. Ferromagnetism in oxide semiconductors. Materials Today. 2006;**9**:28

[23] Edmonds KW, Wang KY, Campion RP, Neumann AC, Farley NRS, Gallagher BL, Foxon CT. High-curie-temperature Ga1–xMnxAs obtained by resistance-monitored annealing. Applied Physics Letters. 2002;**81**:4991

[24] Chiba D, Takamura K, Matsukura F, Ohno H. Effect of low-temperatureannealing on (Ga,Mn)as trilayer structures. Applied Physics Letters. 2003;**82**:3020

[25] Linsebigler AL, Lu G, Yates JT Jr. Photocatalysis on TiO2 surfaces: Principles, mechanisms, and selected results. Chemical Reviews. 1995;**95**:735

[26] Wang R, Hashimoto K, Fujishima A, Chikuni M, Kojima E, Kitamura A, Shimohigoshi M, Watanabe T. Light-induced amphiphilic surfaces. Nature. 1997;**388**:431

[27] Zhuang J, Rusu CN, Yates T Jr. Adsorption and photooxidation of CH3CN on TiO2. Journal of Physical Chemistry B. 1999;**103**:6957

[28] Diebold U. The surface science of titanium dioxide. Surface Science Reports. 2003;**48**:53

[29] Matsumoto YJ, Murakami M, Shono TJ, Hasegawa T, Fukumura T, Kawasaki M, Ahmet P, Chikyow T, Koshihara SY, Koinuma H. Roomtemperature ferromagnetism in transparent transition metal-doped titanium dioxide. Science. 2001;**291**:854-856

[30] Matsumoto Y, Takahashi R, Murakami M, Koida T, Fan X-J, Hasegawa T, Fukumura T, Kawasaki M, Koshihara S-Y, Koinuma H. Ferromagnetism in Co doped TiO2 rutile thin films grown by laser molecular beam epitaxy. Japanese Journal of Applied Physics. 2001;**40**:L1204

[31] Shinde SR, Ogale SB, Das Sarma S, Simpson JR, Drew HD, Lofland SE, Lanci C, Buban JP, Browning ND, Kulkarni VN, Higgins J, Sharma RP, Greene RL, Venkatesan T. Ferromagnetism in laser depositedanatase Ti1-xCoxO2-δ films. Physical Review B. 2003;**67**:115211

[32] Kim DH, Yang JS, Lee KW, Bu SD, Noh TW, Oh S-J, Kim Y-W, Chung J-S, Tanaka H, Lee HY, Kawai T. Formation of co nanoclustersinepitaxial Ti0.96Co0.04O2 thin films and their ferromagnetism. Applied Physics Letters. 2002;**81**:2421

[33] Yamada Y, Toyosaki H, Tsukazaki A, Fukumura T, Tamura K, Segawa Y, Nakajima K, Aoyama T, Chikyow T, Hasegawa T, Koinuma H, Kawasaki M. Epitaxial growth and physical properties of a room temperature ferromagneticsemiconductor: Anatase phase $Ti1\text{-}xCo_xO_2$. Journal of Applied Physics. 2004;**96**:5097

[34] Stampe PA, Kennedy RJ, Xin Y, Parker JS. Investigation of the cobaltdistribution in TiO2:Co thin films. Journal of Applied Physics. 2002;**92**:7114

[35] Higgins JS, Shinde SR, Ogale SB, Venkatesan T, Greene RL. Hall effect in cobalt-doped TiO2-δ. Physical Review B. 2004;**69**:073201

[36] Yang HS, Choi J, Craciun V, Singh RK. Ferromagnetism of anatase Ti1–XCoxO2-δ films grown by ultraviolet-assisted pulsed laser deposition. Journal of Applied Physics. 2003; **93**:7873

[37] Kim J-Y, Park J-H, Park B-G, Noh H-J, Oh S-J, Yang JS, Kim D-H, Bu SD, Noh T-W, Lin H-J, Hsieh H-H, Chen CT. Ferromagnetism induced by clustered co in co-doped anatase TiO2 thin films. Physical Review Letters. 2003;**90**:017401

[38] Murakami M, Matsumoto Y, Hasegawa T, Ahmet P, Nakajima K, Chikyow T, Ofuchi H, Nakai I, Koinuma H. Cobalt valence states and origins of ferromagnetism in co doped TiO2 rutile thin films. Journal of Applied Physics. 2004;**95**:5330

[39] Toyosaki H, Fukumura T, Yamada Y, Nakajima K, Chikyow T, Hasegawa T, Koinuma H, Kawasaki M. Anomalous hall effect governed by electron doping in aroom-temperature transparent ferromagnetic semiconductor. Nature Materials. 2004;**3**:221

[40] Seong NJ, Yoon SG, Cho CR. Effects of co-doping level on themicrostructural and ferromagnetic properties of liquid-delivery metalorganicchemical-vapor-deposited Ti1–xCoxO2 thin films. Applied Physics Letters. 2002;**81**:4209

[41] Chambers SA, Thevuthasan S, Farrow RFC, Marks RF, Thiele JU, Folks L, Samant MG, Kellock AJ, Ruzycki N, Ederer DL, Diebold U. Epitaxial growth and properties of ferromagnetic co-doped TiO2 anatase. Applied Physics Letters. 2001;**79**:3467

[42] Soo YL, Kioseoglou G, Kim S, Kao YH, Sujatha DP, Parise J, Gambinoand RJ, Gouma PI. Local environment surrounding magnetic impurity atoms in a structural phase transition of co-doped TiO2 nanocrystal ferromagnetic semiconductors. Applied Physics Letters. 2002;**81**:655

[43] Calderón MJ, Das Sarma S. Theory of carrier mediated ferromagnetism in dilute magnetic oxides. Annals of Physics. 2007;**322**:2618

[44] Zhao T, Shinde SR, Ogale SB, Zheng H, Venkatesan T, Ramesh R, Das Sarma S. Electric field effect in diluted magnetic insulator anatase co:_TiO2. Physical Review Letters. 2005;**94**:126601

[45] Jaffe JE, Drouban TC, Chambers SA. Oxygen vacancies and ferromagnetismin CoxTi1-xO2-x-y. Journal of Applied Physics. 2005;**97**:073908

[46] Kaminski A, Das Sarma S. Polaron percolation in diluted magneticsemiconductors. Physical Review Letters. 2002;**88**:247202

[47] http://repositorio.ul.pt/bitstream/10451/1635/11/19491_ulsd_re481_CHAPTER_2.pdf

[48] Tada H, Kiyonaga T, Naya S. Chemical Society Reviews. 2009;**38**:1849-1858

[49] Vaseem M, Umar A, Hahn YB. American Science Publications. 2010;**5**:1

[50] Schlamp MC, Peng XG, Alivisatos AP. Journal of Applied Physics. 1997;**82**:5837

[51] Sundar VC, Lee J, Heine JR, Bawendi MG, Jensen KF. Advanced Materials. 2000;**12**:1102

[52] Kazes M, Lewis DY, Ebenstein Y, Mokari T, Bannin U. Advanced Materials. 2002;**14**:317

[53] Britt J, Ferekides C. Applied Physics Letters. 1993;**62**:2851

[54] Wu YL, Lim CS, Fu S, Tok AIY, Lau HM, Boey FYC, Zeng XT. Nanotechnology. 2007;**18**:5604

[55] Baruah JM, Narayan J. Journal of Optics. 2017. https://doi.org/10.1007/s12596-017-0417-y

[56] Makkar M, Viswanatha R. Current Science. 10 APRIL 2017;**112**(7)

[57] Saha A, Shetty A, Pavan A, Chattopadhyay S, Shibata T, Viswanatha R. Uniform doping in quantum-dots-based dilute magnetic semiconductor. Journal of Physical Chemistry Letters. 2016;**7**:2420-2428

[58] Pradhan N, Goorskey D, Thessing J, Peng X. An alternative of CdSe nanocrystal emitters: Pure and tunable impurity emissions in ZnSe nanocrystals. Journal of the American Chemical Society. 2005;**127**:17586-17587

[59] Peng X, Wickham J, Alivisatos A. Kinetics of II–VI and III–V colloidal semiconductor nanocrystal growth: 'Focusing' of size distributions. Journal of the American Chemical Society. 1998;**120**:5343-5344

[60] Sharma P et al. Ferromagnetism above room temperature in bulk and transparent thin films of Mn-doped ZnO. Nature Materials. 2003;**2**:673-677

[61] Jana S, Srivastava BB, Jana S, Bose R, Pradhan N. Multifunctional doped semiconductor nanocrystals. Journal of Physical Chemistry Letters. 2012;**3**:2535-2540

[62] Viswanatha R, Naveh D, Chelikowsky JR, Kronik L, Sarma DD. Magnetic properties of Fe/cu co-doped ZnO nanocrystals. Journal of Physical Chemistry Letters. 2012;**3**:2009-2014

[63] Sanders G, Musfeldt J, Stanton C. Tuning *g*-factors of core-shell nanoparticles by controlled positioning of magneticimpurities. Physical Review B. 2016;**93**:075431

[64] MacKay J, Becker W, Spaek J, Debska U. Temperature and magnetic-field dependence of the Mn2+4T1(4 G) □ 6A1(6 S)photoluminescence band in Zn0.5Mn0.5Se. Physical Review B. 1990;**42**:1743

[65] Viswanatha R, Pietryga JM, Klimov VI, Crooker SA. Spin-polarized Mn2+ emission from Mn-doped colloidal nanocrystals. Physical Review Letters. 2011;**107** 067402

[66] Pandey A, Brovelli S, Viswanatha R, Li L, Pietryga J, Klimov VI, Crooker S. Long-lived photoinduced magnetization in copper-doped ZnSe–CdSe core-shell nanocrystals. Nature Nanotechnology. 2012;**7**:792-797

[67] Beaulac R, Schneider L, Archer PI, Bacher G, Gamelin DR. Light-induced spontaneous magnetization in doped colloidal quantum dots. Science. 2009;**325**:973-976

[68] Zheng W, Strouse GF. Involvement of carriers in the size dependent magnetic exchange for Mn : CdSe quantum dots. Journal of American Chemical Society. 2011;**133**:7482-7489

[69] Ochsenbein ST, Feng Y, Whitaker KM, Badaeva E, Liu WK, Li X, Gamelin DR. Charge-controlled magnetism in colloidal doped semiconductor nanocrystals. Nature Nanotechnology. 2009;**4**:681-687

[70] Shanker GS, Tandon B, Shibata T, Chattopadhyay S, Nag A. Doping controls plasmonics, electrical conductivity, and carrier-mediated magnetic coupling in Fe and Sncodoped In2O3 nanocrystals: Local structure is the key. Chemistry of Materials. 2015;**27**:892-900

[71] Tandon B, Yadav A, Nag A. Delocalized electrons mediated magnetic coupling in Mn–Sncodoped In2O3 nanocrystals: Plasmonics shows the way. Chemistry of Materials. 2016;**28**:3620-3624

Chapter 12

Mn-Doped ZnSe Quantum Dots as Fluorimetric Mercury Sensor

Sundararajan Parani, Ncediwe Tsolekile,
Bambesiwe M.M. May, Kannaiyan Pandian and
Oluwatobi S. Oluwafemi

Additional information is available at the end of the chapter

http://dx.doi.org/10.5772/intechopen.70669

Abstract

Quantum dots (QDs), because of their exciting optical properties, have been explored as alternative fluorescent sensors to conventional organic fluorophores which are routinely employed for the detection of various analytes via fluorometry. QD probes can detect toxic metal ions, anions, organic molecules with good selectivity and sensitivity. This chapter investigates the synthesis of Mn-doped ZnSe QDs using nucleation-doping strategy. The as-synthesized QDs were characterized by various analytical tools such as ultraviolet-visible (UV-vis) absorption, photoluminescence (PL) spectroscopy, X-ray diffractometry (XRD) and transmission electron microscopy (TEM). It was found that Mn doping of QDs significantly increases the PL intensity. The PL of the resulting QDs was examined in the presence of different metal ions to check its selective response. Among the various metal ions, Hg^{2+} exhibits a drastic quenching of the QD's emission intensity. A Stern-Volmer plot of $[Hg^{2+}]$ sensing using the as-synthesized QDs showed linearity in the range of $0–30 \times 10^{-6}$ ML^{-1} with the regression coefficient $R^2 = 0.99$. The detection limit was found to be 6.63×10^{-7} ML^{-1}. Thus, the present Mn-doped ZnSe QDs represent a simple, non-toxic fluorescent probe for the qualitative and quantitative detection of mercury ions in aqueous samples.

Keywords: quantum dots, ZnSe, fluorimetry, doping, heavy metal detection

1. Introduction

Continuous and high-concentration exposure of heavy metals can cause various physiological and biochemical problems to the environment and human being. Thus, the detection of these harmful metal ions has become very important from industrial, environmental and biological point of view. This is a challenging subject for analytical chemists due to the sensitivity, detection limits and acceptable toxicity levels set by global regulation bodies [1, 2]. In addition, similar chemistry of heavy metals is fastidious with respect to selectivity of the determination method. A variety of standardized analytical methods for the detection of metal ions are available. However, only some of them have found application in routine analysis. Recommended procedures for the detection of heavy metals in water samples include photometric methods, flame or graphite furnace atomic absorption spectroscopy (AAS), inductively coupled plasma emission or mass spectrometry (ICP-ES, ICP-MS), total reflection X-ray fluorimetry (TXRF) and anodic-stripping voltammetry (ASV) [3–5]. These methods offer good limits of detection and wide linear ranges, but they are time consuming, require high cost analytical instruments trained operating personals and high maintenance which is a financial burden to analytical laboratories. Furthermore, the required sample pretreatment and preparation time introduce systematic errors in the analysis. However, the development of fast, economical and portable devices for metal ion detection has grown tremendously over the past 10 years. Particularly, chemo-sensors, which offer the ability for both on-line and field monitoring, has attracted many industries in the detection of metal ions in water [6]. This has allowed for quick and continuous supervision monitoring of drinking or ground water and lentic or lotic watercourses. An ideal sensor should provide good sensitivity, high selectivity towards the target analyte, mathematical relationship of signal output to the amount of analyte, fast response time, good signal-to-noise ratio and long-term stability [7]. A variety of sensors have been developed, and these include DNAzymes sensors, optical sensors, electrochemical, colorimetric and fluorescent sensors [1–10] just to mention a few. This chapter aims to introduce the reader to the use of quantum dots (QDs) as metal ion sensors.

1.1. Optical sensors

A chemical sensor can be defined as 'a portable miniaturized analytical device, which can deliver real-time and on-line information in the presence of specific compounds or ions in complex samples' [11]. Chemical sensors can be categorized into electrochemical, optical, mass-sensitive and heat-sensitive, according to the types of transducer. Of these classifications, optical sensors have been the most widely used as contact-less detectors, counting or positioning of parts. An optical sensor device consists of the following components: (a) the recognition element, where specific interaction and identification of the analyte takes place; (b) the transducer element that converts the recognition process into a measurable optical signal; (c) an optical device (process unit) which consists of at least a light source and finally (d) a detector which detects and converts the change of optical properties and amplifies the signal into a unit readout. The optical properties measured can be absorbance, reflectance, luminescence, light polarization, Raman and others. Optical sensors have found many applications in various fields, including biomedical, clinical, environmental monitoring and process controlling [12–18]. They are an attractive analytical tool, whenever continuous monitoring and real-time information is desired. They can track sources of contamination in an industrial process, follow the formation and movement of environmental pollutants and can raise the alarm when a toxic species exceeds an expected level of exposure. For environmental analysis, single-use test strips for various ions, including heavy metals, are commercially available [19], which have their limitations in accuracy and reversibility. In recent years, activities applying optical sensors for the determination of heavy metals increased [20]. The most significant methods are the application of quenchable fluorophores or indicator dyes.

1.2. Quantum dots

Semiconductor quantum dots (QD) are nanocrystalline materials that confines the motion of the charge carriers in three spatial directions. These crystals are intermediate state of matter that display properties present in neither bulk nor molecular systems. The physical and electronic properties of QDs are strongly dependent on size (number of atoms). Their importance was recognized by the 2000 Nobel Prize in Physics awarded to Alferov and Kroemerin for their basic work on information and communication technology using the semiconductor heterostructures. QDs are generally made as binary semiconductor nanocrystals from groups II–VI (e.g. CdSe, ZnS, etc.), groups IV–VI (e.g. PbTe, PbS) or groups III–V (e.g. InAs, InP) in the periodic table [21]. Moreover, QDs of Si, Ge, Ag, also of carbon and graphene [22–24] and ternary QDs (from I–VI) have been reported [25, 26]. QDs have received much attraction because of their stable, tunable, bright and narrow photoemission, high chemical and photo bleaching stability, processability and surface functionality and they offer many advantages in comparison with conventional fluorophore. Thus, a new generation of QDs based sensor containing the unique optical properties of QDs has been constructed for sensing molecules and ions in ultratrace level.

Colloidal synthesis of QDs widely employs a 'bottom-up' approach where the crystals are nucleated and grown from the precursor materials dissolved in a suitable solvent in the presence of a stabilizing agent. This approach was pioneered by L. Brus, an American scientist when he was at Bell laboratories, New Jersey, in the late 1970s and carried over by some of his post docs notably Moungi Bawendi and Paul Alivisatos. Because of its mild preparative conditions, this method is also called as a wet chemical route and has become popular among the scientists and industrialists for their usefulness in the biomedical and analytical fields. Different types of QDs such as alloyed QDs, core/shell QDs, impurity-doped QDs, polymer-QD composites with desired size and desired functional group on the surface can be prepared by colloidal synthesis, which are difficult to obtain by the other methods.

High-quality colloidal quantum dot crystals can be prepared in organic medium. Organic QDs obtained by this method have good degree of monodispersity and high photoluminescence quantum efficiency. This method became familiar after the synthesis of CdSe QD by Murray et al. in 1993 [27, 28]. The precursors for CdSe QDs chosen by Murray were dimethylcadmium (Me_2Cd) and TOPSe (Se dissolved in trioctylphosphine (TOP). The rapid injection of both the precursors together into the hot solution of trioctylphosphine oxide (TOPO) at ~300°C produced yellow/orange CdSe nanocrystallites. However, the use of expensive and/or hazardous organic reagents, harsh reaction conditions, and hydrophobicity of the as-prepared QDs are some of the shortcomings of the organic synthetic routes. To make the QDs water soluble, hydrophobic nature of the QDs surface should be converted into hydrophilic nature by surface encapsulation or ligand exchange. Surface modification processes are tedious, involving multiple steps and usually produce materials with reduced optical properties compared to the parent organic materials [29, 30].

On the other hand, QDs can also be synthesized in aqueous medium directly. For example, in aqueous synthesis of thiol-stabilized CdTe QDs, Cd^{2+} dissolved in water medium would be reacted with a HTe^{-} solution in the presence of water-soluble thiol ligands. Refluxing of the above mixture produces CdTe QDs dispersion. Rajh et al. used a thiol (3-mercapto-l,2-propane-diol) as a stabilizing agent to prepare CdTe QD with 20% photoluminescence quantum yield (PLQY) [31]. Afterwards, numerous thiols were investigated as stabilizing agents [32–34]. Compared with organic phase synthesis, aqueous synthesis involves less toxic precursors, inexpensive and produces excellent water-soluble and biocompatible products.

1.3. Modifications for PL enhancement

Most of the QDs, which are prepared in aqueous conditions, have low stability and low PLQY. As the surface of QDs is highly reactive, they have a high possibility to aggregate in the presence of heat, light, air or some ions. This may cause surface, which further reduced the PLQY. A number of techniques have been used to improve the particle stability, PL efficiency and biocompatibility of the QDs. These include photo-irradiation [35, 36], ultrasonic irradiation [37], doping with transition metals [38–40] and inorganic passivation [41]. Among them, inorganic passivation and doping techniques are the most widely investigated.

Doped semiconductor nanocrystals have been studied extensively in the past two decades since Bhargava et al. [42] reported on Mn-doped ZnS (Mn:ZnS). They stated that it could be possible to obtain efficient emission from the dopant centres even if the host nanocrystals were not of high quality. The PL of doped QDs is purely dopant-specific. Based on this, different colour-emitting (blue to red) QDs doped with metals (Al^{3+}), transition metals (Cu^{+}, Mn^{2+}) and halides (F^{-}) have been reported in the literature [43–45]. Doped nanocrystalline II–VI semiconductors incorporating rare earth (RE) ions such as Tb^{3+}, Eu^{3+} and Er^{3+} have also been reported [46, 47]. However, due to the dissimilar chemical properties (e.g. ionic radius, valence state) between the RE ion and host cation (Cd^{2+}, Zn^{2+}), efficient doping of RE ions into II–VI semiconductor host is not favourable.

In contrast to RE ions, the chemical properties of Mn^{2+} are very similar to those of Cd^{2+} (or Zn^{2+}); thus, incorporating Mn^{2+} into II–VI semiconductor host is much easier. Mn-doped semiconductors are potential luminescent and spintronic materials. The Mn^{2+} ion, used in many luminescent materials, has a d^5 configuration. The Mn^{2+} ion exhibits a broad emission peak, whose position depends strongly on the host lattice due to changes in crystal field strength with host. The emission colour can vary from green to deep red, corresponding to a $^4T_1$–$^6A_1$ transition. Since this transition is spin-forbidden, the typical luminescent relaxation time of this emission is of the order of milliseconds.

Doping of the impurity in the host material can be carried out via nucleation-doping or growth-doping processes [48] as shown in **Figure 1**. In a former process, both host and dopant materials are subjected to nucleation at the same time followed by growth of the host

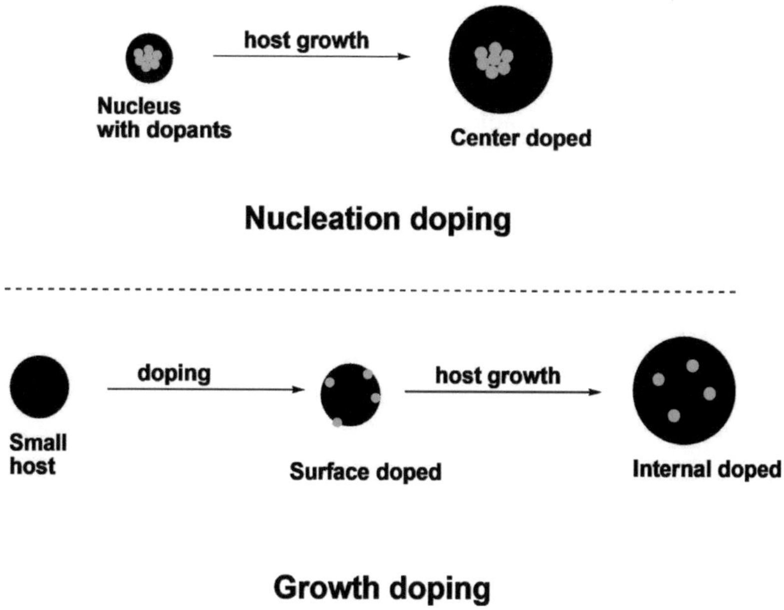

Figure 1. Schematic representations of nucleation- and growth-doping techniques.

material, whereas in the growth-doping process, the host material alone is nucleated and grow for certain time, followed by doping and re-growth of host material again.

1.4. QDs as sensor

QDs possess high surface-to-volume ratio and any change occurring at the surface can influence its surface-dependent properties; hence, luminescence of QDs is sensitive to surface states. Any species that interact directly with the QDs surface by physical or chemical means interferes with radiative recombination of the exciton leading to 'quenching' or enhancement of QDs fluorescence. Following this approach, QDs fluorescent probes can be designed by modifying their surface with suitable ligands so that they can selectively recognize the target analyte.

QD-based sensors are most frequently used to detect metal cations. Generally, metal ion quenches the QDs fluorescence via three different types of interaction: (i) by exchanging the metal cation of QDs, (ii) by displacing the capping ligand from the QDs surface and (iii) by binding with surface ligands [48]. The first two types are based on the competitive interaction between the analyte metal ion and the surface metal ion of QDs to bind with counterpart anion or surface-capping ligand, respectively, thus forming a stable lower solubility product. For instance, CuSe has a lower solubility than CdSe. As a result, surface Cd^{2+} ions in CdSe QDs can be easily exchanged by analyte Cu^{2+} ions to form CuSe particles on the surface of QDs. Similarly, in the case of glutathione (GSH)-capped CdSe QDs, Pb^{2+} ion binds with the thiol group of the capping GSH followed by displacement from QDs surface due to its higher binding affinity than Cd^{2+}. However, the quenching mechanism in both cases is different. The electron and/or hole transfer from the CdSe to the CuSe energy levels (case i) is much faster than the process of fluorescence generation in the CdSe moiety, thus leading to quenching of QD fluorescence. The CuSe particles chemisorbed on CdSe QDs alters the bandgap energy level and crystal lattice. Hence, the shift in absorption, emission and X-ray diffractometry (XRD) patterns of the QDs after the interaction with the analyte metal ions is the indicative of this metal ion exchange process. By contrast, removing the capping ligand from the QDs surface by analyte metal ion (case ii) depassivates the QDs surface and creates drastic imperfections on the QDs surface, resulting in fluorescence quenching. In the ion-binding type (case iii), divalent analyte metal ions adsorb to the QD surface by electrostatic interaction with the capping carboxylic ligands, thus coordinating several QDs together, leading to the formation of closely packed QD aggregates. This causes a decrease in the luminescence intensity due to self-quenching mechanism. The quenching by this interaction is sensitive to pH and ionic strength of the medium. Under acidic conditions, quenching by metal ions is very low due to non-availability of carboxylate for the interaction with metal ions.

Apart from the metal ions, molecules such as phenolic compounds, H_2O_2 [49], 2,4,6-trinitrotoluene (TNT) [50, 51] and glucose [52] can also be detected by fluorometric titration with QDs. The electron-accepting nature of phenolic compounds and TNT shuttled the electron from the conduction band to the valence band of the excited QDs, whereas H_2O_2 oxidizes the surface of QD and destroys its lattice structure resulting in the PL quenching. Glucose can be indirectly detected knowing that glucose can produce H_2O_2 on catalytic oxidation by glucose oxidase.

Most of the QDs that were investigated for fluorometric sensing are based on cadmium QDs; however, a major drawback for their application is the toxicity of cadmium ion. Less toxic particles like doped ZnS or ZnSe QDs may be interesting alternatives for biological imaging and other applications. Mn^{2+}-doped ZnS quantum dots have been extensively investigated in various fields [53]. Fang et al. synthesized high-quality water-dispersible Mn^{2+}-doped ZnSe core/ZnS shell (Mn:ZnSe/ZnS) nanocrystals directly in aqueous media with MPA as the capping ligand [54]. They observed that there was almost no dopant Mn emission in the Mn:ZnSe d-dots and bright Mn luminescence was observed after overcoating the ZnS shell around the Mn:ZnSe dots. In the present work, Mn-doped ZnSe (Mn:ZnSe) QDs have been synthesized by a wet chemical method using 3-mercaptopropionic acid (3-MPA) as capping agent and characterized by various analytical tools. The nucleation-doping method was adopted because it would form a structure similar to core-shell (MnSe/ZnSe) with a diffuse interface. The PL of the resulting QDs was examined in the presence of different metal ions to check its selective response.

2. Methodology

2.1. Materials

All chemicals were of analytical grade and were used without further purification. All solutions were prepared using doubly distilled water. Zn (CH3COO) $_2$•$2H_2O$ ($ZnAc_2$) and Mn

http://dx.doi.org/10.5772/intechopen.70669

$(CH_3COO)_2 \bullet 4H_2O$ ($MnAc_2$) were obtained from Sisco Research Laboratories (SRL) Pvt. Ltd., Mumbai. Selenium powder (99.99%) 3-mercaptopropionic acid and sodium borohydride (NaBH4) were the AR reagents from Sigma Aldrich, Bangalore. Phosphate buffer at a pH of 7.0 was prepared according to literature. The chloride solutions of different metal ions were prepared at the concentration of 1×10^{-4} M using doubly distilled (DD) water.

2.2. Synthesis of Mn-doped ZnSe QDs

Aqueous colloidal solution of Mn:ZnSe QDs was synthesized using MPA as capping agent as depicted in **Figure 2**. Typically, 0.25 mmol of selenium powder and 0.6 mmol of $NaBH_4$ were loaded in a round-bottomed flask containing 10 mL of N_2-purged DD water. The reaction mixture was heated at 100°C for 15 min under N_2 flow until the black selenium powder disappeared completely to give a clear colourless solution. On the other hand, 0.5 mmol of $ZnAc_2$ and 0.0015 or 0.003 mmol of $MnAc_2$ were dissolved in 25 mL of DD water followed by the addition of MPA. The pH of the reaction mixture was adjusted to 10.3 by adding 1 M NaOH solution and purged with N_2 for 20 min. Then, freshly prepared NaHSe solution was added followed by stirring at 50°C for 2 h. The molar ratio of Zn to Se to MPA was set at 1:0.5:2, whereas Zn to Mn was 1:0.03 or 1:0.06. The as-synthesized nanoparticles were purified by precipitation with ethanol, followed by centrifugation and vacuum drying.

2.3. Characterizations

To investigate the optical, crystal structure and morphological properties of QDs, they were characterized by various analytical techniques. ultraviolet-visible (UV-vis) absorption spectra were recorded with a Shimadzu, (Model UV-1800) UV-visible spectrophotometer, Japan. The samples were dispersed in doubly distilled water and loaded in a 4.5-mL precleaned quartz cuvette with 1-cm optical path. The entire spectrum was scanned against the background spectrum of water. PL measurements of the samples were performed in a 4.5-mL quartz cuvette at ambient conditions on a Perkin Elmer LS5B spectrofluorimeter. For a given sample, the excitation wavelength was identified from the absorption spectrum

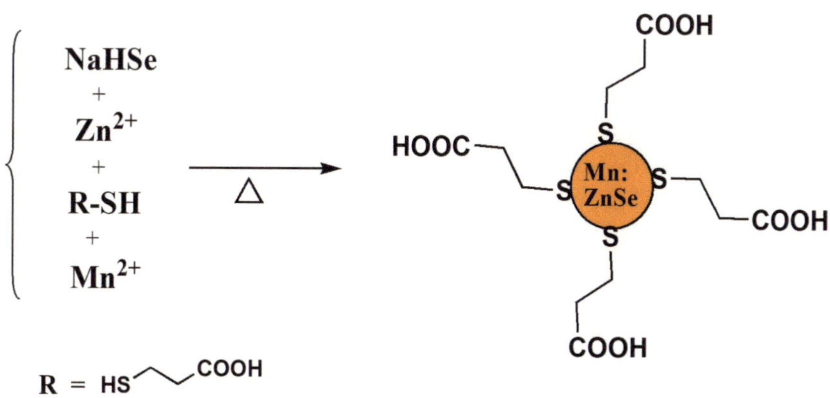

Figure 2. Synthesis of Mn-doped ZnSe QDs.

and it was fixed to scan the emission wavelength. X-ray powder diffraction patterns of the samples were recorded at ambient conditions by using PANanalytical X'Pert PRO diffractometer with monochromatic Cu-Kα1 radiation (λ = 1.5418 Å), 2θ ranging from 10 to 80° in steps of 0.017°/s. The accelerating voltage was set at 40 kV and the current flux was 30 mA. Transmission electron microscopy (TEM) images of QDs were obtained from the FEI Tecnai G2 (T-30) instrument with the operating voltage of 250–300 kV. A small amount of sample for TEM analysis was ultra-sonicated in ethanol or water for a few minutes and then dropped on carbon-coated copper grids. The sample grid was then kept in vacuum desiccators prior to the analysis.

2.4. Metal ion sensing

Fluorescence sensitivity of the QDs towards different metal ions was carried out on a fluorescence microplate reader (Turners Biosystems-Modulus Microplate Multimode Reader-9300-010). Stock solutions of different metal ions (Li^+, Na^+, K^+, Mg^{2+}, Ca^{2+}, Ba^{2+}, Al^{3+}, Mn^{2+}, Fe^{2+}, Co^{2+}, Ni^{2+}, Cu^{2+}, Zn^{2+}, Ag^+, Cd^{2+}, Hg^{2+} and Pb^{2+}) were prepared by dissolving their respective nitrate or chloride salts. Aqueous solution of QDs with the OD = 0.1 was prepared by diluting the stock solution of QDs. The solution of 50 μL was dropped into each well of a 96-well plate followed by filling with different metal solutions to obtain the final volume of 200 μL. The excitation wavelength was selected according to the absorption spectrum of QDs and the relative fluorescence intensity was measured with the plate reader. The standard deviations were calculated from at least six measurements.

3. Results and discussion

3.1. Absorption spectra

Figure 3a shows the absorption spectra of 3 and 6% Mn:ZnSe QDs along with undoped QDs. The Mn content is represented here as the mole % against the host metal (Zn) which is considered as 100% according to the experiment. Compared with bulk ZnSe having the bandgap of ~2.8 eV, the absorption band edge of both doped and undoped QDs is blue-shifted (400–500 meV) indicating quantum confinement of the particles. Furthermore, the bandgap of Mn:ZnSe is blue-shifted with respect to undoped ZnSe QDs under the same experimental conditions, which reveal the formation of smaller-sized particles. This is consistent with the results by Mahamuni et al [55]. The synthesis process itself is affected by Mn additive. In addition, the spectrum of 3% Mn:ZnSe QDs is slightly blue-shifted from that of 6% Mn:ZnSe QDs. Because of lower Mn/Se content, the former has relatively smaller-sized nuclei formed in the nucleation step which leads to a slight blue shift.

3.2. PL spectra

The PL spectra of the 3 and 6% Mn:ZnSe QDs along with undoped QDs (λexc = 365 nm) are shown in **Figure 3b**. The spectrum of undoped ZnSe QDs shows broad emission band centred

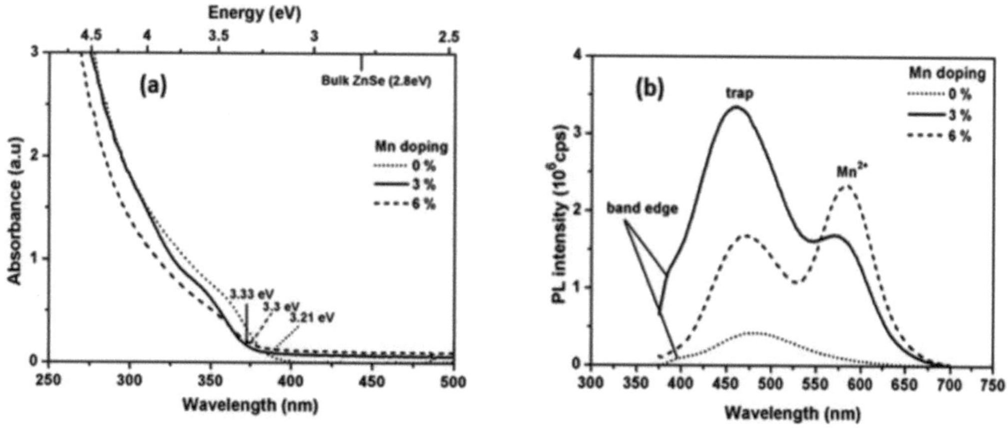

Figure 3. (a) Absorption and (b) PL spectra of undoped and Mn-doped ZnSe QDs.

at 480 nm and a small shoulder at 398 nm assigned to trap-state emission and band-edge emission, respectively. These two emissions appearing together are often observed in CdSe, ZnSe nanocrystals and their sulphur analogues. As discussed by Denzler et al [56], the bulk defects such as vacancies (Schottky defects) and interstitials (Frenkel defects) are the main source of trap states in the aqueous ZnSe QDs. With cubic zinc blende structure (by XRD), ZnSe usually has Schottky defects predominant over Frenkel defects. Therefore, the observed photoluminescence of QDs could be ascribed to a recombination of electrons at the selenium vacancy energy levels because of low Se/Zn ratio synthetic conditions. There are many possible recombination paths through many trap-state emissions, each with different emission energy, causing the relatively wide emission peak. The PL spectrum of as-prepared ZnSe QDs has the full width at half maximum of about 110 nm.

The PL spectrum of 3% Mn:ZnSe QDs exhibits a new band at 579 nm in addition to trap- and band-edge emissions. The appearance of this new emission in the Mn-doped QDs is attributed to Mn^{2+}-related $^4T_1 \rightarrow {}^6A_1$ emission. This transition is spin-forbidden but is allowed because of crystal field effects [55]. Fang et al [54] notified that if the dopant ions are adsorbed on the surface of the host nanocrystal instead of being incorporated into its lattice, no dopant emission is observed and further the host emission is drastically quenched because a loosely adsorbed dopant ion can easily act as a surface trap that quenches the host PL [57]. However, in the present system, after doping with 3% Mn, the luminescence of host emission is enhanced by 7.7 times along with the appearance of Mn emission. This indicates the successful incorporation of Mn^{2+} in the host ZnSe lattice. The enhancement of host PL is mainly attributed to the reduction of non-radiative energy centres by Mn^{2+} doping. **Table 1** shows the PL-integrated intensity ratios calculated from the spectra.

With further increase in the Mn concentration from 3 to 6%, the resultant doped QDs show increased Mn emission followed by subtle red-shift indicating the increase in Mn content in host. However, the host-trap emission is decreased and no band-edge emission is significantly observed. Furthermore, instead of the expected two-fold increase in the Mn emission

PL-integrated intensity ratios of Mn:ZnSe QDs									
Mn doping	Band-edge		Trap		Mn		Cumulative	Trap/Band edge	Trap/ Mn
	Peak (nm)	$I/I_{0\%}$	Peak (nm)	$I/I_{0\%}$	Peak (nm)	$I/I_{3\%}$	$I/I_{0\%}$	I_t/I_b	I_t/I_{Mn}
0%	397	1.0	480	1.0	–	–	1.0	82.2	–
3%	386	6.7	461	7.7	579	1.0	10.1	95.7	3.2
6%	–	–	473	3.1	584	1.4	6.5	–	0.9

*t = trap, b = band edge. (PL peak position and the corresponding integrated intensity are obtained by Gaussian fitting.)

Table 1. PL properties of undoped and Mn-doped ZnSe QDs.

(from 3% doped), only 1.4-fold increase is observed. The reason might be the combination of the following three processes: (i) non-radiative energy transfer between neighbouring Mn^{2+}-dopant ions which quenches Mn emission [58], (ii) adsorption of some Mn ions on the surface instead of incorporation into the host lattice which quenches both host and Mn emission and (iii) initially formed larger MnSe. As the environment of the doping ions in larger nuclei is not as uniform as that in smaller ones, the PL emission performance of doped QDs with larger nuclei will not be satisfactory as those with smaller nuclei [57]. Overall, the PL peak position of the host is in tune with the corresponding band-edge absorption. The 3% doped sample has the highest emission peak intensity, followed by the 6% and then the 0% sample. The above results are interesting that the ratio of the dual-colour emissions (blue and orange) of the Mn-doped ZnSe QDs could be controlled by tuning the Mn-doping levels in the QDs.

3.3. Structural and morphological analysis

X-ray diffraction pattern of the 3% Mn:ZnSe QDs is presented in **Figure 4**. Broad diffraction peaks are observed and are attributed to the nanocrystalline nature of the material. The XRD peaks are close to the characteristic peaks corresponding to the (111), (220) and (311) planes of cubic zinc blende ZnSe. Cubic structures are often obtained in the low-temperature aqueous synthesis of ZnSe QDs. This indicates that the incorporation of Mn^{2+} into the host ZnSe does not bring any obvious change in the crystal lattice and the structure [37, 54]. TEM image of the 3% Mn:ZnSe QDs (**Figure 5a**) shows that QDs are spherical in nature with the average diameter of ~4 nm. Some of the aggregates of the particles are also seen. The energy-dispersive X-ray spectroscopy (EDS) spectrum (**Figure 5b**) confirms the presence of Mn, Zn and Se and the purity of the sample.

3.4. Effect of metal ions on the PL intensity

The fluorescence titrations of the 3% Mn:ZnSe QDs (PL, λ_{max} = 461 nm) with various metal ions are shown in **Figure 6a**. The fluorescence intensity (F) is measured with excitation at 365 nm. The PL of blank QDs is used as a control (F_0). From the figure, it is observed that QDs show maximum response to mercury ions (Hg^{2+}), little response to Ni^{2+}, Cu^{2+} and Pb^{2+} ions and

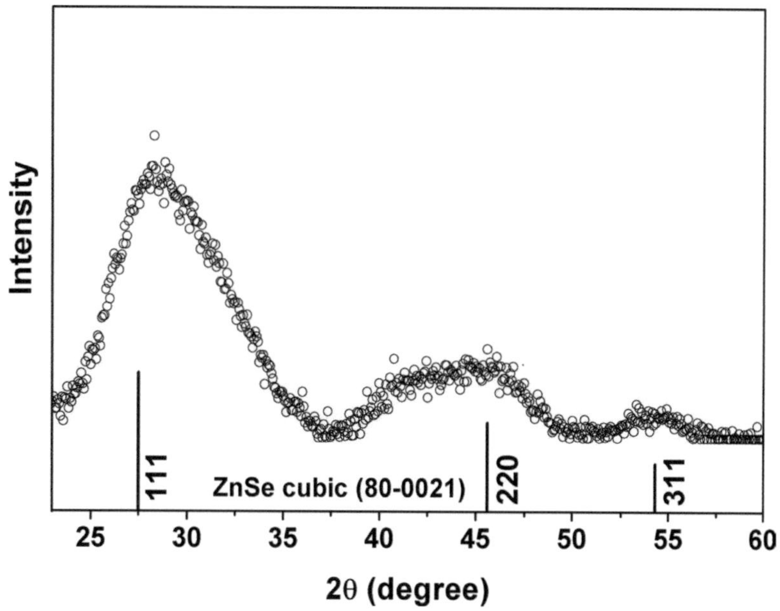

Figure 4. XRD patterns of 3% Mn-doped ZnSe QDs.

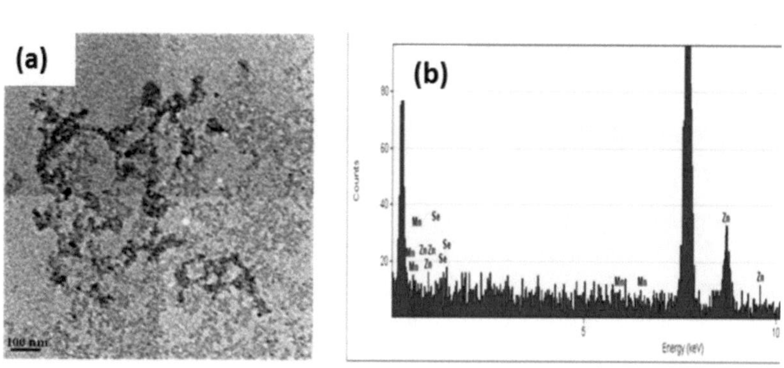

Figure 5. (a) TEM image and (b) EDS of 3% Mn-doped ZnSe QDs.

almost no response to other metal ions. About 74% of the PL intensity is quenched after the addition of 30 μM L^{-1} of Hg^{2+} ions.

The influence of [Hg^{2+}] ion on the QDs fluorescence was studied. **Figure 6b** shows the quenching behaviour of Hg^{2+} ions on the PL intensity of QDs. The PL intensity is quenched drastically and then slightly with the increase in [Hg^{2+}] ions. The fluorescence quenching with respect to the concentration of quencher was analysed using the Stern-Volmer equation. The plot of F_0/F versus [Hg^{2+}] as shown in **Figure 6c** exhibits a good linear relationship up to 30 μM L^{-1} with a correlation coefficient R^2 = 0.9918. The limit of detection (LOD) was calculated according to the following equation:

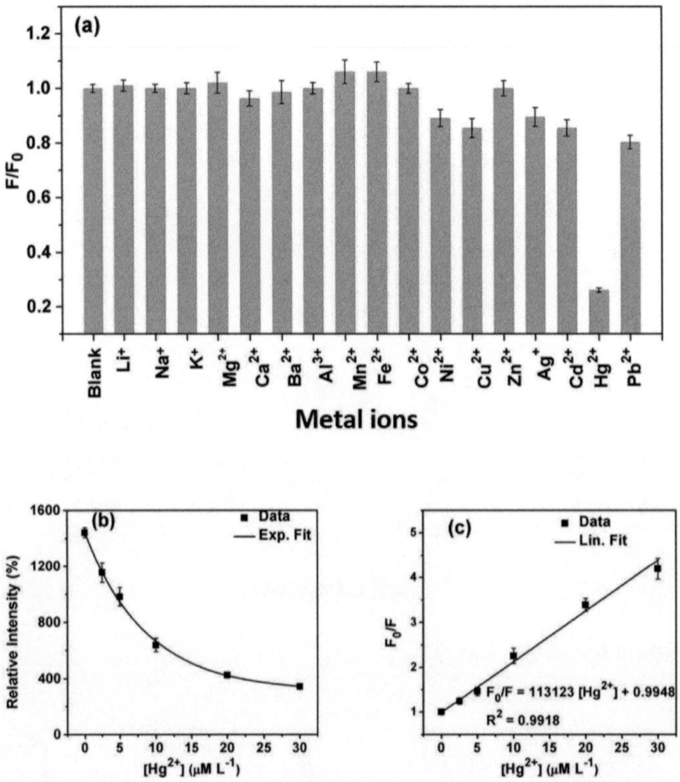

Figure 6. (a) Effect of metal ions (30 μM L^{-1}) on the PL intensity of 3% Mn-doped ZnSe QDs at pH 10.8. (b) The plot of PL intensity of 3% Mn-doped ZnSe QDs versus [Hg^{2+}] ions and (c) the corresponding Stern-Volmer relationship.

$$\mathrm{LOD} = \frac{3\,S_0}{K_{sv}} \tag{1}$$

where 3 is the signal-to-noise ratio, S_0 is the standard deviation of blank measurements (n = 6) and K_{sv} is the slope of calibration graph. LOD of the present probe towards [Hg^{2+}] under the experimental conditions is found to be 6.63×10^{-7} M L^{-1}.

3.5. Mechanism of sensing by quenching

The fluorescence quenching can be explained in terms of strong binding of quencher metal ions (Hg^{2+}) on the surface of QDs. It is known that the solubility HgSe ($K_{sp} = 2 \times 10^{-53}$) is extremely lower than that of ZnSe ($K_{sp} = 3.6 \times 10^{-26}$). The low solubility product is always preferred in a solution and hence the quencher metal ions (Hg^{2+}) displace the Zn on the surface of QDs and form a lower solubility product (HgSe) which deposit on the surface of the QDs

$$Zn_m Se_n + xHg^{2+} \rightarrow Zn_{m-x} Hg_x Se_n + xZn^{2+} \tag{2}$$

The formed particles of HgSe, both isolated and aggregated, can quench the luminescence of QDs by facilitating non-radiative annihilation of charge carriers, which act as electron-hole

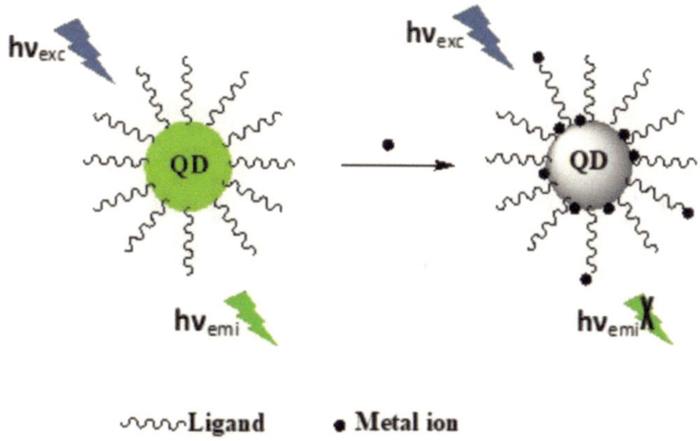

Figure 7. Schematic representation of fluorescence quenching of QDs by metal ions.

recombination centres [59]. However, mere K_{sp} values are not the sole factors in fluorescence quenching. In addition, surface ligands have profound effects on the fluorescence response of QDs to metal ions [60]. They play a critical role in metallic ion selectivity [61]. Also, the quenching of luminescence of the QDs can occur partly through ion binding followed by photoinduced electron transfer process from the thiol ligand to Hg^{2+} ions on the surface of QDs. The schematic representation of fluorescence quenching is shown in **Figure 7**. Theoretical calculation and further study are in progress to gain an insight into the mechanism of fluorescence quenching.

4. Conclusions

In summary, water-soluble MPA-capped Mn-doped ZnSe QDs were synthesized using nucleation-doping method. The absorption spectra of the as-synthesized QDs were blue-shifted in relation to the bulk counterparts due to quantum confinement. The QDs exhibited dual-colour emission (blue and orange). The intensity ratio of the dual-colour emission can be varied by tuning the Mn-doping percentage. It was found that 3% Mn doping in the ZnSe crystal lattice increases the fluorescence (blue) of ZnSe QDs by 10-fold due to the reduction of non-radiative energy centres. Furthermore, Mn^{2+}-related $^4T_1 \rightarrow ^6A_1$ (orange) emission characteristic of Mn doped in the ZnSe host was observed. The as-prepared QDs exhibited a cubic crystal structure according to XRD patterns. TEM images showed aggregates of tiny spherical particles with the average diameter of ~4 nm. The as-prepared $Mn_{3\%}$:ZnSe QDs were treated with different metal ions and were found to be highly selective to Hg^{2+} ions and exhibit pleasing LOD. The possible mechanism of sensing by quenching was also proposed. These studies on metal ion sensing by fluorescent QDs have demonstrated their potential as chemical sensor which can be developed for industrial applications.

Author details

Sundararajan Parani[1,2], Ncediwe Tsolekile[1,2,3], Bambesiwe M.M. May[1,2], Kannaiyan Pandian[4] and Oluwatobi S. Oluwafemi[1,2]*

*Address all correspondence to: oluwafemi.oluwatobi@gmail.com

1 Department of Applied Chemistry, University of Johannesburg, Doornfontein, Johannesburg, South Africa

2 Centre for Nanomaterials Science Research, University of Johannesburg, Johannesburg, South Africa

3 Department of Chemistry, Cape Peninsula University of Technology, Cape Town, South Africa

4 Department of Inorganic Chemistry, University of Madras, Chennai, India

References

[1] McGhee CE, Yong Loh K, Lu Y. DNAzyme sensors for detection of metal ions in the environment and imaging them in living cells. Current Opinion in Biotechnology [Internet]. 2017;**45**:191-201 Available from: http://dx.doi.org/10.1016/j.copbio.2017.03.002

[2] Barone C, Bolzoni L, Caruso G, Montanari A, Parisi S, Steinka I. Food Packaging Hygiene. 2015. Available from: http://link.springer.com/10.1007/978-3-319-14827-4

[3] Zhong WS, Ren T, Zhao LJ. Determination of Pb (Lead), Cd (Cadmium), Cr (Chromium), Cu (Copper), and Ni (Nickel) in Chinese tea with high-resolution continuum source graphite furnace atomic absorption spectrometry. Journal of Food and Drug Analysis [Internet]. 2016;**24**(1):46-55 Available from: http://dx.doi.org/10.1016/j.jfda.2015.04.010

[4] McComb JQ, Rogers C, Han FX, Tchounwou PB. Rapid screening of heavy metals and trace elements in environmental samples using portable X-ray fluorescence spectrometer: A comparative study. Water, Air, & Soil Pollution [Internet]. 2014;**225**(12):2169 Available from: http://link.springer.com/10.1007/s11270-014-2169-5

[5] Barón-Jaimez J, Joya MR, Barba-Ortega J. Anodic stripping voltammetry – ASV for determination of heavy metals. Journal of Physics: Conference Series [Internet]. 2013;**466**:12023 Available from: http://stacks.iop.org/1742-6596/466/i=1/a=012023?key=crossref.8f5785f72a4a87ee677967514ca4e165

[6] Verma R, Gupta BD. Fiber optic SPR sensor for the detection of 3-pyridinecarboxamide (vitamin B3) using molecularly imprinted hydrogel. Sensors and Actuators B: Chemical [Internet]. 2013;**177**:279-285 Available from: http://dx.doi.org/10.1016/j.snb.2012.10.135

[7] Deng G. Principles of chemical and biological sensors. Materials and Manufacturing Processes [Internet]. 1999;**14**(4):623-625 Available from: http://www.tandfonline.com/doi/abs/10.1080/10426919908907570

[8] Neupane LN, Oh ET, Park HJ, Lee KH. Selective and sensitive detection of heavy metal ions in 100% aqueous solution and cells with a fluorescence chemosensor based on peptide using aggregation-induced emission. Analytical Chemistry. 2016;**88**(6):3333-3340

[9] Verma R, Gupta BD. Detection of heavy metal ions in contaminated water by surface plasmon resonance based optical fibre sensor using conducting polymer and chitosan. Food Chemistry [Internet]. 2015;**166**:568-575 Available from: http://dx.doi.org/10.1016/j.foodchem.2014.06.045

[10] Vázquez-González M, Carrillo-Carrion C. Analytical strategies based on quantum dots for heavy metal ions detection. Journal of Biomedical Optics [Internet]. 2014;**19**(10):101503 Available from: http://biomedicaloptics.spiedigitallibrary.org/article.aspx?doi=10.1117/1.JBO.19.10.101503

[11] Chen Y, Li S, Huang L, Pan D. Low-cost and gram-scale synthesis of water-soluble Cu–In–S/ZnS core/shell quantum dots in an electric pressure cooker. Nanoscale. 2014;**6**:1295-1298 Available from: http://pubs.rsc.org/en/Content/ArticleLanding/2014/NR/C3NR05014A#!divAbstract

[12] Szulczyński B, Gębicki J. Currently commercially available chemical sensors employed for detection of volatile organic compounds in outdoor and indoor air. Environments [Internet]. 2017;**4**(1):21 Available from: http://www.mdpi.com/2076-3298/4/1/21

[13] Mehrotra P. Biosensors and their applications—A review. Journal of Oral Biology and Craniofacial Research [Internet]. 2016;**6**(2):153-159 Available from: http://dx.doi.org/10.1016/j.jobcr.2015.12.002

[14] Ramasamy R, Gopal N, Kuzhandaivelu V, Murugaiyan S. Biosensors in clinical chemistry: An overview. Advanced Biomedical Research [Internet]. 2014;**3**(1):67 Available from: http://www.advbiores.net/text.asp?2014/3/1/67/125848

[15] Yeh P, Yeh N, Lee CH, Ding TJ. Applications of LEDs in optical sensors and chemical sensing device for detection of biochemicals, heavy metals, and environmental nutrients. Renewable & Sustainable Energy Reviews [Internet]. 2017;**75**:461-468 Available from: http://dx.doi.org/10.1016/j.rser.2016.11.011

[16] Long F, Zhu A, Shi H. Recent advances in optical biosensors for environmental monitoring and early warning. Sensors (Basel). 2013;**13**(10):13928-13948

[17] Weichert F, Bachmann D, Rudak B, Fisseler D. Analysis of the accuracy and robustness of the Leap Motion Controller. Sensors (Switzerland). 2013;**13**(5):6380-6393

[18] Mathews CB, Libish TM, Kaushalkumar B, Vivek V, Prabhu R, Radhakrishnan P. A fiber optic biosensor for the detection of cholesterol levels based on chitosan coated long period grating. Optoelectronics Letters. 2016;**12**(1):23-26

[19] Daou TJ, Pourroy G, Bégin-Colin S, Grenèche JM, Ulhaq-Bouillet C, Legaré P, et al. Hydrothermal synthesis of monodisperse magnetite nanoparticles. Chemistry of Materials. 2006;**18**(18):4399-4404

[20] Oehme I, Wolfbeis OS. Optical sensors for determination of heavy metal ions. Microchimica Acta. 1997;**126**:177-192

[21] Oluwafemi SO, Revaprasadu N, Ramirez AJ. A novel one-pot route for the synthesis of water-soluble cadmium selenide nanoparticles. Journal of Crystal Growth. 2008;**310** (13):3230-3234

[22] Wang R, Lu K-Q, Tang Z-R, Xu Y-J. Recent progress on carbon quantum dots: Synthesis, properties and applications in photocatalysis. Journal of Materials Chemistry A [Internet]. 2017;**5**:3717-3734 Available from: http://pubs.rsc.org/en/Content/ArticleLanding/2017/TA/C6TA08660H

[23] Kang X, Yang Y, Huang L, Tao Y, Wang L, Pan D. Large-scale synthesis of water-soluble $CuInSe_2$/ZnS and $AgInSe_2$/ZnS core/shell quantum dots. Green Chemistry [Internet]. 2015;**17**(8):4482-4488 Available from: http://dx.doi.org/10.1039/C5GC00908A

[24] Raeyani D, Shojaei S, Kandjani SA, Wlodarski W. Synthesizing graphene quantum dots for gas sensing applications. Procedia Engineering [Internet]. 2016;**168**:1312-1316 Available from: http://dx.doi.org/10.1016/j.proeng.2016.11.356

[25] Chen Y, Li S, Huang L, Pan D. Green and facile synthesis of water-soluble Cu-In-S/ZnS core/shell quantum dots. Inorganic Chemistry. 2013;**52**(14):7819-7821

[26] Chuang P, Lin CC, Liu R. Emission-tunable $CuInS_2$/ZnS quantum dots: Structure, optical properties, and application in white light-emitting diodes with high color rendering index. Applied Materials and Interfaces. 2014;**6**(17):15379-15387

[27] Malik P, Singh J, Kakkar R. A review on CdSe quantum dots in sensing. Advanced Materials Letters. 2014;**5**(11):612-628

[28] Murray CB, Norris DJ, Bawendi MG. Synthesis and characterization of nearly monodisperse CdE (E = sulfur, selenium, tellurium) semiconductor nanocrystallites. Journal of the American Chemical Society [Internet]. 1993;**115**(19):8706-8715 Available from: http://pubs.acs.org/doi/abs/10.1021/ja00072a025

[29] Draaisma GJJ, Reardon D, Schenning APHJ, Meskers SCJ, Bastiaansen CWM. Ligand exchange as a tool to improve quantum dot miscibility in polymer composite layers used as luminescent down-shifting layers for photovoltaic applications. Journal of Materials Chemistry C [Internet]. 2016;**4**(24):5747-5754 Available from: http://xlink.rsc.org/?DOI=C6TC01261B

[30] Karakoti AS, Shukla R, Shanker R, Singh S. Surface functionalization of quantum dots for biological applications. Advances in Colloid and Interface Science [Internet]. 2015;**215**:28-45 Available from: http://dx.doi.org/10.1016/j.cis.2014.11.004

[31] Rajh T, Micic OI, Nozik AJ. Synthesis and characterization of surface-modified colloidal cadmium telluride quantum dots. The Journal of Physical Chemistry. 1993;**97**:11999-12003 Available from: http://pubs.acs.org/doi/abs/10.1021/j100148a026

[32] Oluwafemi OS, Daramola OA, Ncapayi V. A facile green synthesis of type II water soluble CdTe/CdS core shell nanoparticles. Materials Letters [Internet]. 2014;**133**:9-13 Available from: http://www.sciencedirect.com/science/article/pii/S0167577X14012014

[33] Ncapayi V, Parani S, Songca SP, Kodama T, Oluwafemi OS. Simple green synthesis of amino acid functionalised CdTe/CdSe/ZnSe core-multi shell with improved cell viability for cellular imaging. Materials Letters [Internet]. 2017;**189**:168-171 Available from: http://dx.doi.org/10.1016/j.matlet.2016.12.006

[34] Parani S, Tsolekile N, Pandian K, Oluwafemi OS. Thiolated selenium as a new precursor for the aqueous synthesis of CdSe/CdS core/shell quantum dots. Journal of Materials Science: Materials in Electronics. 2017;**28**:11151. DOI: 10.1007/s10854-017-6902-x

[35] Krivenkov V, Tretyachenko A, Samokhvalov PS, Chistyakov AA, Nabiev I. Controllable photo-brightening/photo-darkening of semiconductor quantum dots under laser irradiation. Proceedings of SPIE [Internet]. 2016;**9884**:98843L Available from: http://proceedings.spiedigitallibrary.org/proceeding.aspx?doi=10.1117/12.2228815

[36] Luo C, Li B, Peng H, Tang X, Wang Y, Travas-Sejdic J. The effect of photo-irradiation on the optical properties of thiol-capped CdTe quantum dots. Journal of Nanoscience and Nanotechnology [Internet]. 2012;**12**(4):2998-3005 Available from: http://www.ncbi.nlm.nih.gov/pubmed/22849057

[37] Wang X, Kong X, Yu Y, Sun Y, Zhang H. Effect of annealing on upconversion luminescence of ZnO:Er 3+ nanocrystals and high thermal sensitivity. The Journal of Physical Chemistry C [Internet]. 2007;**111**(41):15119-15124 Available from: http://pubs.acs.org/doi/abs/10.1021/jp0686689

[38] Saleh R, Djaja NF. Transition-metal-doped ZnO nanoparticles: Synthesis, characterization and photocatalytic activity under UV light. Spectrochimica Acta Part A: Molecular and Biomolecular Spectroscopy [Internet]. 2014;**130**:581-590 Available from: http://dx.doi.org/10.1016/j.saa.2014.03.089

[39] Schelonka D, Tolasz J, Štengl V. Doping of zinc oxide with selected first row transition metals for photocatalytic applications. Photochemistry and Photobiology. 2015;**91**(5):1071-1077

[40] Aleksandrova OA, Mazing DS, Musikhin SF, Nikiforova AV. Synthesis of transition metal doped zinc selenide nanoparticles for bioimaging. Photonics North. 2015:1 doi: 10.1109/PN.2015.7292477

[41] Mathew S, Bhardwaj BS, Saran AD, Radhakrishnan P, Nampoori VPN, Vallabhan CPG, et al. Effect of ZnS shell on optical properties of CdSe-ZnS core-shell quantum dots. Optical Materials (Amsterdam). 2015;**39**:46-51

[42] Bhargava RN, Gallagher D, Hong X, Nurmikko A. Optical properties of manganese-doped of ZnS. Physical Review Letters. 1994;**72**(3):1-4

[43] Rajesh C, Phadnis C, Sonawane KG, Mahamuni S. Generation of white light from co-doped (Cu and Mn) ZnSe QDs. Journal of Experimental Nanoscience [Internet]. 2014;**10**(14):1082-1092 Available from: http://www.tandfonline.com/doi/abs/10.1080/17458080.2014.964340

[44] Rajesh C, Phadnis CV, Sonawane KG, Mahamuni S. Synthesis and optical properties of coppper-doped ZnSe quantum dots. Physica Scripta [Internet]. 2015;**90**(1):15803 Available from: http://stacks.iop.org/1402-4896/90/i=1/a=015803?key=crossref.d67be3a40db02beaa895c20d8952b09f

[45] Aphesteguy JC, Kurlyandskaya GV, de Celis JP, Safronov AP, Schegoleva NN. Magnetite nanoparticles prepared by co-precipitation method in different conditions. Materials Chemistry and Physics [Internet]. 2015;**161**:243-249. Available from: http://www.sciencedirect.com/science/article/pii/S0254058415301073

[46] Wang Y, Liang X, Liu E, Hu X, Fan J. Incorporation of lanthanide (Eu(3+)) ions in ZnS semiconductor quantum dots with a trapped-dopant model and their photoluminescence spectroscopy study. Nanotechnology [Internet]. 2015;**26**(37):375601 Available from: http://apps.webofknowledge.com/full_record.do?product=UA&search_mode=GeneralSearch&qid=23&SID=V176ODJebApDXuKoNAp&page=1&doc=6&cacheurlFromRightClick=no

[47] Sureshkumar S, Jothimani B, Sridhar TM, Venkatachalapathy B. Synthesis and characterization of gadolinium-doped ZnSe quantum dots for fluorescence imaging of cancer cells. RSC Advances [Internet]. 2016;**6**(19):16081-16086 Available from: http://xlink.rsc.org/?DOI=C5RA18773G

[48] Guria AK, Pradhan N. Doped or not doped: Ionic impurities for influencing the phase and growth of semiconductor nanocrystals. Chemistry Materials. 2016;**28**(15):5224-5237

[49] Li Y, Li B, Zhang J. H2O2- and pH-sensitive CdTe quantum dots as fluorescence probes for the detection of glucose. Luminescence. 2013;**28**(5):667-672

[50] Shamirian A, Ghai A, Snee PT. QD-based FRET probes at a glance. Sensors. 2015;**15**(6):13028-13051

[51] Xu S, Lu H, Li J, Song X, Wang A, Chen L, et al. Dummy molecularly imprinted polymers-capped CdTe quantum dots for the fluorescent sensing of 2,4,6-trinitrotoluene. ACS Applied Materials & Interfaces. 2013;**5**(16):8146-8154

[52] Ding L, Zhang B, Xu C, Huang J, Xia Z. Fluorescent glucose sensing using CdTe/CdS quantum dots–glucose oxidase complex. Analytical Methods [Internet]. 2016;**8**(14):2967-2970 Available from: http://xlink.rsc.org/?DOI=C5AY03205A

[53] Subha R, Nalla V, Yu JH, Jun SW, Shin K, Hyeon T, et al. Efficient photoluminescence of Mn^{2+}-doped ZnS quantum dots excited by two-photon absorption in near-infrared window II. Journal of Physical Chemistry C. 2013;**117**(40):20905-20911

[54] Fang Z, Wu P, Zhong X, Yang Y. Synthesis of highly luminescent Mn:ZnSe/ZnS nanocrystals in aqueous media. Nanotechnology. 2010;**21**(30):305604

[55] Mahamuni S, Lad AD, Patole S. Photoluminescence properties of manganese-doped zinc selenide quantum dots. Journal of Physical Chemistry C. 2008;**112**(7):2271-2277

[56] Denzler D, Olschewski M, Sattler K. Luminescence studies of localized gap states in colloidal ZnS nanocrystals. Journal of Applied Physics [Internet]. 1998;**84**(5):2841-2845 Available from: http://aip.scitation.org/doi/10.1063/1.368425

[57] Pradhan N, Battaglia DM, Liu Y, Peng X. Efficient, stable, small, and water-soluble doped ZnSe nanocrystal emitters as non-cadmium biomedical labels. Nano Letters. 2007;**7**(2):312-317

[58] Sooklal K, Cullum BS, Angel SM, Murphy CJ. Photophysical properties of ZNS nanoclusters with spatially localized Mn^{2+}. Journal of Physical Chemistry [Internet]. 1996;**100**(11):4551-4555 Available from: http://pubs.acs.org/doi/abs/10.1021/jp952377a

[59] Isarov AV, Chrysochoos J. Optical and photochemical properties of nonstoichiometric cadmium sulfide nanoparticles: Surface modification with copper(II) ions. Langmuir. 1997;**13**(12):3142-3149

[60] Chen Y, Rosenzweig Z. Luminescent CdS quantum dots as selective ion probes. Analytical Chemistry. 2002;**74**(19):5132-5138

[61] Gattás-Asfura KM, Leblanc RM. Peptide-coated CdS quantum dots for the optical detection of copper(II) and silver(I). Chemical Communications [Internet]. 2003;**2003**(21):2684-2685 Available from: http://www.ncbi.nlm.nih.gov/pubmed/14649810

Chapter 13

Quantum Dots and Fluorescent and Magnetic Nanocomposites: Recent Investigations and Applications in Biology and Medicine

Anca Armăşelu

Additional information is available at the end of the chapter

http://dx.doi.org/10.5772/intechopen.70614

Abstract

This chapter presents a comprehensive and updated review on the ongoing research area of nanostructures with a focus on quantum dots (QDs), fluorescent and magnetic nanocomposites, and their applications in biological and medical field. The study includes the essential characteristics of QDs and fluorescent and magnetic nanocomposites, their structure, properties, and methods that are utilized for their characterization. Some interesting qualities of CdSe/ZnS QDs with reference to the research of the microorganism are emphasized. The bioimaging applications of QDs and fluorescent and magnetic nanocomposites and their role as nanoprobes and as contrast enhancing agents are discussed. So, in this work, an overview is exhibited including the case of the most commonly studied QD-based hybrid NPs, which are called MQDs, such as a dual "two-in-one" fluorescent-magnetic nanocomposite materials, that blend both fluorescent and magnetic properties in a unique concept and show the feasibility for clinical diagnostics, drug delivery, and therapy.

Keywords: quantum dots, magnetic quantum dots, nanocomposites, fluorescence property, magnetic property, microorganism labeling, biological imaging, drug delivery

1. Introduction

Fast progresses in nanotechnology and nanoscience have offered a diversity of nanoscale materials possessing very controlled and distinctive optical, electrical, magnetic, or catalytic properties. The diversity of the composition (organic or inorganic, semiconductors or metals), form (particles, rods, wires, cubes, or triangles), and the availability for the functionalization of the surface (physical, chemical, or biological) allowed the production of the different functional nanoscale tools [1, 2]. The scientists have expanded new types of nanoscale tools that could be utilized for forensic science, biology, medicine, electronic technology, environmental science, computer fabrication, and food industries. The researchers in biological and medical fields already used these nanodevices in an assortment of uses varying from the diagnosis of disease to gene therapies. The combination of biomaterials (proteins, peptides, nucleic acids with semiconductor quantum dots (QDs), and metal nanoparticles is expected to generate important advances in molecular biology, bioengineering, medical, and therapeutic diagnostics.

The current progresses comprise the evolution of the functional nanoparticles (electronic, optical, and magnetic) which are conjugated to biological molecules such as peptides, proteins, and nucleic acids. Today, the magnetic nanoparticles (MNPs) are considered to be primary components in therapies and screening methods that are gradually included in many areas of medical practice. Due to the dimension-dependent properties and dimensional similitudes with biomolecules, the magnetic nanoparticles and their bioconjugates are highly appropriate for intracellular tagging [3, 4] and for image contrast-improving agents in magnetic resonance imaging (MRI) [5–8], magnetic separation [7, 9], targeted drug delivery [7, 8, 10], and for usage in hyperthermia [7, 10–12].

QDs are a unique class of fluorescent nanoparticles that are crystalline semiconductors of variable sizes (1–100 nm) and consist of only a few hundred to a few thousand atoms, in spite of the fact that QDs exhibit the same crystal structure as the bulk semiconductor material. The highest-quality QDs are typically composed of atoms from groups II and VI or groups III and V or groups IV and VI of the periodic table [1, 13, 14].

These nanoparticles, compared to their bulk, have smaller exciton Bohr radius which characterizes their definition [15]. This thing establishes what is described as quantum confinement when distinctive optical and electronic properties are created [15]. The nanometric dimension of QDs determines the quantum-confinement effect, which results in unique optical and electronic properties. Due to the effects of quantum confinement, QDs possess distinct photophysical properties that give QDs tremendous advantages over the conventional organic fluorophores [16–18]. Traditional organic dyes exhibit chemical and photophysical limitations such as pH dependence, susceptibility to photo-bleaching, narrow absorption windows of wavelengths, asymmetric emission spectra broadened by a red tail, small Stokes shifts, and short excited state fluorescent lifetimes [17–19].

The researchers in the field have found many production techniques for QDs, from photolithography to wet chemical synthesis. The most utilized QD construction consists of two materials from group II–VI materials, namely a CdSe core with a thin, protective shell of ZnS. Colloidal QDs are fabricated utilizing surfactant micelles, coprecipitation or organic solvent synthesis at high temperature. The last technique is used to manufacture the highest-quality materials [1, 14].

In contrast with the organic dyes and fluorescent proteins, QDs have distinct electronic and optical properties that comprise high quantum yield, broad absorption, large effective Stokes shifts, the ability to size-tune fluorescent emission as a function of core dimension, simultaneous excitation of multiple fluorescence colors, and high resistance against photo-bleaching

http://dx.doi.org/10.5772/intechopen.70614

and against photo- and chemical degradation [19, 20]. QDs have great potential in many applications such as solar cells, light-emitting devices, and photobio-labeling technologies. The unique optical properties of QDs make them appealing *in vivo* and *in vitro* in different biological and clinical applications, such as cell labeling [17, 20], cell tracking, *in vivo* imaging, and DNA detection [19, 21].

The combination of dual-natured parts of optical and magnetic properties on nanometer system can bring new advances in molecular imaging and medical theranosis, which are fundamental for early detection and rapid disease treatment. QDs represent an exciting and versatile category of fluorophores with a bright future, thus increasing the interest in blending the advantages of QDs with those of other materials to obtain composites with multifunctional properties [22, 23].

Nanocomposites, which comprise fluorescent and magnetic particles, represent a basis for multiplexed nanoprobe designs. The area of nanocomposite applications is still in a state of development [24]. The magnetic materials constitute one of the most frequent materials, which when amalgamated with QDs form a captivating class of new materials for bioimaging. In this sense, QD biosensing can be further improved by combination with MNPs (e.g., superparamagnetic iron oxide nanoparticles, SPIONs) or ions (e.g., gadolinium). The fluorescent property of the QDs allows visualization, while the magnetic property of the composite allows imaging, magnetic separation and can bring therapeutic advantages [23]. In this paper, actual investigations using only QDs or MNPs will be reviewed in situations where the applications can be expanded to nanocomposites.

This review examines the properties of QDs and magnetic QDs (MQDs) comprising the applications of these materials. Because the properties of these materials continue to enhance, QDs and MQDs possess the capacity to considerably determine biological imaging, diagnosis, and treatment. The application of QDs for combined targeting and delivery of diagnostic and therapeutic agents can be further developed by a combination with magnetic separation techniques via the recent evolution of MQDs. In this chapter, among many benefits of MQDs-based separation to current procedures, one should also mention the small dimension of MQDs, which are small enough to possibly interact with single-cell biomarkers/cell surface receptors which results in corresponding quantification of the results [22, 25]. These multifunctional fluorescent and magnetic nanoparticles of small dimension, which are MQDs, can target any biomolecule and can be separately controlled by engineering magnetic fields. Some authors reported very useful research papers which refer to a combined result of both types of fluorescent and magnetic properties to approach important biological issues [24, 26, 27]. This work provides a survey of the different application of biosensing technologies which are based on QDs and their MQD correspondents.

2. Properties of quantum dots

In the last decades, NPs were produced as interesting materials, with outstanding results for many applications. As a type of NPs, QDs represent the excellent competitors for optical bioanalysis by virtue of their electronic and optical properties, which can be adjusted by modifying the dimension, morphology, and composition of NPs [18, 28]. A new biological labeling material, QDs are considered that they can bring several important advantages in their applications in comparison with the organic fluorophores [29]. In this paper, these tremendous advantages are considered that determine in special, the fluorescent label comportment and hence the utilization in different cases, which comprise the spectral position, the width of the excitation spectrum, the width of the excitation spectrum, the Stokes shift, the molar absorption coefficient, the fluorescence quantum yield, the photo stability, and the decay lifetime.

The most valuable feature influencing the optical properties is the dimension of the QDs. QDs of varying dimensions change the color emitted or absorbed by the crystal thanks to the energy levels of the crystal [30]. QDs present an electronic structure analogous to atoms, due to the tight confinement of charge carriers in them [30, 31]. So, the discrete size-dependent energy levels of QDs represent the effect of the confinement of the charge carriers (electrons, holes) in three dimensions [18, 20, 31, 32]. As a result, the energy difference between excited and ground state (the bandgap energy) of a QD is a function of the QD size and composition: the smaller the bandgap of QD, the larger the QD [18, 20, 33, 34]. This means that the fluorescence wavelength is a function of the bandgap and therefore a function of the QD size [18, 20, 31]. By modifying the dimension, coating, and composition of the QDs, the emission wavelength can be adjusted from the ultraviolet (UV) to the infrared range of the spectrum such that smaller dots emit higher-energy light that is in the blue range and the larger dots emit lower-energy light that is in the red and near-infrared (NIR) region [15]. Because the dimension of QD is inversely proportional to the bandgap energy level, the frequency light emitted changes and an effect on the color occurs [30, 35].

Many of the conventional fluorophores (organic dyes and protein- based fluorophores) exhibit narrow excitation spectra which necessitate excitation by light of a particular wavelength, which fluctuate between certain fluorophores and present broad red-tailed emission spectra which suggest that the spectra of various conventional fluorophores may overlay to a large extent [20, 32]. Three important properties of QDs are considered to be of interest to specialists in biology, namely the capability of QDs to size-tune the fluorescent emission depending on core size, the broad excitation spectra of QDs, which permit excitation of mixed QDs at a single wavelength [36] and the long luminescent life of QDs, which allows their usage for dynamic imaging of living cells [37].

The narrow emission spectra of QDs permit the multicolor excitation involving the potentiality for simultaneous usage of various functionalized QDs for a number of biological targets at the same time. This fact is suited for the usage of QDs in multiplex immunohistochemistry tests [38–40].

QDs are defined by broadband excitation wavelength, very bright fluorescence even when irradiated only with a light-emitting diode (LED) flashlight [2, 17]. These characteristics of QDs, which have been mentioned earlier, allow simultaneous imaging of many entities in a unique biological experiment. This fact represents a difficult mission with common fluorophores since their relatively narrow excitation and broad emission spectra many times lead to the overlapping of the spectra [41]. QDs exhibit high resistance to physical and chemical degradation (suitable for long-term imaging) high quantum yields and high molar extinction coefficients which are 10–50 times greater than that of organic dyes, which make them much brighter in photon-limited *in vivo* conditions [14].

QDs present photoluminescence (PL), when photons are utilized to excite QDs and another photon of lower frequency is released [42]. QDs are famous for eye-catching photos of differently dimensioned QDs under ultraviolet lighting which exhibit a shining rainbow of photoluminescence [33]. This brilliant PL is obtained like a consequence of high quantum yields ($\varphi = 0.1–0.9$) combined with substantial molar extinction coefficients ($10^5–10^7$ M^{-1} cm^{-1}) [32, 33, 42]. The values of the molar extinction coefficients of QDs are 10–100 times greater than those for most organic fluorophores [41]. In contrast with the organic dyes, another beneficial characteristic of QDs is represented by the very large two-photon action cross section [43].

The majority of QD applications in biology utilize this feature for cellular/molecular tracking and imaging [42]. If a photon excites a QD, but the energy is collected as electricity, QD is utilized as photovoltaic material and represents a good possible choice for the case of the current applications in solar cells [41, 42]. In addition, QD blends the conveniences of inorganic and organic materials. In many QD-based solar cells, QDs do not only help like a light-collecting material but also have numerous purposes in order to assist in load separation and transportation [44]. QDs have also been extensively utilized in solar cells for sensitization. Quantum dot-sensitized solar cells (QDSSCs) have lately captivated a lot of interest due to their benefits over the dye-sensitized solar cells (DSSCs), comprising higher molar extinction coefficient of QDs, tunable energy gaps, and multiple exciton generation [45, 46]. It is also worth mentioning the case in which the higher voltage electrical energy can also segregate electrons from holes to form excitons, and when energy is released in the form of light, QDs are utilized in a new light-emitting diode variation, the QD-LEDs [30, 42].

Compared to organic fluorophores, QDs are about 10–100 times brighter and about 100–1000 times more stable against photo-bleaching [2, 47, 48]. Due to these properties, QDs are excellent for single molecule or, more accurately, particle measurement [33].

The fluorescence time is defined like the average time in which a fluorophore will stay in its excited state before it emits light to return to its ground state [43, 49]. The usual values of the fluorescence lifetimes are from 1 to 10 ns for organic dyes and 10 to 100 ns for QDs [43, 49]. An important property that is used for the diminution of the auto-fluorescence of the biological samples is the possibility to choose any wavelength shorter than the wavelength of fluorescence. This quality of QDs can be obtained by electing the most suitable excitation wavelength for which the auto-fluorescence is reduced to a minimum. Previous studies show that the pushing of the emission wavelength into the near-infrared (650–950-nm) range led to the enhancement of the tissue penetration depth and the decrease of the fluorescence at these wavelengths [14, 50].

QDs have shown other many remarkable advantages compared to traditional fluorophores, such as organic dyes, fluorescent proteins, and lanthanide chelates [15, 32]. One of the most significant properties of QD is the red shift of the emission spectra, called the QD Stokes shift. QD Stokes shift, which can be as large as 300–400 nm, depending on the wavelength of the excitation light, can be utilized for *in vivo* imaging [14, 36, 50]. The high Stokes shift values diminish the phenomenon of auto-fluorescence that enlarges the sensitivity [41].

Considering these advantageous optical properties shown earlier, it is not surprising that QDs have been and are still studied as some alternative fluorophores to conventional organic dyes.

2.1. Quantum dots as fluorescent probes for the bioimaging applications

These benefits of QDs are used in an emerging area of science and technology which blends the biological chemical and engineering sciences and guarantees the achievement of nanometric scale methods in order to study the biological systems in the field of health and disease [51]. The best available QDs for biological applications comprise a semiconductor core (e.g., CdSe, CdS, and CdTe) overcoated with a shell of a semiconductor material with a wider bandgap than the material of the core (e.g., ZnS, CdS) in order to obtain a considerable enhancement in the quantum efficiency [17, 18, 33, 52]. The leader QD material, which is used in almost all biological applications, is certainly represented by CdSe/ZnS core-shell, whose celebrity is imputed to well-determined synthetic protocols, emissions that can be scattered in the visible/NIR region and commercial accessibility [33, 53].

In [17, 20, 54], it has also been shown that CdSe/ZnS core-shell QDs represent an excellent substitute to fluorescent fluorophores utilized to label the microbial cells. In this context, the fluorescence of the CdSe/ZnS core-shell QDs, dispersed in toluene with long-chain amine-capping agents, was studied [17, 20, 54, 55]. The different semiconductor nanocrystals, which were utilized in various research works [17, 20, 54, 55], were purchased from EVIDENT Technologies. The sizes of these QDs are in the field of (3–5) nm and their emission is situated in the domain (490–600) nm. The emission properties of diverse QDs were examined and estimated utilizing Fourier transform visible spectroscopy [17, 18, 54, 55] for two excitation sources (a UV laser or a blue LED). **Figure 1** presents the case of the fluorescence of CdSe/ZnS core-shell QDs with the dimensions of 3.2, 3.8, and 5.0 nm [17].

In this section of the chapter, some recent important bioimaging applications of QDs, which have to do with the study of the microorganisms and toxin detection, are reviewed.

QDs have proven to be convenient in the morphological examination of microorganisms in order to show their shape, position, evolution, number, and so on [17, 20, 54, 56]. The QD-based technologies, which are used for the operations of labeling with QDs, are very relevant in biotechnology medical diagnosis and food safety [17, 20, 54, 56].

Numerous authors described successfully linked QDs to biorecognition molecules such as peptide, antibodies, nucleic acids, or small-molecule ligands for further applications as fluorescent nanoprobes [57, 58], whereas few researchers have reported results obtained in using CdSe/ZnS core-shell QDs in the field of the microbial labeling, for both pure cultures of cyanobacteria (Synechocystis PCC 6803) and mixed cultures of phototrophic and heterotrophic microorganisms [17, 20, 54, 59, 60]. In these last works, the labeling of the biological samples, comprising the cultures of the microorganisms with QDs of 0520 Evidot suspension type, which were incubated in darkness at room temperature, was described. The natural samples including filamentous cyanobacterial cells were microscopically studied (B-352 LD2) by

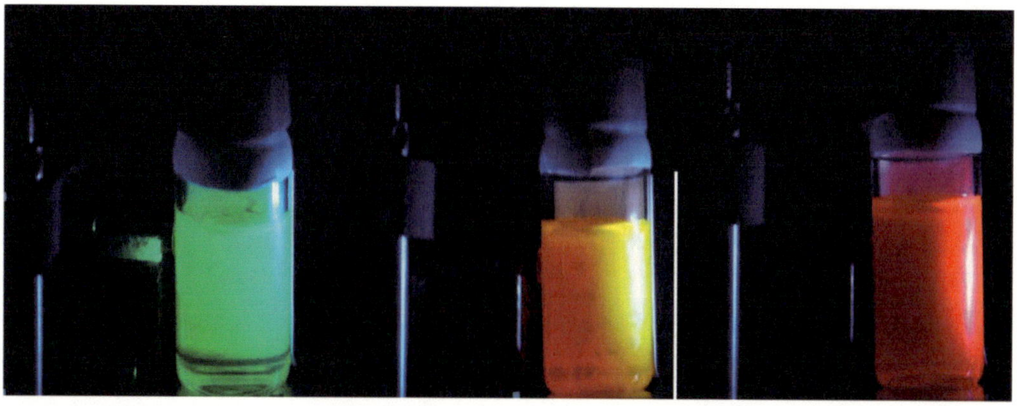

Figure 1. The fluorescence of CdSe/ZnS core-shell, suspended in toluene with long-chain amine-capping agent, under ultraviolet illumination. The various colors are correlated with the various wavelengths of the fluorescence maxim according to the sizes of the QD in the suspension. From left to right, according to the specifications of the EVIDENT Technologies catalog three kinds of CdSe/ZnS core-shell QDs were used: 0490 QDs fluid (with the crystal diameter of 3.2 nm and the color of Lake Placid Blue type), 0520 QDs fluid (with the crystal diameter of 3.8 nm and the color of Hops Yellow type) and 0600 QDs fluid (with the crystal diameter of 5.0 nm and the color of Fort Orange type) [17].

transmission in white light or by epifluorescence, with the help of the blue and green filters [17, 54]. **Figures 2** and **3** show the easy visualization of the individual cyanobacterial cells due to the green fluorescence of QDs. These two figures present the case of the individual filamentous cells in two types of natural samples indicated in the visible microscopy (A), the case of microscopic aspect of the QD-labeled cyanobacterial cells for the blue filter (B) and the case of the microscopic aspect of the same samples for the green filter (C) [17, 54].

Another research paper examined the nonspecific labeling of cyanobacteria in natural samples and enriched cultures with CdSe/ZnS core-shell and the impact of CdSe/ZnS core-shell QDs on the global color of epifluorescence microscopy images [20]. The same paper [20] exhibited the use of the digital color analysis method for the study of the epifluorescence microscopy images, demonstrating the color transformation of the epifluorescence images of filamentous cyanobacteria and showing in this way the potential toxic effects of QDs on cyanobacteria.

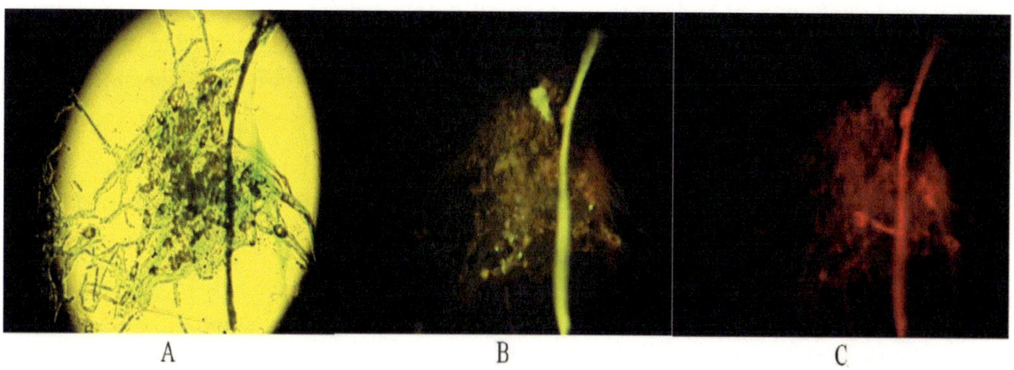

Figure 2. Microscopic aspect of the labeling of the cyanobacterial cells with QDs in the case of the sample 1: A: Transmission in white light; B: Epifluorescence utilizing a blue filter; C: Epifluorescence utilizing a green filter [54].

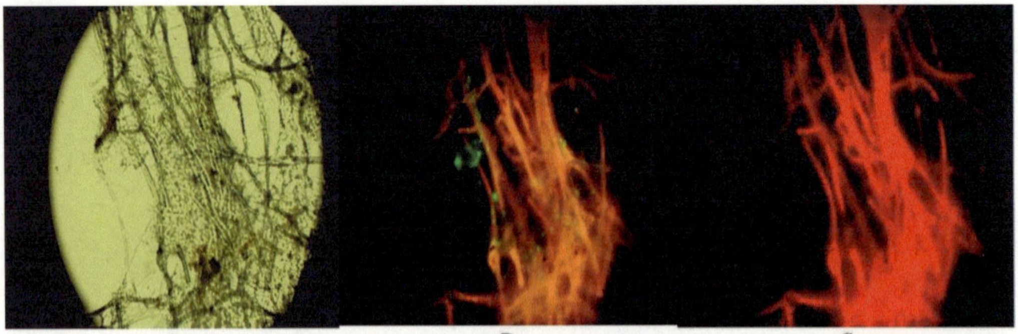

Figure 3. Microscopic aspect of the labeling of the cyanobacterial cells with QDs in the case of the sample 2: A: Transmission in white light; B: Epifluorescence utilizing a blue filter; C: Epifluorescence utilizing a green filter [54].

3. Fluorescent and magnetic nanocomposites as nanoprobes for multimodal imaging applications

QDs represent a new category of molecular imaging instruments which created a significant effect in biological and medical research, thus helping to further develop new applications. Using new layers, QD appears as a fundamental element for further manufacture of multifunctional nanostructures and nanodevices which can be manufactured by integrating QDs with NIR emission, paramagnetic or superparamagnetic nanomaterials [15]. In the next part of the paper, the latest applications of the nanocomposites are exhibited, which comprise fluorescent and magnetic particles and help for the construction of the novel multiplexed nanoprobe models.

Fluorescent-magnetic nanocomposites comprise a diversity of materials which integrate silica-based, dye-functionalized MNPs and QDs-MNPs composites. Different papers have described various types of techniques to fabricate composites of fluorescent semiconductor QDs and MNPs, such as the mixing of the two materials for the construction of a single heteromeric particle with optical and magnetic properties, the enclosing of separately synthesized fluorescent and magnetic particles in a polymer or silica matrix, the enclosing of the single particles in a polymer or silica gel, magnetically doped QDs and ionic aggregates, which are composed of a magnetic core and fluorescent ionic composites [24, 61].

These multifunctional fluorescent magnetic nanocomposites can be utilized in a range of biological and biomedical applications in nanobiotechnology, such as imaging and therapy, cell tracking and sorting, separation, and drug delivery.

Corr et al. [61] emphasized the fact that the merging of a magnetic and fluorescent entity offers novel two-in-one multifunctional nanomaterials, with a wide gamut of feasible applications, for two reasons: the first reason is that multimodal magnetic-fluorescent tests would be an advantage for *in vitro* and *in vivo* bioimaging applications (magnetic resonance imaging and fluorescence microscopy) and the second reason is that these multifunctional fluorescent magnetic nanocomposites can be used as instruments in the nanomedicine realm. Thus, fluorescent magnetic nanocomposites are utilized to visualize and synchronously treat diverse diseases [61].

Koole et al. described four diverse techniques of obtaining a single nanoparticle, which comprises the fluorescence and magnetic properties and represents new sensitive bimodal contrast instrument for two extremely powerful and highly complementary imaging methods: fluorescence imaging and MRI [27].

In some research papers [21, 53], new approaches of the therapeutic procedures and imaging modalities (such as the technique of the correlation between MRI and ultrasensitive optical imaging) are exhibited in order to be developed and to become mainstream clinical methods for the visual detection of the microscopic tumors during an operation and the complete elimination of the diseased cells and tissues. The authors of these research works show that medical imaging methods can detect the illness, but do not furnish a visual template during a certain surgery. This matter can be resolved with the help of some MQD probes. MQDs are a form of magnetic contrast agents in MRI. In this regard, paramagnetic and superparamagnetic agents Gd(III) and different forms of iron oxide (Fe_2O_3) in molecular form and in nanoparticle form are attached to QDs to utilize in a variety of MRI applications with the scope of the improving image contrast [21, 53].

Many researchers have investigated paramagnetic QDs (pQDs) [62–65]. So, in Refs. [62, 64], a model of multifunctional fluorescent magnetic nanocomposites, comprising silica-coated Fe_3O_4 and TGA-capped CdTe QDs, was communicated. This type of nanocomposite has been used for the labeling and imaging of HeLa cells in a magnetic separation. In Refs. [63, 65], pQDs were improved by coating CdSe/ZnS core-shell QDs with a PEGylated phospholipid and a Gd lipid, making the particles biocompatible and MRI active. The pQDs were also conjugated by maleimide to cyclic RGD peptides for targeting angiogenic vascular endothelium as proved by *in vitro* investigations with human umbilical vein endothelial cells (HUVECs).

Ahmed et al. [66] developed a novel technique for the manufacture of QDs enclosed MNPs based on layer-by-layer (LbL) self-assembly method in order to be used for cancer cells imaging. In a research study, Park's group [67] reported long-circulating, micellar hybrid nanoparticles (MHNs) which include MNs, QDs, and the anticancer drug doxorubicin (DOX) in a single poly(ethylene glycol) (PEG)-phospholipid micelle and furnish the first models of concomitant targeted drug delivery and dual-mode near-infrared fluorescence imaging and MRI of diseased tissue *in vitro* and *in vivo* [24, 67].

A very useful review study, which includes the mention of the specialized literature on the applications of MQDs, which represent one of the most currently explored QD-based hybrid NPs, was realized by Wegner and his colleague [68]. In this study, a principal result is indicated, which was obtained by Qiu et al. [69]. These researchers blended QDs, superparamagnetic iron oxide nanoparticles (SPIONs), and gold (Au) NPs in a single poly(lactic-co-glycolic acid) (PLGA) NP. This type of particle is significant for imaging, tracking, and manipulating neutrophils and is used for *in vivo* applications and localized photothermal treatment.

4. Conclusions

QDs and fluorescent and magnetic nanocomposites, which have essential physicochemical characteristics, represent some excellent candidates for numerous applications in bioanalysis and bioimaging. In this review, the important properties of these nanoprobes are summarized and the recent developments of the applications of the QD-based techniques in biomedical uses (biological imaging, cell tracking, magnetic bioseparation, and bio- and chemo-sensoring) are highlighted.

Author details

Anca Armăşelu

Address all correspondence to: anca_armaselu@yahoo.com

Department of Electrical Engineering and Applied Physics, Faculty of Electrical Engineering and Computer Science, Transilvania University of Braşov, Braşov, Romania

References

[1] Wang Y, Tang Z, Kotov NA. Bioapplication of nanosemiconductors. Materials Today. 2005;**8**(5):20-31. DOI: 10.1016/S1369-7021(05)00892-8

[2] Aguilar Z. Nanomaterials for Medical Applications. New York: Elsevier; 2012. p. 544. ISBN: 978-0-12-385089-8

[3] Zborowski M, Chalmers JJ, Lowrie WG. Magnetic cell manipulation and sorting. In: Lee W, Tseng P, Di Carlo D, editors. Microtechnology for Cell Manipulation and Sorting. Switzerland: Springer International Publishing; 2017. p. 15-55. DOI: 10.1007/978-3-319-44139-9_2

[4] Kolosnjaj-Tabi J, Wilhelm C, Clément O, Gazeau F. Cell labeling with magnetic nanoparticles: Opportunity for magnetic cell imaging and cell manipulation. Journal of Nanobiotechnology. 2013;**11**(1):S7. DOI: 10.1186/1477-3155-11-S1-S7

[5] La Conte L, Nitin N, Bao G. Magnetic nanoparticle probes. Materials Today. 2005;**8**(5):32-38. DOI: https://doi.org/10.1016/S1369-7021(05)00893-X

[6] Wang Y, Zhang Y, Du Z, Wu M, Zhang G. Detection of micrometastases in lung cancer with magnetic nanoparticles and quantum dots. International Journal of Nanomedicine. 2012;7:2315-2324. DOI: 10.2147/IJN.S30593

[7] Thanh NTK. Magnetic Nanoparticles: From Fabrication to Clinical Applications. Taylor & Francis: Boca Raton, London, New York: CRC Press; 2012. p. 616. DOI: http://doi.org/10.1201/b11760

[8] Veiseh O, Gunn JW, Zhang M. Design and fabrication of magnetic nanoparticles for targeted drug delivery and imaging. Advanced Drug Delivery Reviews. 2010;**62**(3):284-304. DOI: 10.1016/j.addr.2009.11.002

[9] Dušak P. Magnetic Nanoparticles for Selective Magnetic Separation in Biotechnology [Thesis]. Jožef Stefan International Postgraduate School; 2015 Available from: http://www.marin-berovic.si/znanost/files/doktorat_peter_dusak.pdf

[10] Arruebo M, Fernández-Pacheco R, Ibarra MR, Santamaria J. Magnetic nanoparticles for drug delivery. Nano Today. 2007;**2**:22-32. DOI: 10.1016/S1748-0132(07)70084-1

[11] Khandhar AP, Ferguson RM, Simon JA, Krishnan KM. Tailored magnetic nanoparticles for optimizing magnetic fluid hyperthermia. Journal of Biomedical Material Research A. 2012;**100A**(3):728-737. DOI: 10.1002/jbm.a.34011

[12] Obaidat IM, Issa B, Hail Y. Magnetic properties of magnetic nanoparticles for efficient hyperthermia. Nanomaterials. 2015;**5**:63-89. DOI: 10.3390/nano5010063

[13] Bhatia S. Nanoparticles types, classification, characterization, fabrication methods and drug delivery applications. In: Bhatia S, editor. Natural Polymer Drug Delivery Systems: Nanoparticles, Plants, and Algae. Switzerland: Springer International Publishing Switzerland; 2016. p. 33-93. DOI: 10.1007/978-3-319-41129-3_2

[14] Rhyner MN, Smith AM, Gao X, Mao H, Yang Z, Nie S. Quantum dots and multifunctional nanoparticles: new contrast agents for tumor imaging. Nanomedicine. 2006;**1**(2):209-217. DOI: 10.2217/17435889

[15] Ghaderi S. Development of Fluorescent Nanoparticles "Quantum Dots" for Biomedical Application [Thesis]. UCL Center for Nanotechnology and Regenerative Medicine Division of Surgery & Interventional Sciences University College London; 2012 Available from: http://discovery.ucl.ac.uk/1370637/5/1370637_Shirin%20Thesis.pdf

[16] Sortino S. Light-Responsive Nanostructured Systems for Applications in Nanomedicine. Top. Curr. Chem 370. Switzerland: Springer International Publishing Switzerland; 2016. p. 263. DOI: 10.1007/978-3-319-22942-3

[17] Damian V, Ardelean I, Armăşelu A, Apostol D. Fourier transform spectra of quantum dots. Proceedings of the SPIE. 2010;**7469**:7460E. DOI: 10.1117/12.862053

[18] Armăşelu A. Determination of quantum dot size by Fourier transform visible spectroscopy. Optoelectronics and Advanced Materials, Rapid Communications. 2015;**9**(3):531-536 Print: ISSN 1842-6573. On line: ISSN 2065-3824

[19] Wang J, Han S, Ke D, Wang R. Semiconductor quantum dots surface modification for potential cancer diagnostic and therapeutic applications. Journal of Nanomaterials. 2012;**2012**(2012):1-8. DOI: 10.1155/2012/129041

[20] Armăşelu A, Popescu A, Apostol I, Ardelean I, Damian V, Iordache I, Sarchizian I, Apostol D. Passive nonspecific labeling of cyanobacteria in natural samples using quantum dots. Optoelectronics and Advanced Materials, Rapid Communications. 2011;**5**(10):1084-1090 Print: ISSN 1842-6573. On-line: ISSN 2065-3824

[21] Zhang X. Synthesis of CdTe and PbS Semiconductor Quantum Dots and their Biological and Photochemical Applications [Thesis]. The University of Texas at Arlington; 2010 Available from: https://uta-ir.tdl.org/uta-ir/bitstream/handle/10106/4889/Zhang_uta_2502M_10608.pdf?sequence=1

[22] Winter JO. (182 c) cellular and molecular biosensing using magnetic quantum dots. In: AIChE Annual Meeting: Proceedings: Sensor: Topical Plenary: Advances in Biosensing (Invited Talks). The Ohio State University; 2016. Available from: https://www.aiche.org/conference/aiche-annual-meeting/2016/proceeding/paper/182c-cellular-and-molecular-biosensing-using-magnetic-quantum-dots

[23] Mahajan KD, Fan Q, Dorcéna J, Ruan G, Wiinter JO. Magnetic quantum dots in biotechnology – synthesis and applications. Biotechnology journal. 2013;**8**(12):1424-1434.DOI: 10.1002/biot.201300038

[24] Thakur D. Fluorescent and Magnetic Nanocomposites for Multimodal Imaging [thesis]. The Ohio State University; 2010.Available from: https://etd.ohiolink.edu/rws_etd/document/get/osu1274630209/inline

[25] Makkar M, Viswanatha R. Recent advances in magnetic ion-doped semiconductor quantum dots. Current Science. 2017;**112**(7):1421-1429. DOI: 10.18520/cs/v112/i07/1421-1429

[26] Roullier V, Grasset F, Boulmedais F, Artzner F, Cador O, Marchi-Artzner V. Small Bioactivated Magnetic Quantum Dot Micelles. Chem. Matter. 2008;**20**(21):6657-6665. DOI: 10.1021/cm801423r

[27] Koole R, Mulder WMJ, Van Schooneveld MM, Strijkers GJ, Meijerink A, Nicolay K. Magnetic quantum dots for multimodal imaging. Wiley Interdisciplinary Reviews Nanomedicine and Nanobiotechnology. 2009;**1**(5):475-491. DOI: 10.1002/wnan.14

[28] Fahmi MZ, Chang JY. Potential application of oleylamine-encapsulated $AgInS_2$-ZnS quantum dots for cancer cell labeling. Procedia Chemistry. 2016;**18**(2016):112-121. DOI: 10.1016/j.proche.2016.01.018

[29] Svenson S, Prud'homme RK. Multifunctional Nanoparticles for Drug Delivery Applications: Imaging, Targeting and Delivery. Springer-Verlag New York: Nanostructure Science and Technology Springer Science + Business Media, LLC; 2012. p. 374. DOI: 10.1007/978-1-4614-2305-8

[30] Bagher AM. Quantum dots applications. Sensors & Transducers. 2016;**198**(6):37-43 Available from: https://www.sensorportal.com/HTML/DIGEST/march_2016/Vol_198/P_2807.pdf

[31] Parak WJ, Pellegrino T, Plank C. Labeling of cells with quantum dots. Nanotechnology. 2005;**16**(2):R9-R25. DOI: 10.1088/0957-4484/16/2/R01

[32] Jamieson T, Bakhshi R, Petrova D, Pocock R, Imani M, Seifalian AM. Biological applications of quantum dots. Biomaterials. 2007;**28**(31):4717-4731. DOI: 10.1016/j.biomaterials.2007.07.014

[33] Petryayeva E, Algar WR, Medintz IL. Quantum dots in bioanalysis: A review of applications across various platforms for fluorescence spectroscopy and imaging. Applied Spectroscopy. 2013;**67**(3):215-253. DOI: 10.1366/12-06948

[34] Krishan B, Garg MR. A literature review on quantum dots. International Journal of Advanced Research in Electrical, Electronics and Instrumentation Engineering. 2015;**4**(9):7857-7862. DOI: 10.15662/IJAREEIE.2015.0409085

[35] Hegazy MA, El-Hameed AMA. Characterization of CdSe-nanocrystals used in semiconductors for aerospace applications: Production and optical properties. NRIAG Journal of Astronomy and Geophysics. 2014;**3**(1):82-87 https://doi.org/10.1016/j.nrjag.2014.05.002

[36] Medintz IL, Uyeda HT, Goldman ER, Mattoussi H. Quantum dot bioconjugates for imaging, labeling and sensing. Nature Materials. 2005;**4**(6):435-446. DOI: 10.1038/nmat1390

[37] Shen L. Biocompatible polymer/quantum dots hybrid materials: Current status and future developments. Journal of Functional Biomaterials. 2011;**2**:355-372. DOI: 10.3390/jfb2040355

[38] Chan P, Yuen T, Ruf F, Gonzales-Maeso J, Sealfon SC. Method for multiplex cellular detection of mRNAs using quantum dot fluorescent in situ hybridization. Nucleic Acids Research. 2005;**33**(18):161. DOI: https://doi:org/10.1093/nar/gni162

[39] Shi C, Zhou G, Zhu Y, Su Y, Cheng T, Zhau HE, Chung LWK. Quantum dots-based multiplexed immunohistochemistry of protein expression in human prostate cancer cells. European Journal of Histochemistry. 2008;**52**(2):127-134. DOI: 10.4081/1202

[40] Anceno AJ, Bonduush I, Dutta J, Stuetz RM, Shipin OV. Of quantum dots and microbes: Smart materials for fluorescence based characterization of environmental microflora. In: Proceedings of the International Conference on Future Trends in Structural, Civil, Environmental and Mechanical Engineering: FTSCEM 2013; 13-14 July 2013; Bangkok, Thailand: SEEK Digital Library; 2013. p. 61-65. DOI: 10.3850/978-981-07-7021-1_53

[41] Fu A, Gu W, Larabell C, Alivisatos AP. Semiconductor nanocrystals for biological imaging. Current Opinion in Neurobiology. 2005;**15**(5):568-575. DOI: 10.1016/j.comb.2005.08.004

[42] Fan Q. Designing photo-switchable quantum dots for super resolution imaging [thesis]. In: . Graduate Program in Chemical Engineering. The Ohio State University; 2015 Available from: https://etd.ohiolink.edu/!etd.send_file?accesion=osu1446217324&disposition=inline

[43] Resch-Genger U, Grabolle M, Cavaliere-Jaricot S, Nitschke R, Nann T. Quantum dots versus organic dyes as fluorescent labels. Nature Methods. 2008;**5**(9):763-777. DOI: 10.1038/NMETH.1248

[44] Hu MZ, Zhu T. Semiconductor nanocrystal quantum dot synthesis approaches towards large-scale industrial production for energy applications. Nanoscale Research Letters. 2015;**10**(1):469-483. DOI: 10.1186/s11671-015-1166-y

[45] Kong E-H, Chang Y-J, Jang HM. Hierarchically nanostructured photoelectrodes for quantum dot-sensitized solar cells. In: Wang ZM, Neogi A, editors. Nanoscale Photonics and Optoelectronics, Lecture Notes in Nanoscale Science and Technology. Vol. 15. New York: Springer; 2014. p. 39-66. DOI: https://doi.org/10.1007/978-1-4614-8148-5_2

[46] Sudhagar P, Juárez-Pérez EJ, Kang YS, Mora-Seró I. Quantum dot-sensitized solar cells. In: Lin Z, Wang J, editors. Low-Cost Nanomaterials. Green Energy and Technology. London: Springer; 2014. p. 89-136. DOI: https://doi.org/10.1007/978-1-4471-6473-9_5

[47] Zhang Y, Clapp A. Overview of stabilizing ligands for biocompatible quantum dot nanocrystals. Sensors. 2011;**11**(12):11036-11055. DOI: 10.3390/s111211036

[48] Malik S, Sachan M, Nara S. Nano-strategies for neuro-imaging and diagnostics. In: Pillay V, Choonara YE, editors. Advances in Neurotherapeutic Delivery Technologies. Foster City: OMICS Group International; 2015. p. 193. DOI: http://dx.doi.org/10.4172/978-1-63278-036-2-037-009

[49] Lens P, Virkute J, Jegatheesan V, Al-Abed S. Nanotechnology for Water and Wastewater Treatment. Integrated Environmental Technology Series. Vol. 12. London SW1H 0QS, UK: IWA Publishing; 2013. p. 500. DOI: 10.2166/9781780404592

[50] Smith AM, Mancini MC, Nie S. Second window for in vivo imaging. Nature Nanotechnology. 2009;**4**(11):710-711. DOI: 10.1038/nnano.2009.326

[51] Vu TQ, Rajan SS. Quantum dot imaging of neural cells and tissue. In: Silva GA, Parpura V, editors. Nanotechnology for Biology and Medicine: At the Building Block Level. New York: Springer-Verlag, Springer Science + Business Media, LLC; 2012. p. 151-168. DOI: 10.1007/978-0-387-31296-5

[52] Vasudevan D, Gaddam RR, Trinchi A, Cole J. Core-shell quantum dots: Properties and applications. Journal of Alloys and Compounds. 2015;**636**:395-404. DOI: http://dx.doi.org/10.1016/j.jallcom.2015.02.102

[53] Mazumder S, Dey R, Mitra MK, Mukherjee S, Das GC. Review: Biofunctionalized quantum dots in biology and medicine. Journal of Nanomaterials. 2009;**2009**:1-17. DOI: 10.1155/2009/815734

[54] Armășelu A, Popescu A, Damian V, Ardeleanu I, Apostol D. Fluorescence properties of quantum dots used in the study of microorganisms. Journal of Optoelectronics and Advanced Materials. 2011, 4;**13**:439-443 ISSN: PRINT: 1454-4164. ON-LINE: 1841-7132

[55] Armășelu A. New spectral applications of the Fourier transforms in medicine, biological and biomedical fields. In: Nikolic GS, Cakic MD, Cvetkovic DJ, editors. Fourier Transforms-High-Tech Application and Current Trends. Rijeka, Croatia: INTECH; 2017. p. 235-252. DOI: 10.5772/66577

[56] Radwan SH, Hassan MES. Quantum dots for nanomedicine. In: Sattler KD, editor. Handbook of Nanophysics: Nanomedicine and Nanorobotics. Boca Raton: CRC Press Taylor & Francis Group; 2011. pp. 22-1-22-23. ISBN-13: 978-1-4200-7549-6

[57] Shao L, Gao Y, Yan F. Semiconductor quantum dots for biomedical applications. Sensors. 2011;**11**:11736-11751. DOI: 10.3390/s111211736

[58] Agrawal A, Wang MD, Nie S. Bioconjugated nanoparticles for ultrasensitive detection of molecular biomarkers and infectious agents. In: Gonsalves KE, Halberstadt CR,

Laurencin CT, Nair LS, editors. Biomedical Nanostructures. NJ: John Wiley & Sons; 2008. pp. 207-222. ISBN: 978-0-471-92552-1

[59] Ardelean I, Sarchizian I, Manea M, Damian V, Apostol I, Cîrnu M, Armăşelu A, Iordache I, Apostol D. CDSe/ZnS quantum dots cytotoxicity against phototrophic and heterotrophic bacteria. In: TANGER Ltd, editor. Proceeding of NANOCON 2011; 21-23 September 2011; Brno, Czech Republic; 2011. pp. 608-617

[60] Ardelean II. Metallic nanoparticles synthesis by cyanobacteria: Fundamentals and applications. In: Sahoo D, Seckbach J, editors. The Algae World. Springer Netherlands: Springer Science + Business Media Dordrecht; 2015. p. 598. DOI: 10.1007/978-94-017-7321-8

[61] Corr SA, Rakovich YP, Gun'ko YK. Multifunctional magnetic-fluorescent nanocomposites for biomedical applications. Nanoscale Research Letters. 2008;**3**:87-104. DOI: 10.1007/S 11671-008-9122-8

[62] Jin S, Hu Y, Gu Z, Liu L, Wu HC. Application of quantum dots in biological imaging. Journal of Nanomaterials Hindawi Publishing Corporation. 2011;2011:1-13. DOI: http://dx.doi.org/10.1155/2011/834139

[63] Swierczewska M, Lee S, Chen X. Inorganic nanoparticles for multimodal molecular imaging. Molecular Imaging. 2011;**10**(1):3-16. DOI: 10.2310/7290.2011.00001

[64] Sun P, Zhang H, Liu C, Fang J, Wang M, Chen J, Zhang J, Mao C, Xu S. Preparation and characterization of Fe_3O_4/CdTe magnetic/fluorescent nanocomposites and their applications in immunolabeling and fluorescent imaging of cancer cells. Langmuir. 2011;**29**(2):1278-1284. DOI: 10.1021/1a9024553

[65] Mulder WJM, Koole R, Brandwijk RJ, Storm G, Chin PTK, Strijkers GJ, De Mello DC, Nicolay K, Griffioen AW. Quantum dots with a paramagnetic coating as a bimodal molecular imaging probe. Nano Letters. 2006;**6**(1):1-6. DOI: 10.1021/nI051935m

[66] Ahmed SR, Dong J, Yui M, Kato T, Lee J, Park EY. Quantum dots incorporated magnetic nanoparticles for imaging colon carcinoma cells. Journal of Nanobiotechnology. 2013;**11**(28):1-9. DOI: 10.1186/1477-3155-11-28

[67] Park JH, Von Maltzahn G, Ruoslahti E, Bhatia SN, Sailor MJ. Micellar hybrid nanoparticles for simultaneous magnetofluorescent imaging and drug delivery. Angewandte Chemie International Edition England. 2008;**47**(38):7284-7288. DOI: 10.1002/anie.200801810

[68] Wegner KD, Hildebrandt N. Quantum dots: bright and versatile *in vitro* and *in vivo* fluorescence imaging biosensors. Chemical Society Review. 2015;**44**(14):4792-4834. DOI: 10.1039/c4cs00532e

[69] Qiu Y, Palankar R, Echeverria M, Medvedev N, Moya SE, Delcea M. Design of hybrid multimodal poly(lactic-co-glycolic acid) polymer nanoparticles for neutrophil labeling, imaging and tracking. Nanoscale. 2013;**5**(24):12624-12632. DOI: 10.1039/c3nr04013e